Preface

In this book we give a mathematical account of some of the methods of data simplification which are involved in or suggested by the practice of biological taxonomy. The computable methods derived are offered as potentially useful tools for taxonomists, rather than as substitutes for their activities.

Superficially similar problems of data simplification arise in pattern recognition and in the various sciences which make substantial use of classificatory systems: biological taxonomy, ecology, psychology, linguistics, archaeology, sociology, etc. But more detailed examination shows that the kinds of classification used and the kinds of data on which they are based differ widely from science to science. Whilst we recognize that some of the methods described here, particularly in Part II of the book, are more widely applicable, we have deliberately limited discussion to biological taxonomy. The emphasis throughout is on the clarification of the mathematical properties of methods of automatic classification and of the conditions under which their application is valid, so that anyone who wishes to apply the methods in other fields shall be aware of their limitations, and of the lines along which they may profitably be developed and modified.

The historical development of methods of automatic classification is described only briefly. For the bibliography of the subject the reader is referred to Sokal and Sneath's book *Principles of Numerical Taxonomy* (Freeman, San Francisco, 1963) and to the sources cited in Appendix 8.

The first part of the book, entitled *The measurement of dissimilarity*, deals with methods for deriving dissimilarity coefficients on a set of populations, given as data descriptions of members of each population. The mathematical basis lies in information theory, for which background may be found in Kullback's book *Information Theory and Statistics* (Wiley, New York, 1959). The second part of the book, entitled *Cluster analysis*, gives a general treatment of methods for the construction of classificatory systems from data in the form of a dissimilarity coefficient. The mathematics used involves some elementary set theory and some ideas of continuity in a general context. For this background the reader is referred to one of the many textbooks available, for example Kolmogorov and Fomin's book *Elements of the Theory of Functions and Functional Analysis (Vol. 1)* (English translation: Graylock, New York, 1957). The third part of the book, entitled *Mathematical and biological taxonomy*, deals with both theoretical and practical aspects of the application in biological taxonomy of the methods developed in Parts I and II. No sophisticated knowledge of biology is assumed, but Davis and Heywood's book *Principles of Angiosperm Taxonomy* (Oliver and Boyd, Edinburgh, 1963), Mayr's book *Principles of Systematic Zoology* (McGraw-Hill, New York, 1969), Simpson's book *Principles of Animal Taxonomy* (Columbia University Press, New York, 1961), and Briggs and Walters' book *Plant Variation and Evolution* (Weidenfeld and Nicolson *World University Library*, London, 1969) are useful complementary reading. Occasional mention is made of various methods of multivariate analysis. The reader is referred to Kendall and Stuart's book *The Advanced Theory of Statistics (Vol. 3, 3-volume ed.)* (2nd ed., Griffin, London, 1968) and to Rao's book *Advanced Statistical Methods in Biometric Research* (Wiley, New York, 1952). Occasional mention is made also of methods used in pattern recognition. For up-to-date coverage of the field the reader is referred to Watanabe (ed.), *Methodologies of Pattern Recognition* (Academic Press, London and New York, 1969) and to Mendel and Fu (eds.), *Adaptive, Learning, and Pattern Recognition Systems–Theory and Applications* (Academic Press, London and New York, 1970).

Tables of various statistics and details of algorithms and programs are given in Appendices 1 to 6. Examples of the application of the new methods developed here are given in Appendix 7. Chapters and sections are headed on the decimal system, starting with 1.1. Thus a reference to 'Chapter *m.n*' refers to Section *n* of Chapter *m*. References are given separately for the introduction and each of the three parts of the book, and are given according to the Harvard system, thus: 'Sokal and Sneath (1963)'.

We have profited greatly from the advice and encouragement of Professor D. G. Kendall, Dr S. M. Walters, Dr P. H. A. Sneath, and Dr R. M. Needham. We have had many useful conversations with Dr K. Spärck Jones, M. I. C. Lerman, M. M. Roux, Dr M. B. Hesse, and Mr J. D. Holloway. We wish to thank also all those who attended the informal conference *Mathematical and Theoretical Aspects of Numerical Taxonomy* held at King's College, Cambridge in September 1969, and sponsored by the United Kingdom Science Research Council and by King's College Research Centre. The discussion at the conference was very helpful to us. We acknowledge

Mathematical Taxonomy

NICHOLAS JARDINE and ROBIN SIBSON
Fellows, King's College, Cambridge

John Wiley & Sons Ltd

London *New York* *Sydney* *Toronto*

Library of Congress Catalog Card Number 70-149578.

ISBN 0 471 44050 7

Printed in Great Britain by
William Clowes & Sons Limited, London, Colchester and Beccles

gratefully the substantial help given by our coworkers Mr C. J. Jardine, Dr J. K. M. Moody, Mr C. J. van Rijsbergen, Mrs E. Bari, Mr L. F. Meintjes, and Mr A. D. Gordon, who have made substantial contributions to the appendices and have helped with the implementation of many of our methods and techniques. We thank especially Mrs H. M. Hunt, who drew the figures, collated the references, and typed and checked much of the text, and we are grateful to Miss A. D. Shrewsbury and Miss A. Kemper, both of whom have done substantial amounts of typing.

We acknowledge with gratitude the support of various institutions and bodies who have sponsored our work. Both authors hold Fellowships at King's College, Cambridge. Extensive facilities have been provided by King's College Research Centre and by Cambridge University Mathematical Laboratory. Much of the work has been sponsored by the United Kingdom Science Research Council through a research grant held by N. Jardine for a project entitled *The Use of Computers in Classification and Diagnosis.* The Royal Society has provided a grant for the purchase of computing equipment, and N. Jardine has held a Royal Society Scientific Information Research Fellowship since October 1968.

Both authors take responsibility for the entire contents of this book; there is no senior author.

King's College, Cambridge, England *N. J., R. S.*
February 1970

Introduction

Terminological confusions and the conceptual confusions which they conceal have loomed large in the development of methods of automatic classification. Partly this has arisen from the parallel development of related methods under different names in biology, pattern recognition, psychology, linguistics, archaeology, and sociology. Partly it has arisen from failure by mathematicians working in the various fields to realize how diverse are the problems included under the headings 'classification', 'taxonomy', and 'data analysis'; and from the concomitant failure of scientists to realize that until they are quite clear about how they wish to represent their data, and what kinds of inference they wish to base upon their representations, they are not in a position to select appropriate methods.

The term *taxonomy* as used in this book covers all the various activities involved in the construction of classificatory systems. Biologists have sometimes used the term 'taxonomy' as a synonym for 'biosystematics', a wider interpretation including all methods for the experimental investigation and description of patterns of variation in organisms. The term *classification* will likewise be restricted to the processes whereby classificatory systems are constructed. This runs counter to its use by some statisticians, who have used it to describe the discrimination of classes or, given a system of classes, the diagnosis or identification of individuals. This ambiguity is present also in ordinary language: compare 'How would you classify (identify) this?' and 'How are these best classified (grouped)?'. The importance of distinguishing classification meaning 'the construction of classificatory systems' from classification meaning 'diagnosis' has been emphasized by many authors including Dagnelie (1966), Kendall (1966), and Whewell (1840). The terms *classificatory system* and *taxonomic system* will be used to describe any set of subsets of a set of objects which in some way conveys information about the objects. The term *object* will be used to describe whatever are the elements of a set which is decomposed into subsets in a classificatory system, whether the elements be individuals or classes of individuals. To avoid subsequent confusion it is as well to note here that the classes of one classificatory system may be the objects of another. Thus if populations of organisms are partitioned into species, the populations are the objects. If subsequently the species are grouped in genera, families, etc., the objects of this classificatory system will be the classes of the previous one.

The problems to which methods of automatic classification have been applied fall roughly into two categories; compare Jardine (1970a). First, there are applications where classificatory methods are applied in attempts to find good solutions to particular problems, and where there is an extrinsic criterion of the 'goodness' of a solution. For example, where the data specify the connections between electrical components in a machine, methods of automatic classification have been successfully applied in attempts to find ways of allocating the components to boxes of given sizes in such a way that the connections between boxes are as few as possible (MacWilliams, 1969). Similarly, where the data is the cost per unit distance of flow of persons and equipment between activities, methods of automatic classification have been used in attempts to find allocations of activities to sites in buildings which reduce the cost of communication between sites; see Levin (1964), and Moseley (1963).

For many such problems it is known that there exists in general no unique optimal solution, and no computationally feasible algorithm is available which is guaranteed to find an optimal solution. The criterion by which the validity of application of a method of automatic classification to such problems must be judged is, simply, 'How well does it work?'. A cavalier attitude to the mathematical properties of such methods is quite in order.

Secondly, there are problems where what is sought is some simplified representation of complex data which will serve as a fruitful source of hypotheses. In such problems there are no simple extrinsic criteria by which the results of classificatory methods can be evaluated. Knowledge of the mathematical properties of methods forms a crucial part of the justification for their use. In order to find appropriate methods for these purposes it is necessary to construct rigorous mathematical models for the process of data simplification. Heuristic geometrical and graph-theoretic models have played a large part in the development of methods of automatic classification. Whilst useful in the development of algorithms they are inadequate to determine the analytic properties of methods, and may actively mislead by providing spurious intuitive justification for methods with undesirable properties. The application of methods of automatic classification in biological taxonomy is of this second kind.

Not all applications of methods of automatic classification can be categorized in this way. For example, their application in document clustering and index term clustering in information retrieval systems appears to be an intermediate case. Although in this field general constraints about the way in which a method should represent data can be laid down, there are many extrinsic constraints which complement and may in some cases override the general constraints.

The strategy which we have adopted in developing methods of automatic classification appropriate to biological taxonomy is as follows. We have treated the construction of classificatory systems as a two-stage process. The first stage is to construct a measure of dissimilarity on a set of populations in such a way as to describe the pattern of variation of the organisms studied with respect to the attributes chosen. The second stage is to represent the dissimilarity coefficient by an appropriate classificatory system. Treatment of the construction of classificatory systems as a two-stage process is widely current amongst both orthodox and numerical taxonomists. It is convenient because it renders the subject mathematically tractable, and leads to methods which appear to work well in practice; but there seems to be no theoretical reason why such treatment is essential. What is beyond doubt is that any approach which does not break up the construction of classificatory systems in this way would be complex and would lead to computationally formidable methods except under very special circumstances. Certain possible alternative approaches are discussed briefly in Chapters 8, 9, and 12.

For each of the two stages a similar strategy is used for finding adequate methods. First, precise characterizations of the form of the data and of the form of the required representation are set up. Any method can then be regarded as a transformation from a structure of one kind to a structure of another kind. Next, criteria of adequacy for the transformation are laid down. These may include specifications of operations on the data under which the transformation is required to be invariant or covariant, specifications of structure in the data which is to be preserved in the result, and optimality conditions. Then it must be ascertained whether there exist methods which satisfy the requirements, and what are the further properties of the methods so determined. Finally, efficient algorithms to implement the methods are sought.

The account of the measurement of dissimilarity between populations leads to a method which satisfies the criteria of adequacy suggested, but no axiomatic treatment is given. In the account of cluster analysis an axiomatic model is set up, and it is shown that under certain circumstances particular methods are uniquely determined by simple and intuitively necessary constraints.

At the outset it will be convenient to list some of the structures which are used as tools in the discussion of classificatory methods. Some of the forms of data which we may wish to represent by classificatory systems are as follows.

(a) When the objects to be classified are individuals we may have a specification of the states of a finite set of attributes for each individual.
(b) When the objects to be classified are populations we may have a specification of a sample of individuals from each population in terms of states of some or all of a finite set of attributes.
(c) We may have a binary symmetric relation on P, where P is the set of objects to be classified. If the states of the relation stand for 'alike' and 'unlike', or for dissimilarity more or less than some threshold value, this is a *nominal* similarity or dissimilarity measure.
(d) We may have a rank-ordering on $P \times P$: that is, an *ordinal* similarity or dissimilarity measure.
(e) We may have a non-negative real valued function on $P \times P$: that is, a *numerical* similarity or dissimilarity coefficient.

The first part of the book is mainly concerned with the construction of an appropriate transformation from (b), to (e).

The kind of classificatory system with which we shall be concerned is a set of subsets which is exhaustive of the objects under consideration. These we shall call *clusterings*. Clusterings may be categorized as follows.

(a) A *simple* clustering is one in which no class includes any other class. Otherwise a clustering is *compound*.
(b) A *partitional* clustering is one in which every two classes are either disjoint or one is included in the other. Otherwise a clustering is *overlapping*. A simple partitional clustering is a *partition*.
(c) A *stratified* clustering is a sequence of simple clusterings on a set of objects in which each member of the sequence is indexed by a numerical or ordinal level. If each member of the sequence is partitional it is a *hierarchic* stratified clustering. If the members of the sequence are not partitional it is a *non-hierarchic* or overlapping stratified clustering.

The taxonomic or Linnaean hierarchy is an ordinally stratified hierarchic clustering. Numerically stratified hierarchic clusterings are often called *dendrograms*.

The second part of the book will be primarily concerned with obtaining stratified clusterings from data of forms (d), and (e). Stratified clusterings are selected for special attention for two reasons. First, they are selected because they have richer structure than other kinds of clustering. Thus one way to obtain various kinds of simple clustering is by further operations on stratified clusterings. The construction of classificatory systems, like other kinds of data simplification, involves throwing away information, and it seems advisable to throw away information in as gradual and controlled a manner as possible. Even when the representation eventually required is a simple clustering it may be better to approach it indirectly. Secondly, they are selected because they are the kind of clustering which, for good reasons, have conventionally been used in biological taxonomy.

Finally, we must counter a possible objection to the whole approach adopted in this book. Many biologists argue that when a data-handling method is to be chosen, the proof of the pudding lies in the eating, and that the kind of theoretical treatment offered here is irrelevant. One value of a theoretical approach is that it can rule out some seemingly luscious puddings as poisonous. It makes it possible to understand the properties of known methods, so that in testing and applying them biologists know how to interpret their results; and, contrary to what is often supposed, it can lead to new methods whose properties render them of potential value.

REFERENCES

Dagnelie, P. A. (1966). A propos des différentes methodes de classification numerique. *Rev. Statist. Appl.*, **14**, pp. 55-75.

Jardine, N. (1970a). Algorithms, methods, and models in the simplification of complex data. *Comput. J.*, **13**, pp. 116-117.

Kendall, M. G. (1966). Discrimination and classification. In P. R. Krishnaiah, (Ed.), *Multivariate Analysis,* Academic Press, London and New York, pp. 165-184.

Levin, P. H. (1964). Use of graphs to decide the optimum layout of buildings. *Architects' J. Inf. Library,* 7 October 1964, pp. 809-815.

MacWilliams, J. (1969). Paper given at the Department of Pure Mathematics and Mathematical Statistics, Cambridge University.

Moseley, L. (1963). A rational design for planning buildings based on the analysis and solution of circulation problems. *Architects' J. Inf. Library,* 11 September 1963, pp. 525-537.

Whewell, W. (1840). *The Philosophy of the Inductive Sciences,* Vol. 1, 1st ed. Parker, London.

Contents

5. Other Measures of Similarity and Dissimilarity

PART II CLUSTER ANALYSIS

6. The Form and Analysis of the Data

7. Hierarchic Cluster Methods

8. Non-Hierarchic (Overlapping) Stratified Cluster Methods

9. An Axiomatic Approach to Cluster Analysis

Part I

THE MEASUREMENT OF DISSIMILARITY

'Two added to one—if that could but be done',
It said, 'with one's fingers and thumbs!'
Recollecting with tears how, in earlier years,
It had taken no pains with its sums.

LEWIS CARROLL, *The Hunting of the Snark*

CHAPTER 1

Kinds of Similarity and Dissimilarity

1.1. TERMINOLOGY

At the outset of a discussion of the various concepts of similarity and dissimilarity, we establish a consistent terminology for the description of individuals, and classes of individuals, on which measures of similarity and dissimilarity are based. We need to distinguish the following.

(a) Descriptive terms applicable to individuals, or to parts of individuals; for example, *petals red, leaf length 2 cm.*

(b) Sets of mutually exclusive descriptive terms of this kind; for example, *PETAL COLOUR* to which *petals red, petals yellow* belong. The former we shall call *attribute states*; the latter, *attributes.* The states of any attribute are jointly exhaustive of whatever individuals are under consideration, except when the attribute in question is conditionally defined. An attribute is said to be *conditionally defined* when its states can only meaningfully describe individuals which show particular states of some other attribute or attributes. For example, the attribute *HAIR LENGTH* is conditionally defined on the state *hairy* of the attribute *HAIR PRESENCE* = {*hairy, hairless*}. The case where, owing to conditional definition, no state of an attribute can conceivably describe an individual should be distinguished from the case where, as a matter of fact on a particular occasion, no state describes it because of inadequate observation or some other contingency. Another distinction which is needed is between the following.

(c) Probability measures over an attribute; that is, over a set of attribute states which form an attribute.

(d) Sets of such probability measures. The former may be called *character states*; the latter, *characters.* This use of the terms 'character state' and 'character' runs counter to much previous usage by biologists. We have adopted it deliberately because the terms have usually been used by biologists to denote the kinds of description of populations of individual organisms which serve as a basis for the construction of classificatory systems, although there has been much dispute and confusion about what kinds of descriptions are appropriate. In this chapter we shall argue that it is precisely probability measures over sets of attribute states which form the basis for the measurement of dissimilarity between classes of individuals, and hence for the construction of classificatory systems in biological taxonomy.

There has been extensive and persistent confusion in the literature of numerical taxonomy between the concepts defined here. We shall maintain the terminology expounded above, with the exception that in the informal discussion we shall sometimes use the term 'property' for 'attribute state'. When an attribute is invariant within a class of individuals, confusion between the attribute and the associated character arises naturally. Thus we habitually speak of ravens being black; but the state *feathers black* of the attribute *FEATHER COLOUR* describes individual ravens, and not the class of individuals *Corvus corax*. If it is true that all ravens are black, then what describes *Corvus corax* is the character state which assigns probability 1 to *feathers black* and probability 0 to the union of all other states of the attribute *FEATHER COLOUR*. When attributes vary within classes of individuals then the distinction becomes obvious. Thus, *leaf length 2 cm* is an attribute state, whereas *leaf length normally distributed, mean 2 cm, variance 0.09* cm^2 is a character state.

1.2. KINDS OF SIMILARITY AND DISSIMILARITY

The first stage in the construction of a classificatory system is often the calculation of a measure of similarity or dissimilarity. All such measures have something in common; they are in some way derived from information about the objects to be classified, whether they be individuals or classes of individuals. The terms 'similarity' and 'dissimilarity' are ambiguous, for they can cover at least the following three distinct concepts.

(a) *A-similarity* (*Association*)
(b) *I-distinguishability*
(c) *D-dissimilarity*

A-similarity. This kind of similarity is properly applied either to individuals, or to classes of individuals each of which does not vary with respect to the attributes considered. It is similarity by virtue of shared attribute states, and increases as the number or proportion of shared attribute states increases. Sometimes we may restrict the attributes considered to those which are relevant, in the sense that they are shared by some, but not all, the objects under consideration. In other cases, for example in classifying bacteria using a fixed list of standard tests, this restriction may not be applicable. A-similarity measures may be converted into dissimilarity measures for any fixed list of attributes, but not otherwise.

Examples of such measures are the coefficients listed in Dagnelie (1960), Goodman and Kruskal (1954, 1959), and Sokal and Sneath (1963, Chap. 5) under the heading 'measures of association'; perhaps the most familiar of those listed by Sokal and Sneath are Jaccard's Coefficient and the Simple Matching Coefficient.

Two kinds of A-similarity measure can be distinguished. Some measures, such as Jaccard's coefficient, depend on a preliminary labelling of the states of each attribute. An example of such labelling is the identification of one state of each binary attribute as positive, denoting presence of some part or constituent, and the other as negative, denoting its absence. Other measures, such as the Simple Matching Coefficient, are independent of any such preliminary labelling of attribute states.

I-distinguishability is the extent to which classes of individuals can be distinguished. It may be thought of in terms of the probability of correct reassignment of an individual from one of the classes on a basis of information about its attribute states. A measure of I-distinguishability is generally a dissimilarity coefficient, but is often bounded above by its value in the case of complete discrimination. The variation distance between two probability measures (character states) is an example of I-distinguishability; it is bounded above by the value 2. The Pearson (1926) Coefficient of Racial Likeness for two normal populations with the same dispersion, and Mahalanobis' D^2 statistic (Mahalanobis, 1930; Rao, 1952) for normal populations with the same dispersion, are measures of I-distinguishability.

D-dissimilarity is a rather more subtle concept. If X is a class of individuals known to be one of A, B, then its identification as A or as B gives information about the character states which describe it, further to the information given by knowing simply that X is one of A, B, but not which one. The amount of information given in this way is not well defined since it may differ for the two outcomes of the identification. D-dissimilarity is a suitable typical value of the expected information gain for such identification.

Some of the mathematical relations between the three kinds of measure will be investigated later. All such coefficients can be regarded as functions from $P \times P$ (where P is the set of objects) to the real numbers. A *dissimilarity coefficient* is a function d from $P \times P$ to the non-negative real numbers such that

$$d(A, B) \geqslant 0 \qquad \text{for all } A, B \in P$$

$$d(A, A) = 0 \qquad \text{for all } A \in P$$

$$d(A, B) = d(B, A) \quad \text{for all } A, B \in P.$$

A similar characterization of a similarity coefficient can be given; see Harrison (1968), Lerman (1970).

Many of the processes of automatic classification and scaling can be regarded as methods whereby similarity or dissimilarity coefficients satisfying different constraints are converted into one another; compare Hartigan (1967). It will be seen, for example, in Chapter 9 that stratified hierarchic clustering can be regarded as the conversion of an arbitrary dissimilarity coefficient into a dissimilarity coefficient satisfying a stronger constraint. A second example is given by the techniques of non-metric multidimensional scaling outlined briefly in Chapter 14.6 which are methods for the transformation of an ordinal or numerical similarity or dissimilarity measure into a metric array of distances representable in a small number of euclidean dimensions.

1.3. SIMILARITY AND DISSIMILARITY BETWEEN INDIVIDUALS

Of the kinds of similarity and dissimilarity described, only A-similarity measures are appropriate for individuals; as mentioned earlier, an A-similarity measure can be converted into a dissimilarity measure only when a fixed list of attributes is used. In constructing measures of A-similarity between individuals we do not face the difficulties raised by variability and correlation which arise in the measurement of dissimilarity between classes of individuals, but we do face the problems raised by conditional definition amongst attributes. Furthermore we face certain problems which can be avoided in measuring dissimilarity between classes of individuals, notably the problems of coding continuously variable attributes into discrete states, and of scaling discrete-state attributes with ordered states along the real line. It is extremely difficult to see how such coding and scaling can be performed in a non-arbitrary manner.

One line of argument which yields a unique A-similarity measure is to stipulate that the dissimilarity between individuals a and b should be the same as the D-dissimilarity between a class A composed entirely of individuals identical to a and a class B composed entirely of individuals identical to b.

This will, as shown later, yield a coefficient related to the Simple Matching Coefficient. This argument would have the advantage that it yielded a very simple method for dealing with conditional definition amongst attributes. It would be dealt with by precisely the form of weighting used in measuring *D*-dissimilarity between classes for attributes which are invariant within the classes but show conditional definition. The line of argument has, unfortunately, a serious defect. For if applied to attributes whose states are ordered, or can be mapped onto the real line, it would result in the same contribution to dissimilarity from any attribute whose states differ in the two objects, regardless of the magnitude of the difference between the states. The failure of this line of argument points to a basic difficulty in constructing measures of *A*-similarity with respect to many attributes. This is the difficulty in finding ways of scaling attribute states so as to render the contributions to *A*-similarity from different attributes strictly comparable so that they may meaningfully be combined.

Sometimes more appropriate methods for scaling attribute states can be devised. For example, suppose that the individuals are described by a number of quantitative attributes. It might be reasonable to dissect the range of each attribute by estimating percentiles of the distribution for the entire set of individuals for each attribute. Dissimilarity between individuals would then be measured for each attribute by the number of percentiles difference between their values, and the dissimilarities for each attribute somehow combined to obtain an overall measure of dissimilarity with respect to all the attributes. It should be noted that when attributes have been scaled in this way *addition* of dissimilarities for each attribute to obtain an overall measure of dissimilarity may not be meaningful.

The choice of a measure of association between individuals may sometimes be partially determined by the nature of the classificatory problem. Thus in measuring similarity between sites in ecology given as data lists of the species present in each site it is reasonable to use one of the *A*-similarity measures which depend on identification of positive and negative states of binary attributes. In measuring similarity between individual organisms measures of similarity which are independent of any initial labelling of attribute states are needed, since the states of attributes describing the morphology of organisms cannot reasonably be partitioned into positive and negative states; much confusion has arisen from failure to appreciate this. When clustering methods which commute with monotone transformations of the data are used, the choice reduces to a choice between equivalence classes of mutually monotone *A*-similarity measures. The same is true if the cluster methods described in Chapter 12 which depend only on the complete or partial rank-ordering of the dissimilarities are used. The monotonicity relations between some of the various association measures listed in Sokal and Sneath (1963, Chap. 5) are investigated by Lerman (1970).

In many cases, however, it is likely that the construction of classificatory systems based on individuals *via* a dissimilarity measure is misguided, for the necessity to code or scale attributes in the calculation of an association measure renders the process arbitrary, except in those cases where the particular classificatory problem imposes sufficient special constraints to indicate how this should be done.

1.4. DISSIMILARITY BETWEEN CLASSES OF INDIVIDUALS

The situation which arises when we come to consider the measurement of dissimilarity between classes is apparently more complex. We have to consider not only the problems raised by conditional definition of attributes, but also the problems raised by the variability and correlation of attributes within classes.

In order that a measure of overall dissimilarity with respect to many attributes be suitable for taxonomic purposes, we require that it allow for classes which are already completely discriminated by some attribute to vary in the extent to which they are dissimilar. Men are doubtless completely discriminated from both contemporary mice and contemporary monkeys, but we still require for taxonomic purposes that mice should be considered to differ more from men than men do from monkeys. This eliminates measures of I-distinguishability as they stand. Nevertheless, the combination of measures of I-distinguishability for the character states exhibited by classes can, under certain circumstances, lead to an overall measure of dissimilarity appropriate for use in taxonomy. Such a combination is not in itself a measure of I-distinguishability, because a measure of I-distinguishability with respect to many attributes will be obtained by consideration of the joint distributions of each class for the set of attributes, rather than the marginal distributions. We shall require also that the introduction or elimination of irrelevant attributes, that is, attributes which do not vary over the entire system of classes, shall make no difference to the measure of dissimilarity. Many of the coefficients of association do not have this property. The Simple Matching Coefficient can be converted into a coefficient with this property as follows. Let S be the Simple Matching Coefficient, and let n be the number of binary attributes, constant in each class, used in its calculation; then $n(1 - S)$ is a metric dissimilarity coefficient with the property. As will be shown, this dissimilarity measure is the special case of a general measure of D-dissimilarity.

In the next two chapters we show that it is possible to construct an information-theoretic measure of *D*-dissimilarity which satisfies the above conditions and which can handle the problems raised by combining continuously variable and discrete attributes, and can deal with variability and conditional definition amongst the attributes. Several authors have emphasized the importance of finding a non-arbitrary method for combining contributions from different attributes in order to obtain an overall measure of dissimilarity; see, for example, Carmichael, Julius, and Martin (1965), Colless (1967), and Sokal and Sneath (1963, Chap. 5). Some authors have attempted to render the contributions of continuously variable attributes comparable with those of discrete-state attributes by dissecting the range of the former, so as to be able to use an association measure (Kendrick, 1964, 1965; Talkington, 1967). Others have attempted to arrange the states of the latter along the real line, so as to be able to use one of the various measures of taxonomic distance (Carmichael, Julius, and Martin, 1965, Rohlf and Sokal, 1965; Sokal and Sneath, 1963, Chap. 6); both of these approaches will be shown to be unnecessary when used in the construction of an overall measure of dissimilarity between classes of objects; their use in constructing measures of similarity or dissimilarity between individuals may often be valid. The information-theoretic measure of dissimilarity which we shall construct avoids such arbitrary devices.

CHAPTER 2

Information Theory

2.1. BACKGROUND

We now introduce the information theory used in constructing a mathematical form for *D*-dissimilarity. For those readers who are unfamiliar with information theory the books by Feinstein (1958) and Kullback (1959) will prove to be useful sources. This account is based on Sibson (1969), prefaced by some introductory remarks. Information theory in its modern form was initiated by Shannon (1948), although Fisher (1925) had introduced the term 'information' into statistics. Fisher's 'information' is not the same as the Shannon version, although there are links between them. Shannon's particular concern was with communication theory, and much of the development of information theory since then has been in this direction. Despite a paper by Lindley (1956) and Kullback's book (1959), statisticians have been comparatively slow to appreciate that an information-theory approach to problems can prove profitable, although there has recently been much important work done along these lines, especially by Rényi and other workers in eastern Europe.

In this chapter, we shall find it appropriate and convenient to work in terms of a general probabilistic framework; an excellent account of the fundamentals of modern probability theory may be found in Kingman and Taylor (1966). Thus, integrals will be integrals over general measure spaces, and probability densities will be Radon-Nikodým derivatives. Any reader who is unfamiliar with this branch of mathematics will, for the purposes of this book, lose little by considering two special cases, one being that of a finite measurable space, and the other, the real line or euclidean space of some higher dimension. This is because the vast majority of the attributes encountered in taxonomy are either multistate attributes or (possibly multivariate) numerical attributes. In the latter case the character states (probability measures) encountered in practice possess ordinary probability density functions. Very occasionally there may be a numerically scaled attribute for which the character states have one or more atoms; this may happen, for example, for an attribute like *INTERNODE LENGTH,* which may take the value 0 with non-zero probability. In the case of a multistate attribute integration reduces to summation over the states of the attribute, and in the case of a numerical attribute it becomes ordinary (strictly, Lebesgue) integration.

Sibson (1969) follows Rényi (1961) in dealing with information of all orders. In this treatment we confine ourselves to information of order 1, and omit mention of the order; order 1 information is what is generally called information by statisticians and communications engineers. Little use has as yet been made of order α information. We shall omit proofs, giving references instead.

2.2. INFORMATION GAIN

Suppose that $(X,\bar{\mathscr{X}}) = \mathbf{X}$ is a measurable space, and μ, ν are probability measures on $\mathbf{X}$, with $\nu \gg \mu$. The *information gain* $I(\mu|\nu)$ is defined to be

$$I(\mu|\nu) = \int_X \log_2 \frac{d\mu}{d\nu}(x)\,\mu(dx).$$

The integral either exists or diverges to $+\infty$, in the latter case the value of $I(\mu|\nu)$ is taken to be $+\infty$. If λ is a measure on $\mathbf{X}$, and $\lambda \gg \nu$, then, writing $p = d\mu/d\lambda$, $q = d\nu/d\lambda$, we have

$$I(\mu|\nu) = \int_X p(x) \log_2 [p(x)/q(x)]\,\lambda(dx).$$

In the two special cases mentioned above, this reduces to

$$I(\mu|\nu) = \sum_{x \in X} p(x) \log_2 [p(x)/q(x)]$$

and

$$I(\mu|\nu) = \int_{-\infty}^{+\infty} p(x) \log_2 [p(x)/q(x)]\,dx.$$

The condition $\mu \ll \nu$ is precisely the condition that $p(x)/q(x)$ should be everywhere defined.

Informally, we may interpret $I(\mu|\nu)$ as follows. $p(x),q(x)$ are the probability densities of μ, ν at x, or the *likelihoods of* μ, ν at x. $p(x)/q(x)$ is the *likelihood ratio,* and $\log_2\ [p(x)/q(x)]$ is the *information* for rejecting ν for μ given by the observation x. $I(\mu|\nu)$ is the average value of this with respect to μ; that is, it is the expected amount of information for rejecting ν for μ if μ is correct. Thus $I(\mu|\nu)$ may be regarded as a measure of the information gained if evidence allows ν to be replaced by μ. The above argument is not formal because in general $p(x)$, $q(x)$ are defined only as elements of $L_1(\mathbf{X}, \lambda)$, not as individual functions.

Another way of looking at information gain is by way of an axiomatic approach. Rényi, for example, has given simple axioms which in the finite case characterize $I(\mu|\nu)$ (Rényi, 1961). So if we want a measure of information gain satisfying these conditions, it is $I(\mu|\nu)$ which must be used.

$I(\mu|\nu)$ has the property that it is non-negative, and is zero if and only if $\mu = \nu$. It is, however, not symmetrical, and very often it is desirable to have a symmetrical measure of divergence which is meaningful in terms of order 1 information theory. The variation distance

$$\rho(\mu_1, \mu_2) = \int_X |p_1(x) - p_2(x)| \, \lambda(dx),$$

(which does not depend on λ) has no interpretation in order 1 information theory, although it was shown by Sibson (1969) to be interpretable in terms of order ∞ theory. The construction of information radius was originally carried out in this context, to provide a symmetric information-theoretic divergence measure. The symmetrized information gain J has often been used for this purpose. J is defined by

$$J(\mu_1, \mu_2) = \tfrac{1}{2}[I(\mu_1|\mu_2) + I(\mu_2|\mu_1)],$$

and in consequence is only defined for $\mu_1 \equiv \mu_2$; that is, for $\mu_1 \ll \mu_2$ and $\mu_2 \ll \mu_1$. Frequently we wish to compare arbitrary probability measures, for which purpose J is thus inappropriate. The information radius K can, however, compare any two probability measures on $\mathbf{X}$. Further investigation indicated that there is a much more natural context in which to consider information radius, namely that of prior probability measures in the sense of Bayesian statistics. To use this general framework here would introduce unnecessary mathematical complexity, and we shall adopt a middle course consistent with the general approach.

2.3. INFORMATION RADIUS

Suppose that $\mu_1, \ldots, \mu_n$ are probability measures on $\mathbf{X}$, and that $w_1, \ldots, w_n$ are weights; that is, $w_1, \ldots, w_n \geqslant 0$ and $\Sigma w_i > 0$. Let M denote the array

$$\begin{bmatrix} \mu_1, \ldots, \mu_n \\ w_1, \ldots, w_n \end{bmatrix}.$$

Define $K(M|\nu)$, the *information moment* of M about ν, to be

$$K(M|\nu) = \Sigma w_i \, I(\mu_i|\nu)/\Sigma w_i\,, \text{ where } \nu \gg \Sigma w_i \, \mu_i\,.$$

Clearly $K(M|\nu)$ is independent of the order of the μ's, and if $w_1, \ldots, w_n$ are replaced by $cw_1, \ldots, cw_n$ with $c > 0$, then $K(M|\nu)$ is unchanged, and if $w_i = 0$ then that and the associated μ_i may be omitted from the array, and finally if $\mu_i = \mu_j$ then $K(M|\nu)$ depends only on $w_i + w_j$ and not on the ratio $w_i : w_j$. In other words, $K(M|\nu)$ is dependent only on the probability distribution over $\mu_1, \ldots, \mu_n$ specified by $w'_1, \ldots, w'_n$ where $w'_i = w_i/\Sigma w_j$. In the general approach, M is taken to be a probability measure of a more general kind. If we interpret the w'_i as the probabilities of $\mu_1, \ldots, \mu_n$ being correct, then $K(M|\nu)$ is the average gain in information on rejecting ν in favour of a μ_i. We know what $\mu_1, \ldots, \mu_n$ are, so we can regard $\inf_\nu K(M|\nu)$ as the average gain in information on identifying which μ_i is correct, given initially that the probability that μ_i is correct is w'_i. We define the *information radius of M* to be $K(M) = \inf_\nu K(M|\nu)$. The following properties of information radius were given in Section 2 of Sibson (1969), to which the reader is referred for proofs.

Information moment satisfies the formula

$$K(M|\nu) = K(M|\tau M) + I(\tau M|\nu),$$

where $\tau M = \Sigma w_i \mu_i / \Sigma w_i$. It follows that $K(M|\nu)$ is minimal uniquely for $\nu = \tau M$ if it is finite for any ν, since $I(\mu|\nu)$ is positive, and is zero if and only if $\mu = \nu$. Information radius is in consequence given by the following formula.

$$K(M) = \int_X \sum_i \left\{ \frac{w_i \, p_i(x)}{\Sigma_j \, w_j} \log_2 \frac{p_i(x) \Sigma_j \, w_j}{\Sigma_j \, w_j \, p_j(x)} \right\} \lambda(dx).$$

$K(M)$ is zero if and only if all μ_i for which $w_i > 0$ are equal; in other words, when M is a singular probability measure. Otherwise it is positive. If we take $n = 2$, and fix the values of $w_1, w_2 > 0$, then

$$K \begin{bmatrix} \mu_1 & \mu_2 \\ w_1 & w_2 \end{bmatrix}$$

is a measure of divergence between μ_1 and μ_2, and, if $w_1 = w_2$, it is symmetric in μ_1 and μ_2. It is in any case topologically equivalent to the variation distance ρ. $K(M)$ satisfies the boundedness conditions

$$K(M) \leqslant \sum_i \left\{ \frac{w_i}{\Sigma_j \, w_j} \log_2 \frac{\Sigma_j \, w_j}{w_i} \right\} \leqslant \log_2 n$$

Equality is attained in the left-hand inequality if and only if all the μ_i for which $w_i > 0$ are mutually singular, and $K(M) = \log_2 n$ if and only if in addition all the w_i are equal.

2.4. PRODUCT INFORMATION RADIUS

Ordinary information radius is obtained by absolute minimization of $K(M|\nu)$ over ν. This may be relativized by requiring that ν should be some special kind of probability measure. Let $\mathbf{X}_1, \ldots, \mathbf{X}_m$ be measurable spaces, and let $\mathbf{X}$ be their product. If $\mu_1, \ldots, \mu_n$ are probability measures on $\mathbf{X}$, we define their *product information radius* $K^{\times}(M)$ to be inf $\{K(M|\nu) : \nu = \Pi\nu_j\}$ where each ν_j is a probability measure on $\mathbf{X}_j$.

The main interest of product information radius is that it satisfies an additivity theorem, namely that if $\mu_i = \Pi\mu_{ij}$, where each μ_{ij} is a probability measure on $\mathbf{X}_j$, then

$$K^{\times}(M) = \Sigma_j K(M_j)$$

where M_j is the array consisting of the μ_{ij}'s and the w_i's.

2.5. NORMAL INFORMATION RADIUS

Normal information radius is another kind of relative information radius; in this case $\mathbf{X}$ is m-dimensional euclidean space and ν is required to be a multivariate normal probability measure. We shall consider only probability measures which are absolutely continuous with respect to Lebesgue measure, and thus possess density functions. We consider only good probability measures: a probability measure is defined to be *good* if it has a mean vector (β), a positive-definite covariance matrix (Σ), and finite entropy

$$\mathscr{H}(\mu) = -\int_{-\infty}^{+\infty} \log_2 p(x)\mu(dx).$$

The theory of normal information radius is developed in detail in Sibson (1969) Section 3. For the present purposes only one result is of interest, namely that the normal information radius with equal weights on two probability measures, denoted by $N(\mu_1, \mu_2)$, is given by the following formula in the case where both μ_1 and μ_2 are normal distributions with means β_1, β_2 and covariance matrices Σ_1, Σ_2 respectively.

$$N(\mu_1, \mu_2) = \tfrac{1}{2} \log_2 \left\{ \frac{(\det\{\frac{1}{2}(\Sigma_1 + \Sigma_2) + \frac{1}{4}(\beta_1 - \beta_2)(\beta_1 - \beta_2)^{\mathrm{T}}\}}{(\det \Sigma_1)^{\frac{1}{2}} (\det \Sigma_2)^{\frac{1}{2}}} \right\}$$

$$= \tfrac{1}{2} \log_2 \left\{ \frac{(\det[\frac{1}{2}(\Sigma_1 + \Sigma_2)]}{(\det \Sigma_1)^{\frac{1}{2}} (\det \Sigma_2)^{\frac{1}{2}}} \right\}$$

$$+ \tfrac{1}{2} \log_2 \{1 + \tfrac{1}{4}(\beta_1 - \beta_2)^{\mathrm{T}} [\tfrac{1}{2}(\Sigma_1 + \Sigma_2)]^{-1} (\beta_1 - \beta_2)\}$$

If $\Sigma_1 = \Sigma_2 = \Sigma$, then this reduces to the following.

$$N(\mu_1, \mu_2) = \tfrac{1}{2} \log_2 \{1 + \tfrac{1}{4}(\beta_1 - \beta_2)^{\mathrm{T}} \Sigma^{-1} (\beta_1 - \beta_2)\}$$

$$= \tfrac{1}{2} \log_2 \{1 + \tfrac{1}{4} D^2\}$$

where D^2 is Mahalanobis' D^2 statistic. Thus in the equal covariance case the normal information radius between two multivariate normal distributions reduces essentially to D^2; it follows that N may be regarded as providing a generalization of D^2 appropriate to the unequal-covariance case.

CHAPTER 3

K-Dissimilarity

3.1. INFORMATION RADIUS AND D-DISSIMILARITY FOR ONE ATTRIBUTE

In Chapter 2 we showed how it was possible to interpret information gain and information radius in terms of the information change involved in such statistical processes as rejection of a hypothesis, or discrimination. Recall that $I(\mu|\nu)$ measures the information obtained on rejection of ν in favour of μ, and $K(M|\nu)$ the average information gained on rejection of ν in favour of one of $\mu_1, \ldots, \mu_n$ with weights $w_1, \ldots, w_n$. $K(M)$, the infimum of $K(M|\nu)$, has been shown to be obtained as $K(M|\tau M)$; that is, if ν is chosen so as to incorporate as much as possible of what is known about which probability measure $\mu_1, \ldots, \mu_n$ should be employed, then $K(M)$ measures the remaining deficit in information. But this is, for the special case of a set of probability measures with weights, just what was described in Chapter 1 as D-dissimilarity. Thus $K(M)$ measures the D-dissimilarity of $\mu_1, \ldots, \mu_n$ with weights $w_1, \ldots, w_n$, and in particular $K(\mu_1, \mu_2)$ measures the D-dissimilarity of μ_1 and μ_2 with equal weights and

$$K\begin{bmatrix} \mu_1 & \mu_2 \\ w_1 & w_2 \end{bmatrix}$$

their D-dissimilarity with weights w_1 and w_2. This discussion provides the link between the formal mathematical reasoning in Chapter 2 and the conceptual discussion in Chapter 1, and shows that K may be regarded as a model for D-dissimilarity within a single character. The weights represent the prior probabilities of the two outcomes of the rejection of ν in favour of one of μ_1, μ_2.

3.2. THE CONSTRUCTION OF K-DISSIMILARITY

The next step is to construct the corresponding model for D-dissimilarity based on many attributes. Suppose first that no characters which arise from conditionally defined attributes are included. Corresponding to each OTU there will be a system of character states, one for each character. Two such systems may be denoted by $\mu_{11}, \ldots, \mu_{1n}; \mu_{21}, \ldots, \mu_{2n}$. The information for rejecting $\nu_1, \ldots, \nu_n$ together in favour of $\mu_{i1}, \ldots, \mu_{in}$ given by an observation $(x_1, \ldots, x_n)$ is

$$\log_2\{\Pi_j[d\mu_{ij}/d\nu_j](x_j)\}$$

which is equal to

$$\Sigma_j \log_2\{[d\mu_{ij}/d\nu_j](x_j)\},$$

both Σ_j and Π_j being taken over the range $j = 1, \ldots, n$. The expectation value of this on hypothesis $\mu_{i1}, \ldots, \mu_{in}$ is $\Sigma_j I(\mu_{ij}|\nu_j)$, and the weighted average value of this for $i = 1, 2$ with weights w_1, w_2 is

$$\Sigma_j K\begin{bmatrix} \mu_{1j} & \mu_{2j} & |\nu_j \\ w_1 & w_2 & \end{bmatrix}.$$

On minimization this yields

$$\Sigma_j K\begin{bmatrix} \mu_{1j} & \mu_{2j} \\ w_1 & w_2 \end{bmatrix}.$$

In the majority of situations there will be no need to introduce unequal object weighting, and we shall take $w_1 = w_2$. Thus, the combination of dissimilarities from attributes which are not conditionally defined is carried out simply by adding the values of K. The resultant sum is called the K-dissimilarity. Reference to Chapter 2.4 shows that the K-dissimilarity is the product information radius of the marginal distributions over the individual attributes. This illustrates how the use of K-dissimilarity as a model for D-dissimilarity, by working only on marginal distributions, makes a formal independence assumption between different characters. The consequences of this will be explored in more detail when we discuss correlation and redundancy of attributes in the next chapter.

3.3. CONDITIONAL DEFINITION

Now suppose that the list of characters contains some which arise from conditionally defined attributes. It is necessary that if an attribute which is conditionally defined is included, then any attributes on which it is conditional should also be included. Thus, if *PETAL COLOUR* is included, and if some individuals have flowers without petals, then the attribute *PETAL PRESENCE* must also be included. If all plants have petaloid flowers, then the attribute *PETAL PRESENCE* is constant throughout, and can be omitted, since its inclusion will make no difference even though *PETAL COLOUR* would be conditional on it if varied.

Suppose first that there are just two attributes to be considered, and that the second is conditional upon the first taking some state in a subset $S \subseteq X_1$ of the set of all states of the first attribute. Then two kinds of observation are possible: (x_1, x_2) if $x_1 \in S$, and (x_1) if $x_1 \notin S$. If we take the probability measures on X_1 to be $\mu_{11}, \mu_{21}, \nu_1$ and those on X_2 (conditional upon the second attribute being defined) to be $\mu_{12}, \mu_{22}, \nu_2$ in the notation employed earlier, then the information for discrimination is

$$\log_2\{[d\mu_{i1}/d\nu_1](x_1)\,[d\mu_{i2}/d\nu_2](x_2)\} \text{ if } x_1 \in S$$

and

$$\log_2\{[d\mu_{i1}/d\nu_1](x_1)\} \text{ if } x_1 \notin S,$$

and the expectation is

$$I(\mu_{i1}|\nu_1) + \mu_{i1}(S)\,I(\mu_{i2}|\nu_2)\,.$$

The average value of this for $i = 1, 2$ with weights w_1, w_2 is

$$K\begin{bmatrix} \mu_{11} & \mu_{21} \\ w_1 & w_2 \end{bmatrix}|\nu_2 + \frac{w_1\,\mu_{11}(S) + w_2\,\mu_{21}(S)}{w_1 + w_2}\,.\,K\begin{bmatrix} \mu_{12} & \mu_{22} \\ w_1\,\mu_{11}(S) & w_2\,\mu_{21}(S) \end{bmatrix}|\nu_2,$$

and on minimization this gives

$$K\begin{bmatrix} \mu_{11} & \mu_{21} \\ w_1 & w_2 \end{bmatrix} + \frac{w_1\,\mu_{11}(S) + w_2\,\mu_{21}(S)}{w_1 + w_2}\,.\,K\begin{bmatrix} \mu_{12} & \mu_{22} \\ w_1\,\mu_{11}(S) & w_2\,\mu_{21}(S) \end{bmatrix}.$$

Again, the weights will usually be taken to be equal, and the reduced form is then

$$K(\mu_{11}, \mu_{21}) + \tfrac{1}{2}[\mu_{11}(S) + \mu_{21}(S)]\,K\begin{bmatrix} \mu_{12} & \mu_{22} \\ \mu_{11}(S) & \mu_{21}(S) \end{bmatrix}.$$

It is important to observe that in the case of conditional definition the term contributed by the conditionally defined attribute consists of two factors, and even in the case $w_1 = w_2$ the information radius will have unequal weights in it, unless it happens fortuitously that $\mu_{11}(S) = \mu_{21}(S)$. In the general case these are $w_i\mu_{i1}(S) (i = 1, 2)$ where the w_i are the overall weights. The two factors of the term contributed by a conditionally defined attribute are called respectively the *occurrence factor* and the *information factor.* It is easy to see that this term exhibits appropriate behaviour in various boundary cases. Suppose that $w_1 = w_2$. If $\mu_{11}(S) = \mu_{21}(S) = 1$, then the second attribute is defined except for a set of measure zero under μ_{11}, μ_{12} in the first attribute, and hence, at least for this pair of OTU's, is effectively not conditional upon it. In this case the term contributed by the attribute which was treated as conditionally defined reduces to a term of the type obtained in the case of non-conditional definition. If just one of $\mu_{11}(S)$, $\mu_{21}(S)$ is zero, then the information factor has unequal weighting, and one of the weights is zero. Thus the information factor vanishes and so in consequence does the entire term. This is to be expected, because if for one of the two OTU's under consideration the second attribute is never available for comparison, that attribute cannot be the basis for any non-zero contribution to the overall dissimilarity between those two OTU's. Note that this is a point which would be very difficult to discuss without a precise terminology; if confusion between 'attribute' and 'attribute state' is allowed to creep in here, confusion between the above situation and the discussion of presence/absence attributes is almost inevitable. If both $\mu_{11}(S)$ and $\mu_{21}(S)$ are zero, then the information factor is undefined. The occurrence factor is zero, and because information radius is uniformly bounded the whole term can be taken as zero; again, this is what is wanted.

It is easy to see that any pattern of conditional definition may be analysed simply by observing what the function giving the information-for-discrimination is for each type of observation and over what regions integration of each of these functions is to be carried out. The result is a sum of terms, one for each attribute, and those which arise from conditionally defined attributes have two factors, an occurrence factor and an information factor, as above. The individual terms are called *K-terms*; their sum, the overall measure of dissimilarity whose construction is the main object of this part, we have already called *K-dissimilarity*. In general patterns of conditionality the weights introduced by the conditionality are more complicated, but still depend only on the character states over those attributes on which the attribute being considered is conditional. The general term is

$$\tfrac{1}{2}[m_{1j} + m_{2j}]K\begin{bmatrix}\mu_{1j} & \mu_{2j}\\ m_{1j} & m_{2j}\end{bmatrix}$$

where μ_{1j}, μ_{2j} are the character states in the j'th character, and m_{1j}, m_{2j} are the products

$$\Pi \, \{\mu_{li}(S_{ji}) : j\text{'th attribute conditional on states } S_{ji} \text{ of } i\text{'th}\}, \text{ where } l = 1, 2.$$

The actual construction of *K*-dissimilarity is now complete, and it can be seen to satisfy the various criteria for adequacy of a general taxonomic measure of dissimilarity.

CHAPTER 4

Correlation and Weighting of Attributes

4.1. INTRODUCTION

Character correlation, character weighting, and the relations between them have long been central issues in the discussion of taxonomic methods by biologists (Cain, 1959b). Many taxonomists have objected to methods of automatic classification on the grounds that they ignore the correlations of attributes and that by giving equal weight to all attributes they fail to take account of the information on which all successful taxonomy depends. We shall show that both 'correlation' and 'weighting' are ambiguous concepts. Certain of the kinds of weighting which taxonomists use intuitively are in fact incorporated in the calculation of K-dissimilarity, and there is a precise sense in which the application of cluster analysis to such a dissimilarity coefficient can be considered as a method for the analysis of character correlations. However, certain of the other kinds of weighting and correlation which taxonomists have discussed will be shown to be relevant to the selection of attributes, rather than to the calculation of dissimilarity and analysis of dissimilarity coefficients once attributes have been selected.

4.2. WEIGHTING OF ATTRIBUTES

One of the axioms of the approach to taxonomy adopted by the pioneers in the use of numerical methods was that '*a priori* every character is of equal weight in creating natural taxa' (Sokal and Sneath, 1963, p. 50). Much of the resultant dispute has stemmed from failure to comprehend the alleged distinction between *a priori* weighting and *a posteriori* weighting. This is understandable since it is hard to imagine that anyone could rationally use truly *a priori* weighting; that is, differential weighting of attributes based on no empirical information. In practice the dispute has concerned the question of what forms of *a posteriori* weighting should be used in taxonomy.

Some of the many disputed grounds for differential weighting which have been suggested are listed.

(a) Should attributes be differentially weighted according to their relative complexities?

(b) Should attributes which are highly variable within OTU's be accorded less weight than attributes which are less variable within OTU's?

(c) Should more weight be given to attributes which are good discriminators between OTU's than to attributes which are poor discriminators?

(d) Should sharing of rare attribute states be considered more important than the sharing of widespread attribute states?

(e) Should attributes be differentially weighted according to their diagnostic importance?

(f) Should attributes be differentially weighted according to their biological or functional importance?

(g) Should attributes be differentially weighted according to their known or supposed importance as indicators of evolutionary relationships?

(h) Should redundant or correlated attributes be accorded less weight than those which are independent?

We shall discuss these grounds for differential weighting in order. It will be shown that the model for *D*-dissimilarity given in the previous chapter incorporates *a posteriori* attribute weighting by criteria (a)–(c). We shall argue that differential weighting of attributes by criteria (d)–(g) is misguided. Criterion (h) turns out to involve several distinct issues related to the selection rather than the weighting of attributes.

But before the weighting of attributes is discussed it is necessary to clarify the conditions under which the contributions of different attributes to dissimilarity are strictly comparable; only when the contributions are strictly comparable can they meaningfully be combined to yield an overall measure of dissimilarity between OTU's.

Failure to understand the nature of comparability between attributes stems largely from the confusion between attribute states and character states which was discussed in Chapter 1. When only attributes constant within OTU's are being considered it is an easy step from confusing attribute states with character states to attempting to measure dissimilarity between attribute states rather than between character states. Likewise, confusion between character states and parameters of character states leads to attempts to measure dissimilarity between parameters of character states rather than between character states themselves. For example, the difference between means is often employed in numerical taxonomy as a measure of difference between distributions on the line. Measures of this kind, which are generally called taxonomic distances, are discussed in detail in the next chapter, but it is worth pointing out here that one reason for rejecting them as they stand is that they are dimensional and incorporate in this sense part of the structure of the attribute from which they are derived. Obviously there can be no comparability between such differently dimensioned quantities as length and weight. Likewise there can be no comparability between quantities having the same dimensions when they do not refer to the same attribute. Thus *PETAL LENGTH* and *CALYX LENGTH* cannot be comparable in this way even though they are both length measurements. Much of the discussion about character weighting breaks down at this point, failing as it does to appreciate that under these circumstances there is no basis for comparison at all.

When character states and attribute states are clearly distinguished, the situation alters radically. The measure of dissimilarity which we suggest is information-theoretic, and its units are units of information (*bits*). Thus its values are always comparable and there is a basis on which to begin the discussion of character comparability and weighting. Such modifications as rescaling attributes do not affect the dissimilarity values so obtained.

K-dissimilarity consists of a sum of terms, one for each attribute. These we have called *K-terms*; each such term consists of two factors, the *information factor*, and the *occurrence factor.* In constructing *K*-dissimilarity we have adopted *a priori* equal weighting for the different information-for-discrimination terms arising from each observation. Because of the limited availability of observations in cases where conditional definition is involved, integration to obtain expectation values leads to the introduction of weighting factors (the occurrence factor and the possibly unequal object weights in the information factor). The variation in the information factors of the *K*-terms makes appropriate allowance for the different extents to which attributes vary within OTU's. For example, the information factor for two disjoint probability measures (that is, mutually singular measures) over an attribute which is not conditionally defined takes the value 1; if the probability measures overlap, its value drops sharply. Thus dependence of attributes' contributions to dissimilarity on their discriminant power is basic to *K*-dissimilarity. It is the discriminant power of attributes which is generally referred to when orthodox taxonomists write of 'good' and 'bad' attributes. It seems that conditional definition is closely related to what taxonomists have in mind when writing of 'complexity' of attributes. An attribute such as {*flowers present, flowers absent*} is considered complex because many other attributes are conditional on its taking the state *flowers present.* The way in which *K*-dissimilarity operates emphasizes the importance of investigating the pattern of variation of each attribute within each OTU, and removes the dependence on invariant attributes which is produced by the use of measures of association in numerical taxonomy.

Differential weighting of matches between OTU's in attribute states according to the relative frequency of the state in question amongst the OTU's studied has been suggested by Goodall (1964, 1966), Rogers and Tanimoto (1960), and Smirnov (1966, 1968). This differential weighting is proposed for the construction of measures of *A*-similarity between OTU's and therefore involves only attributes which do not vary within the OTU's studied. The argument given for such weighting is that the weight attached to a match should be related to the probability of the match occurring given some appropriate null hypothesis. The null hypothesis adopted by Goodall, and implicitly adopted by Smirnov, is that the states of each attribute are distributed, with probabilities equal to their observed relative frequencies, at random amongst the OTU's. In the extreme case where an attribute is present in all the OTU's studied it receives zero weight. It is reasonable that a measure of similarity or dissimilarity between OTU's should be independent of the introduction of irrelevant attributes; *K*-dissimilarity has this property Unfortunately the resultant measures of *A*-similarity have the very curious property that the measure of similarity between an OTU and itself varies from OTU to OTU. The counterintuitive results of such weighting suggest that it may be ill-conceived. It seems possible that the idea that rare attribute states should be given greater weight may have arisen from the observation that rare attribute states are more likely to be good diagnostic criteria than are common attribute states.

The idea that some attributes are more important than are others is rather vague. Sometimes what is meant is that the states of some attributes are better diagnostic criteria than are the states of others. This point is well discussed in Sokal and Sneath (1963, p. 267). To some extent this property corresponds to the average value of the *K*-terms arising from the corresponding character being large, and in this sense attributes which provide good diagnostic criteria tend to contribute more to dissimilarity than others; but there seems to be no justification for any further weighting in this direction, and under certain circumstances the possibility of circularity and self-reinforcing argument may arise if attempts to introduce such a weighting are made.

The idea that functional importance should be used to weight attributes is of long standing in biological taxonomy. Cuvier was one of the last great exponents of functional weighting (see Cain, 1959b, for a lucid and detailed discussion). With the advent of evolutionary theory the opposite view has sometimes been canvassed: that functionally important (biological or adaptive) attributes are more likely to be modified relatively rapidly by natural selection than are functionally unimportant (fortuitous) attributes, and hence should be accorded less weight. This point is fully discussed in Davis and Heywood (1963, pp. 121-126). It is hard to see how functional importance of attributes could be objectively measured. Indeed it is doubtful whether it really makes sense to talk of the functional importance of individual attributes; for it is parts or complexes of parts of organisms which perform functions, not individual attributes.

Differential weighting of attributes according to their importance as indicators of evolutionary affinity–*phylogenetic* weighting–has been suggested by many authors. A decision about phylogenetic weighting depends on fundamental views about the aims of biological taxonomy. This issue is discussed in detail in Chapter 13.3. A practical objection is that it is difficult to find objective methods for phylogenetic weighting which are independent of the process of classification itself. Very often the attributes presumed to be of high phylogenetic weight are precisely those which are diagnostic for groups of organisms which are thought to be strictly monophyletic, that is, to comprise all the available descendants of a single common ancestor. There is an obvious danger of circularity here, for the groups of organisms presumed to be monophyletic are often precisely the groups recognized as taxa on the basis of overall resemblance with respect to many attributes.

4.3. CORRELATION AND REDUNDANCY

There are at least five distinct senses in which taxonomists have written of 'correlation' of attributes and characters. We shall attempt to disentangle these senses, and to clarify both the relations between them and their relevance to the measurement of dissimilarity between OTU's.

(a) *Logical correlation of attributes.* This is the phenomenon of conditional definition. There is little doubt that some form of weighting must be introduced into any method for the measurement of similarity or dissimilarity in order to deal with conditional definition. Usually it is easy to decide whether or not an attribute is conditionally defined on some state of another attribute. In certain cases, however, the decision may be awkward. Suppose that in a group of organisms studied substance *X*, when present, is always produced by a biosynthetic pathway *P* in which substance *Y* is a precursor. Should the attribute {*X present, X absent*} be considered conditional on the attribute {*Y present Y absent*} taking the state *Y present,* despite the fact that *X* might be derived from other precursors by some other biosynthetic pathway? In practice such difficulties can usually be eliminated by careful choice of attributes to convey a given body of information. In this case it would be better to treat the attributes as independent and to add a further attribute {*X derived from Y, X not derived from Y*} should other precursors of *X* be discovered. Alternatively, the presence or absence of biosynthetic pathway *P* might be considered as an attribute, and further attributes conditional upon its presence be used to indicate its terminal stage.

(b) *Functional correlation of attributes.* This kind of correlation is that which attributes have when their states can describe parts of organisms which are jointly involved in the performance of some function. Thus the attributes *SIZE OF CANINE TEETH, SIZE OF MASTOID PROCESS, SIZE OF STERNAL PROCESS* were functionally correlated in the sabre-toothed tigers (*Smilodon*), in which the powerful sternomastoid muscles were used to wield the head as a whole in striking prey with the massive canine teeth. Functional correlations may have predictive power, since given a valid inference about function from some attribute states it may be possible to predict the states of other attributes. Many biologists have exaggerated the reliability of such predictions. It is important to be aware that the functional correlations of attributes will in general vary greatly between different populations of organisms adapted to different environments, and that the same part of an organism will usually be involved directly or indirectly in many different functions. Olson and Miller (1958) have shown in detail that the statistical correlations of quantitative attributes within homogeneous populations may often coincide with functional correlations of attributes which can be determined on independent grounds.

(c) *Statistical correlation of attributes within populations.* This is the only use of the term correlation by taxonomists which coincides with ordinary usage by statisticians. It is obvious, but does not seem to have been generally recognized by biologists, that estimation of the pairwise product-moment correlation coefficients between quantitative attributes assumed to be normally distributed is not generally adequate to explore in detail the statistical dependences of attributes within a population. Olson and Miller (1958) have shown that the correlations of attributes within populations may vary very widely between otherwise closely similar populations. Methods for investigation of statistical correlation amongst attributes are discussed by Rao (1952) and Kendall and Stuart (1968, Chaps. 41-44).

(d) *Taxonomic correlation of attributes.* This is the kind of correlation which taxonomists have in mind when they state, for example, that organisms can be classified consistently because there exists correlation between characters; see Briggs and Walters (1969, p. 11). This kind of correlation is quite unrelated to the statistical correlation of attributes. It may be described as follows. Suppose that for two attributes $j = 1, 2$ we estimate probability measures μ_{ij} on the states of each attribute (i.e. character states) for each of a set of OTU's, $i = 1, \ldots, n$. If the character states μ_{i1}, μ_{i2} tend to discriminate the OTU's in similar ways the two attributes are said to be taxonomically correlated. In general the taxonomic correlation of any two sets of attributes can be measured by comparing the K-dissimilarity coefficient on the set of OTU's derived from one set of attributes with that derived from the other set of attributes. Alternatively, their taxonomic correlation with respect to a particular method of cluster analysis might be measured by comparing the clusterings obtained from the two dissimilarity coefficients. Methods for measuring the concordance between dissimilarity coefficients and clusterings on a given set of OTU's are discussed in detail in Chapter 11.3.

(e) *Redundancy.* It is easy to give examples of lists of attributes which everyone would agree to be redundant for taxonomic purposes and hence in need of shortening or simplification. Thus if we are dealing with OTU's whose component plant specimens vary widely in size but not in shape and other attributes, a list of attributes including many lengths and breadths of parts would be redundant. It is, however, hard to conceive of any general procedure for the elimination of redundancy in selections of attributes. It is certainly true that redundant attributes which are both statistically correlated and concordant in a given set of OTU's may cease to be so when further OTU's are considered.

All these various kinds of 'correlation' amongst attributes pose important topics for investigation using methods of automatic classification. The importance of the investigation of concordance between different kinds of attribute, for example morphological and biochemical attributes, is discussed in detail in Chapter 14.4. The rôle of the study of concordance between attributes in the investigation of the stability of classifications is discussed in Chapter 14.3. Such investigations are a prerequisite for deciding how many attributes should be used in any taxonomic study. A further important field for investigation is the relation between functional correlation and statistical and taxonomic correlation. Olson and Miller (1958) showed that functionally correlated attributes often show high statistical correlation. There are grounds for supposing that, at least at low taxonomic levels, attributes within the same functional complex may be more taxonomically correlated than are attributes or sets of attributes from different functional complexes. This situation would arise if, as seems probable, attributes from the same functional complex are more likely to be subjected to the same selection pressures in a given environment than are attributes from different functional complexes. This phenomenon is sometimes called mosaic evolution.

Study of both statistical and taxonomic correlation may play a part in the selection of attributes for use in a numerical taxonomic study. When attributes are both highly correlated statistically and show high taxonomic correlation there may be reason to suspect redundancy. High statistical correlation amongst attributes not accompanied by high taxonomic correlation does not suggest redundancy but may suggest that selected attributes do not convey the available information as efficiently as possible. Thus if the seeds of plants in a given set of OTU's were of more or less constant shape, the attributes *SEED LENGTH* and *SEED WEIGHT* would be highly correlated, and would not discriminate the OTU's as well as would *SEED LENGTH, SEED DENSITY*. Similarly, the sizes of parts often show high statistical correlations, and better discrimination may often be achieved by using ratios of sizes as attributes. Some suggestions about the ways in which attributes should be selected in taxonomy are made in Chapter 15.3. But it should be emphasized that the selection and definition of appropriate attributes in taxonomy is a largely unexplored field which presents many problems outside the scope of this book. For example the substantial problems involved in finding adequate ways of describing the shapes of parts of organisms are not discussed here.

It is important to note that different ways of selecting attributes to convey a given body of observations may lead to alterations in the dissimilarity coefficient on a given set of OTU's. We have pointed out that the selection of attributes requires particular care when conditional definition is involved. The extent to which the relative magnitudes of dissimilarity values are affected in this way is a subject which merits much further investigation. We are indebted to Mr F. A. Bisby for discussions on this topic.

—CHAPTER 5

Other Measures of Similarity and Dissimilarity

5.1. INTRODUCTION

We have already argued that a particular concept of dissimilarity, *D*-dissimilarity, corresponds to the intuitive judgment of dissimilarity between populations of organisms by taxonomists. *K*-dissimilarity is a mathematical model for *D*-dissimilarity. It is certainly not the only possible model, for example, the order α measures of information gain for $\alpha \neq 1$ could be used to provide alternative models. However, the majority of measures of similarity and dissimilarity between populations which have been proposed for use in taxonomy are models for *D*-dissimilarity either under no circumstances or only under special circumstances. It is the purpose of this chapter to find out which of these numerous measures are, or are reasonably closely related to, *D*-dissimilarity. The criticisms which we offer are intended only as criticisms of the specialized use of the various measures in the automatic classification of populations of organisms, not as general criticisms. Thus we criticize the use of most association measures in taxonomy, whilst recognizing that their use in ecology is correct; similarly we criticize the use of an analogue of the product moment correlation coefficient whilst (obviously) implying no criticism of its use as a measure of linear dependence in a bivariate normal distribution. In this respect our treatment of measures of dissimilarity is more parochial than our treatment of cluster methods, for there we are able to show that certain proposed cluster methods have properties which militate not only against their use in biological taxonomy, but also against their usefulness in the majority of applications.

In our discussion of measures of similarity and dissimilarity we shall follow the order adopted in Sokal and Sneath (1963, Chap. 5) which should be consulted for fuller details of the various methods.

5.2. ASSOCIATION MEASURES

The obvious practical objection to the use of association measures in taxonomy is that they can be used only for discrete-state attributes which do not vary within populations. Quantitative attributes can be used only if it is possible to find some way of dissecting the range into intervals such that the range of variation of each population lies entirely within a single interval: this is likely both to be an arbitrary procedure, and rarely to be practicable. Attempts to avoid the apparent arbitrariness involved in dissection of the range of quantitative attributes have been made by Kendrick (1965) and Talkington (1967).

Many association measures for binary discrete-state attributes depend upon identification of the states of different attributes as positive or negative, or as indicating presence or absence of some part or constituent. As pointed out by Sokal and Sneath (1963, p. 127) it is unrealistic to assume that all states of binary attributes of organisms can meaningfully be identified in this way.

Amongst the association measures which do not depend upon such a classification of attribute states is the Simple Matching Coefficient; this coefficient is related to K-dissimilarity as follows. It we have a list of n two-state attributes constant within OTU's, then the Simple Matching Coefficient $S(A, B)$ between two OTU's A, B is the proportion of the total number of attributes for which the two OTU's are described by the same attribute state. Thus $n.S(A, B)$ is the number of attributes where all members of the OTU's are described by the same state, and $n - n.S(A, B) = n(1 - S(A, B))$ is the number of attributes where all members of the one OTU are described by one state, and all members of the other OTU by the other state. The probability measures which describe the OTU's are all of a type which concentrate all probability on one or the other of the two states of each attribute. Thus if members of OTU X are constantly described by state $P_{X,i}$, then the corresponding probability measure $\mu_{X,i}$ will attach probability 1 to $P_{X,i}$ and probability 0 to the other state of the i'th attribute. Now the K-dissimilarity between A and B is given by $\Sigma_i K(\mu_{A,i}, \mu_{B,i})$, and we know from the properties of information radius that this is simply a sum of terms each of which is either 1 or 0 as $P_{A,i} \neq P_{B,i}$ or $P_{A,i} = P_{B,i}$. Thus in this special case K-dissimilarity is the number of attributes in which there is disagreement, that is, $n(1 - S(A, B))$. The Simple Matching Coefficient is thus seen to have, in addition to simplicity, an indirect theoretical justification since it is related to a measure of dissimilarity appropriate to populations whose members are constantly described by binary discrete-state attributes. It must be emphasized, however, that when a measure of similarity or dissimilarity between individuals described by discrete-state attributes is required there are many cases where the Simple Matching Coefficient is inappropriate, and choice of an association coefficient is guided by quite different constraints. For example, in Salton (1968, Chap. 4) the selection of association coefficients suitable for measuring similarity between documents described by index terms is discussed; in Dagnelie (1960) the selection of association measures appropriate for measuring ecological similarity between sites described by species occurrence is discussed.

5.3. CORRELATION COEFFICIENTS

A coefficient similar in mathematical form to the product-moment correlation coefficient has been widely used in numerical taxonomy. It was apparently introduced by Sokal and Michener (1958). Its calculation involves averaging over the states of different quantitative attributes to produce an 'average attribute state' for each OTU. This is absurd, and Eades (1965) and Minkoff (1965) have pointed out other grounds on which the correlation coefficient is unsatisfactory.

5.4. TAXONOMIC DISTANCES

A wide variety of measures have been proposed under the name 'taxonomic distance'; see Bielecki (1962), Penrose (1954), and Sokal (1961). The measures involve the assumption that the attribute states of individuals within each OTU can be used to obtain a disposition of points representing the OTU's in a suitable euclidean space. Williams and Dale (1965) have suggested the use of a model involving non-euclidean spaces, but no-one has yet produced a non-euclidean measure of taxonomic distance. The distance between points representing each pair of OTU's is then used as a measure of pairwise dissimilarity. If discrete-state attributes are used they have to be converted into quantitative form; for example, by disposition along the real line. The notion is widespread that the simple geometrical model available for measures of Taxonomic Distance provides them with some justification, but this justification is specious.

In order to explore the properties of such measures let us consider first a single attribute measured on a numerical scale. This will rarely be constant within OTU's. For two OTU's we must first estimate the distributions which form the character states of the OTU's for the attribute in question. For example, we may decide on the family of normal distributions and then estimate their parameters. If these distributions are widely separated with small dispersion, the difference between means will provide a reasonable measure of dissimilarity between the OTU's with respect to the single attribute, measured in units the same as the scale for that attribute. Various methods of combining such measures to obtain a measure of overall similarity with respect to many attributes have been suggested. Czekanowski's (1932) *durschnittliche Differenz* and Cain and Harrison's (1958) Mean Character Difference take the average distance between means for each pair of OTU's. Sokal's (1961) unstandardized Taxonomic Distance takes the root mean square difference. Both methods of combination are invalid because they involve addition of incommensurables, either blatantly by adding quantities of different physical dimensions or more subtly by combining measurements describing different (that is, in organisms, non-homologous) parts.

Various authors, including Sokal (1961), have tried to avoid this difficulty by standardization so as to express the difference between means in dispersion units. The resultant quantities are then dimensionless and can be combined. This procedure is reasonable only when the distributions involved all have about the same dispersion, in which case the standardized Taxonomic Distance is equivalent to the one-dimensional form of Mahalanobis' Distance (the square root of Mahalanobis' D^2 statistic). This is rarely true of quantitative attributes in biological populations.

Let us examine in more detail the circumstances under which the mean separation is a reasonable measure of divergence between distributions. It seems unquestionable that the divergence between distributions should be zero if and only if they are the same. It follows that the mean separation is an acceptable divergence measure only for a one-parameter family of distributions each specified by the mean. The normal distributions of fixed variance form such a one-parameter family, each distribution being specified by its mean; if the common variance is σ^2, the standardized mean separation is $|\beta_1 - \beta_2|/\sigma$, where β_1, β_2 are the means.

This is the one-dimensional form of Mahalanobis' Distance (Huizinga, 1962; Mahalanobis, 1936; Rao, 1948), and is related to Pearson's (1926) Coefficient of Racial Likeness for a single character. Ali and Silvey (1966) showed that there is a wide class of divergence measures for normal distributions which are related to $|\beta_1 - \beta_2|/\sigma$ by self-homeomorphisms of $[0, \infty)$, that is, by continuous maps $[0, \infty) \to [0, \infty)$ which possess continuous inverses; the information radius K is one of these.

5.5. MEASURES OF *I*-DISTINGUISHABILITY

The one-dimensional form of Mahalanobis' Distance is a measure of *I*-distinguishability, and is, as shown above, related to *K*-dissimilarity for a single character. For multivariate normal distributions *K* is likewise related to Mahalanobis' Distance by a self-homeomorphism of $[0, \infty)$. Pearson's Coefficient of Racial Likeness cannot be generalized to the multivariate case (Fisher, 1936).

Mahalanobis' Distance may be generalized to the case of inconstant covariance in various ways (Chaddha and Marcus, 1968; Reyment, 1961). The generalization adopted will be determined by the nature of the assumptions made in handling pooled populations which are individually multivariate normal. If these are to be treated as being correctly described by the best available distribution of arbitrary type, then the appropriate generalization must be obtained *via* the relationship to *K*, although an explicit expression for this relationship is not available. If the pooled populations are to be treated as being themselves multivariate normal, then *normal information radius* (*N*) provides the required generalization. Details of this and of the explicit relationship between *N* and D^2 have been given in Chapter 2.

Whilst there is a close mathematical relationship between information radius and Mahalanobis' D^2 statistic, their uses in taxonomy are very different. Mahalanobis' D^2 statistic is a measure of *I*-distinguishability. When it is based upon many characters, each normally distributed within OTU's, it operates on the joint distributions. As pointed out in Chapter 1 measures of *I*-distinguishability are in general inappropriate in taxonomy since we require populations which are completely discriminated by some character still to be allowed to vary in the extent of their dissimilarity. It is only in the study of infraspecific variation that taxonomists are faced with populations which are not discriminated by any character; hence only at these low taxonomic levels is the D^2 statistic applicable. Even in these cases its use is generally precluded by the fact that character states are often not normally distributed, and when they are normally distributed very rarely have constant covariance. The case for using Mahalanobis' D^2 rather than *K*-dissimilarity in the cases where the former is applicable rests on the fact that the D^2 statistic, by operating on the joint distributions, takes account of the statistical correlations amongst the attributes. So also does the information radius on the joint distributions, which may, like D^2, be used as a measure of *I*-distinguishability. *K*-dissimilarity involves a formal assumption of independence amongst the attributes. Many authors have felt that, by taking account of covariance within populations, the D^2 statistic eliminates redundancy amongst the selected attributes. We have argued, however, in Chapter 4 that reduction of unnecessary redundancy is something which should be incorporated in the initial selection of attributes, and further that redundancy and statistical correlation of attributes are not directly related.

REFERENCES

Ali, S. M., and S. D. Silvey (1966). A general theory of coefficients of divergence of one distribution from another. *Jl R. statist. Soc. (Ser. B)*, **27**, 131-142.

Bielecki, T. (1962). Some possibilities for estimating inter-population relationships on the basis of continuous traits. *Current Anthrop.*, **3**, 3-8; discussion, 20-46.

Briggs, D., and S. M. Walters (1969). *Plant Variation and Evolution*, Weidenfeld and Nicolson, London.

Cain, A. J. (1959b). Function and taxonomic importance. In A. J. Cain (Ed.), *Function and Taxonomic Importance*. Systematics Association, Publ. No. 3, London. pp. 5-19.

Cain, A. J., and G. A. Harrison (1958). An analysis of the taxonomist's judgement of affinity. *Proc. zool. Soc. Lond.*, **131**, 85-98.

Carmichael, J. W., R. S. Julius, and P. M. D. Martin (1965). Relative similarities in one dimension. *Nature, Lond.*, **208**, 544-547.

Chaddha, R. L., and L. F. Marcus (1968). An empirical comparison of distance statistics for populations with unequal covariance matrices. *Biometrics*, **24**, 683-694.

Colless, D. H. (1967). An examination of certain concepts in phenetic taxonomy. *Syst. Zool.*, **16**, 6-27.

Czekanowski, J. (1932). Coefficient of racial likeness and *'durschnittliche Differenz'*. *Anthrop. Anz.*, **9**, 227-249.

Dagnelie, P. A. (1960). Contribution à l'étude des communautés végétales par l'analyse factorielle. *Bull. Serv. Carte phytogéogr. (Sér. B)*, **5**, 7-71; 93-195.

Davis, P. H., and V. H. Heywood (1963). *Principles of Angiosperm Taxonomy*, Oliver and Boyd, Edinburgh and London.

Eades, P. C. (1965). The inappropriateness of the correlation coefficient as a measure of taxonomic resemblance. *Syst. Zool.*, **14**, 98-100.

Feinstein, A. (1958). *Foundations of Information Theory*, McGraw-Hill, New York.

Fisher, R. A. (1925). Theory of statistical estimation. *Proc. Camb. phil. Soc.*, **22**, 700-725.

Fisher, R. A, (1936). The use of multiple measurements in taxonomic problems. *Ann. Eugen.*, **7**, 179-188.

Goodall, D. W. (1964). A probabilistic similarity index. *Nature, Lond.*, **203**, 1098.

Goodall, D. W. (1966). A new similarity index based on probability. *Biometrics*, **22**, 882-907.

Goodman, L. A., and W. H. Kruskal (1954). Measures of association for cross-classifications. *J. Am. statist. Ass.*, **49**, 732-764.

Goodman, L. A., and W. H. Kruskal (1959). Measures of association for cross-classifications II. Further discussion and references. *J. Am. statist. Ass.*, **54**, 123-163.

Harrison, I. (1968). Cluster analysis. *Metra,* **7**, 513-528.

Hartigan, J. A. (1967). Representation of similarity matrices by trees. *J. Am. statist. Ass.*, **62**, 1140-1158.

Huizinga, J. (1962). From *DD* to D^2 and back. The quantitative expression of resemblance. *Proc. K. ned. Akad. Wet. (Sect. C)*, **65**, 380-391.

Kendall, M. G., and A. Stuart (1968). *The Advanced Theory of Statistics*, Vol. 3, 2nd ed., Griffin, London.

Kendrick, W. B. (1964). Quantitative characters in computer taxonomy. In V. H. Heywood, and J. McNeill (Eds.), *Phenetic and Phylogenetic Classification,* Systematics Association, Publ. No. 6, London. pp. 105-114.

Kendrick, W. B. (1965). Complexity and dependence in computer taxonomy. *Taxon,* **14**, 141-153.

Kingman, J. F., and S. Taylor (1966). *Introduction to Measure and Probability*, Cambridge University Press.

Kullback, S. (1959). *Information Theory and Statistics.* Wiley, New York.

Lerman, I. C. (1970). *Les Bases de la Classification Automatique*, Gauthier-Villars, Paris.

Lindley, D. V. (1956). On a measure of the information provided by an experiment. *Ann. math. Statist,* **27**, 986-1005.

Mahalanobis, P. C. (1930). On tests and measures of group divergence. *J. Asiat. Soc. Beng.*, **26**, 541-588.

Mahalanobis, P. C. (1936). On the generalized distance in statistics. *Proc. nat. Inst. Sci. India,* **2**, 49-55.

Minkoff, E. C. (1965). The effects on classification of slight alterations in numerical technique. *Syst. Zool.*, **14**, 196-213.

Olson, E. C., and R. L. Miller (1958). *Morphological Integration.* Chicago University Press.

Pearson, K. (1926). On the coefficient of racial likeness. *Biometrika,* **18**, 105-117.

Penrose, L. S. (1954). Distance, size, and shape. *Ann. Eugen.*, **18**, 337-343.

Rao, C. R. (1948). The utilization of multiple measurements in problems of biological classification. *J. R. statist. Soc. (Ser. B)*, **10**, 159-193.

Rao, C. R. (1952). *Advanced Statistical Methods in Biometric Research*, Wiley, New York.

Rényi, A. (1961). On measures of entropy and information. In J. Neyman (Ed.), *Proceedings of the 4th Berkeley Symposium in Mathematical Statistics and Probability.* California University Press, Berkeley and Los Angeles. pp. 547-561.

Reyment, R. A. (1961). Observations on homogeneity of covariance matrices in palaeontologic biometry. *Biometrics,* **18**, 1-11.

Rogers, D. J., and T. T. Tanimoto (1960). A computer program for classifying plants. *Science, N.Y.*, **132**, 1115-1118.

Rohlf, F. J., and R. R. Sokal (1965). Coefficients of correlation and distance in numerical taxonomy. *Kans. Univ. Sci. Bull.*, **45**, 3-27.

Salton, G. (1968). *Automatic Information Organisation and Retrieval.* McGraw-Hill, New York.

Shannon, C. E. (1948). A mathematical theory of communication. *Bell Syst. tech. J.,* **27**, 379-423; 623-656.

Sibson, R. (1969). Information radius. *Z. Wahrsch'theorie & verw. Geb.,* **14**, 149-160.

Smirnov, E. S. (1966). On the expression of taxonomic affinity. *Zh. obshch. Biol.,* **27**, 191-195. (In Russian).

Smirnov, E. S. (1968). On exact methods in systematics. *Syst. Zool.,* **17**, 1-13.

Sokal, R. R. (1961). Distance as a measure of taxonomic similarity. *Syst. Zool.,* **10**, 70-79.

Sokal, R. R., and C. D. Michener (1958). A statistical method for evaluating systematic relationships. *Kans. Univ. Sci. Bull.,* **38**, 1409-1438.

Sokal, R. R., and P. H. A. Sneath (1963). *Principles of Numerical Taxonomy.* Freeman, San Francisco and London.

Talkington, L. (1967). A method of scaling for a mixed set of discrete and continuous variables. *Syst. Zool.,* **16**, 149-152.

Williams, W. T., and M. B. Dale (1965). Fundamental problems in numerical taxonomy. *Adv. bot. Res.,* 2, 35-68.

Part II

CLUSTER ANALYSIS

It next will be right
To describe each particular batch:
Distinguishing those that have feathers, and bite,
From those that have whiskers, and scratch.

LEWIS CARROLL, *The Hunting of the Snark*

CHAPTER 6

The Form and Analysis of the Data

6.1. GENERAL CONSIDERATIONS

This part of the book is concerned with the stage of the process of automatic classification which starts with a dissimilarity coefficient and derives from it a classification of some kind. Whilst the measurement of dissimilarity between populations with respect to many attributes may reasonably be regarded as being rather special to taxonomy—at least in the form in which it is discussed in Part I—cluster analysis has many applications outside biological taxonomy, and indeed outside the biological sciences altogether. Because of this, much more attention has been focussed on it than on the measurement of dissimilarity, and a great number of cluster methods have been devised. It is only recently that any serious attempt has been made to provide a theoretical background for cluster analysis, and in consequence it is hardly surprising to find that many widely-used methods are unsatisfactory as methods of data simplification when considered in terms of a general theory.

The process of deriving a dissimilarity coefficient carried through in Part I may be regarded as a process of throwing away information. The initially available information about all the character states for the selected attributes tells us far more about the OTU's than does the final measure of dissimilarity, and the reason for discarding such a large body of fact lies in the belief that, for the particular purpose of describing patterns of overall variation, the measure of dissimilarity provides just the requisite summary of the initially much larger amount of information. Such processes of discarding information are very common in data-handling. A set of scale readings is often reduced to a measure of its location, for example the mean, or to measures of its location and dispersion, for example mean and variance. This can always be done for any finite set of scale readings, but it will only be appropriate to do it, first if it provides the information which is wanted, and secondly if it is a reasonable process to carry out. If the readings cluster closely about two points of the scale, then the mean and variance are not generally useful statistics to calculate. So in circumstances where these statistics are calculated, an assumption is being made about the nature of the underlying probability distribution, namely that it is a roughly bell-shaped distribution which tails away reasonably quickly in either direction. These conditions would certainly be met if the underlying distribution were normal, but the mean and variance are often of value in circumstances where much less is known about the distribution than its precise form. In calculating a statistic it is also necessary to ensure that it transforms appropriately under transformations of the scale: for example, the geometric mean of a set of readings does not transform in a equivariant manner under change of zero-point (that is, under translation operations), whereas the arithmetic mean does.

Thus three considerations arise in the analysis of what may in general be called a data-simplification process. First, relevance of result; secondly, appropriateness to data; and thirdly, what may loosely be called invariance properties. K-dissimilarity is constructed as a step-by-step model for an intuitive process, and certainly provides a relevant result. We have seen in Part I that it is satisfactory from the point of view of invariance properties, and its construction makes it appropriate for all data in the form of lists of character states. Thus the process of measuring K-dissimilarity satisfies these general criteria for a data-simplification process; one of the tasks of this second part will be to discuss methods of cluster analysis in similar terms.

Basic to cluster analysis is the assumption that it is reasonable to seek clusters in the data, these being subsets of the set P of OTU's characterized by possession of the properties of coherence and isolation. Ideal data for cluster analysis would yield clusters so obvious that they could be picked out, at least in reasonably small-scale cases, without the need for complicated mathematical techniques and without making precise what is meant by 'cluster'. It might under these circumstances be possible to observe that the clusters themselves fell into well-marked clusters, and so on. Thus already there are two possible descriptive forms: either the objects can simply be split up into clusters to produce a simple covering or partition of the set P, or some kind of nested system of clusters, a compound clustering of the set P, can be devised. This might start, for instance, with highly coherent but ill-isolated clusters such as single-OTU clusters, and finish with the whole of

P. The term 'cluster method' is widely employed in both contexts, and indeed methods of the two kinds are closely related. Clusters obtained at some stage in the application of a method of the second kind can be viewed in isolation as the output of a method of the first kind, and sometimes clusters obtained by methods of the first kind can be fitted together into a stratified system which may be regarded as the output of a method of the second kind. Which kind of method is selected in any particular case will depend in part on the use to which the resultant system of clusters is to be put. When comparison between different sets of data is an object of the exercise, the extra flexibility of the stratified systems usually results in their being chosen as the appropriate mode of description, and the treatment given here is concerned mainly with methods of this kind.

Despite the assumption that there is a cluster structure underlying the data, it will only rarely happen that all the clusters are laid out plain to see. There may be some very well-marked groupings of indisputable merit. Other rather less well-marked groupings may make themselves apparent differently to different observers. Finally, there may be sections of the data in which no definite cluster-structure of any kind seems to exist. The next stage is clearly to find some objective definition of 'cluster', or at least to find out what is meant by 'a reasonably well-marked cluster'. Numerous definitions of this kind have been suggested in the literature. Some of them are unsuitable for the construction of stratified systems since they contain no built-in notion of the level at which the cluster is defined. The selection of an appropriate definition incorporates further assumptions about the form of structure which is believed to underlie the data, the way in which the data may fail to reflect this properly, and the use to which the resultant description is to be put.

The choice of a precise definition of 'cluster' and the conditions which are imposed on cluster methods in this treatment are designed to incorporate the idea that a grouping together of objects in terms of the measure of dissimilarity always reflects some aspect of the structure, but that there is a possibility of objects becoming separated 'accidentally'. This assumption is designed to correspond to taxonomic practice; thus the treatment given here is intended to be of particular application in biological classificatory problems, this being one of the principal fields of application of cluster methods. Nevertheless, it is hoped on two grounds that the treatment will be of wider interest: first, that the requirements about the way in which classifications should represent data which are here adopted in the context of biological taxonomy may be relevant to classification in other fields, secondly, that the model offered here will provide a pattern for the way to construct a model designed to incorporate different basic assumptions.

6.2. MODELS, METHODS, AND ALGORITHMS

The ideal logical strategy by which methods of data simplification may be developed is as follows. First, precise mathematical characterizations of the data and of the kind of representation wanted are set up, so that methods of simplification can be treated as transformations from a structure of one kind to a structure of another kind. Next, criteria of adequacy for such transformations are laid down. These include specifications of the operations on the data under which the representation should be invariant or covariant, specification of structure in the data which should be preserved in the representation, and optimality conditions of various kinds. Then it must be ascertained whether there exist methods which satisfy the requirements, whether the requirements determine a unique method, and what are the further mathematical properties of the method or methods. The proof of the existence (and uniqueness) of methods having the required properties may not be constructive or may specify an algorithm which is computationally unfeasible. The final stage, therefore, is the quest for efficient algorithms.

The development of clustering methods has not followed this sequence. It has indeed followed an inverse sequence. A variety of algorithms for clustering were proposed in the 1940's and 1950's. It has gradually been realized in the last few years that some of these, despite superficial differences, implement the same method, and that different methods differ very widely in their properties and results. Only very recently have attempts been made to construct any general theory within which the properties of such methods can be analysed so as to make clear the conditions under which their application is valid. The development of a general theory has been hindered by two widespread confusions.

One widespread confusion is between algorithms and the methods which they implement. For example, many authors have distinguished two main kinds of cluster method, *divisive* and *agglomerative*. Divisive methods are said to work by successive partitions of the set of objects, and agglomerative methods are said to work by successive pooling of subsets of the set of objects. Lance and Williams (1967) have suggested as a general theory of hierarchic clustering what is in fact a generalized agglomerative algorithm, for the distinction is correctly applied to algorithms rather than methods. It is easy to show that the single-link method, for example, can be implemented by a divisive algorithm, an agglomerative algorithm, or by an algorithm which belongs to neither category.

Failure to observe the distinction between algorithms and the methods which they implement has arisen partly from the use of clustering and scaling algorithms for purposes other than data simplification. Thus clustering and scaling algorithms have been used in attempts to allocate electrical components to boxes of given sizes in such ways that the connexions between boxes are as few as possible; they have been used in seeking good solutions to the travelling salesman and related problems; and they have been used to tackle the problem of allocating activities to sites in buildings so as to minimize the cost of movement of equipment and people between sites. For most such problems there exists in general no unique optimal solution and no algorithm which is guaranteed to find an optimal solution. In evaluating the applications of clustering and scaling algorithms to such problems the rule is that anything goes if it works well: a cavalier attitude to the mathematical properties of the transformations induced by algorithms is quite in order. In data simplification the aim is to represent the data in a way which will suggest fruitful hypotheses rather than to process the data in such a way as to get an answer to a problem which is already set up. Hence the distinction between algorithms and the methods which they implement becomes crucial, and the knowledge of the mathematical properties of methods is an essential part of the justification for the use of particular algorithms.

Another widespread confusion concerns the role of models in data simplification; see Jardine (1970). The term 'model' has been used in two quite different ways. One use covers the mathematical frameworks within which it is possible to analyse the properties of methods of data simplification. The other use covers descriptions of algorithms in terms of their application to some interpretation of the data. The latter may be called analogue models. Two kinds of analogue model have been widely used in cluster analysis; see Sneath (1969a). First, there are models which treat the objects as points or unit masses in euclidean space; see, for example, Gower (1967), Wishart (1969). Secondly, there are models which treat the objects as vertices of a graph and values of the dissimilarity coefficient less than or equal to some threshold as edges; see, for example, Estabrook (1966), Jardine and Sibson (1968a). Geometrical models can obviously be applied only if the data is metric. Despite the fact that the dissimilarity coefficients which arise in practice are often non-metric and not meaningfully convertible into metric form, the alleged desirability of geometrical models has led several authors to suggest that metric coefficients are to be preferred. Even when the data is naturally metric a plausible geometrical interpretation for an algorithm does not necessarily provide any justification for the method which it implements. Thus the various average-link and centroid algorithms which have simple geometrical interpretations will be shown to suffer from very serious defects. The graph-theoretic models are more generally applicable since any dissimilarity coefficient and any stratified clustering can be characterized by a sequence of graphs. We shall show that all cluster methods of a certain type which are independent of order isomorphisms on the data can be described by separate operations on the graphs for each value of a dissimilarity coefficient, and that this description is useful in suggesting efficient algorithms. But like the various geometrical models this kind of model provides no validation of the methods.

6.3. DISCUSSION OF CLUSTER METHODS

Methods leading to simple clusterings consisting of overlapping clusters have been developed recently. However, most available stratified cluster methods are of hierarchic type, in which the clusters at each level are disjoint. Numerous such cluster methods have been suggested, some of which are discussed in detail in Chapter 7. One of the earliest to appear in the literature was the single-link (nearest-neighbour) method, and more recently procedures based on some kind of weighted average process have become popular. The single-link method leads to OTU's connected by intermediate OTU's being clustered together, average-link processes attempt to avoid this. In Chapter 7 the defects of the average-link methods are pointed out, and such methods are shown to be of dubious computational value. In Chapter 8 some alternative methods are introduced, which, by generalizing to allow stratified systems of overlapping clusters, succeed in avoiding the defects of the average-link methods and also in recovering more information than the single-link method; this is achieved at the cost of greater complexity in the resultant classification. Chapters 9 and 10 are concerned with the setting up of a mathematical theory of stratified clustering and with the demonstration within the theory of the extent to which methods are uniquely specified by constraints placed on them. In Chapter 9 a very general model is set up, and it is shown that under reasonable constraints the cluster method is determined once the type of clustering wanted as output is selected. A further constraint is added in Chapter 10, and is shown to lead to a particularly simple and computationally feasible family of cluster methods. In Chapter 11 we discuss ways in which the results of cluster methods may be compared and evaluated. Finally, in Chapter 12, we discuss certain approaches to cluster analysis not captured within the theoretical framework established in Chapter 9 and their relations to methods which have been so captured.

CHAPTER 7

Hierarchic Cluster Methods

7.1 DC's AND DENDROGRAMS

In Chapters 9 and 10, cluster methods are treated from a general axiomatic point of view. To relate this treatment to actual methods, and to provide examples for discussion, we devote this chapter and the next to the construction of some of the mathematical objects we shall use and the construction of cluster methods in terms of them. We shall be concerned throughout with numerically stratified methods, and shall not bother to put this into the terminology in most cases. This chapter deals with familiar territory, namely the hierarchic methods. In Chapter 8 we go on to develop non-hierarchic methods, that is, methods where clusters at the same level may overlap.

The starting point of cluster methods will be taken to be a dissimilarity coefficient (DC) as defined in Chapter 1. The definition will be repeated here for reference, although we defer detailed discussion of DC's to the beginning of Chapter 9. A *dissimilarity coefficient* (DC) on a set P of OTU's is a real-valued function d on $P \times P$ satisfying the following conditions:

DC1	$d(A, B) \geqslant 0$	for all $A, B \in P$
DC2	$d(A, A) = 0$	for all $A \in P$
DC3	$d(A, B) = d(B, A)$	for all $A, B \in P$

Later it will be necessary to consider various ways of strengthening this definition, but it will be adequate as it stands for this chapter and the next.

We have discussed the construction of DC's in the context of dissimilarity between populations in Part I, but this part is not dependent on the use of any particular method of measuring dissimilarity; our approach applies equally well to any DC, the only restriction being that we do require numerical values rather than, say, ranking of dissimilarities. Often dissimilarities measured in some way may have special properties; for example they might form a *metric* rather than simply a DC, but we shall not find it necessary to assume this. We may informally think of DC values as 'distances',

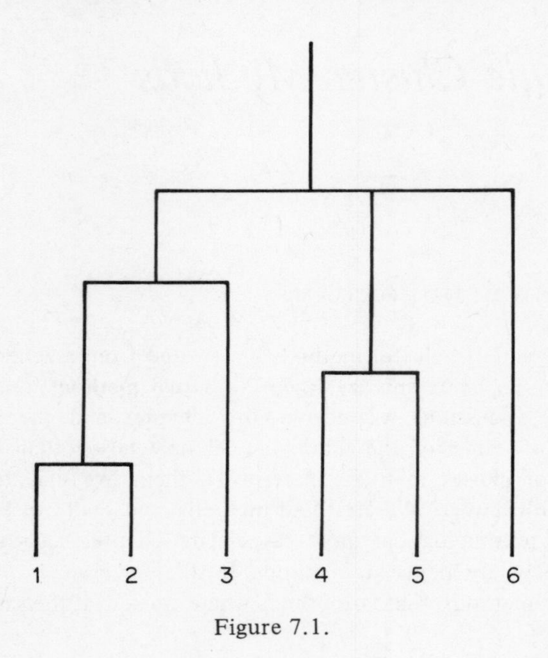

Figure 7.1.

but even if the DC is metric the pattern of distances may be far removed from anything which is realizable in euclidean space.

For hierarchic cluster methods, the endpoint of the process is a dendrogram, or tree diagram; we shall in fact expect a dendrogram to be a special kind of tree diagram in which numerical levels are associated with the branch points. What we usually draw as a tree diagram is a representation of the dendrogram and there are numerous different ways of drawing a tree diagram to describe the same dendrogram. Figures 7.1, 7.2, 7.3, all show the same dendrogram, but it is drawn in different ways: Figure 7.2 is in a different style from Figure 7.1, and in Figure 7.3 the order of the OTU's along the

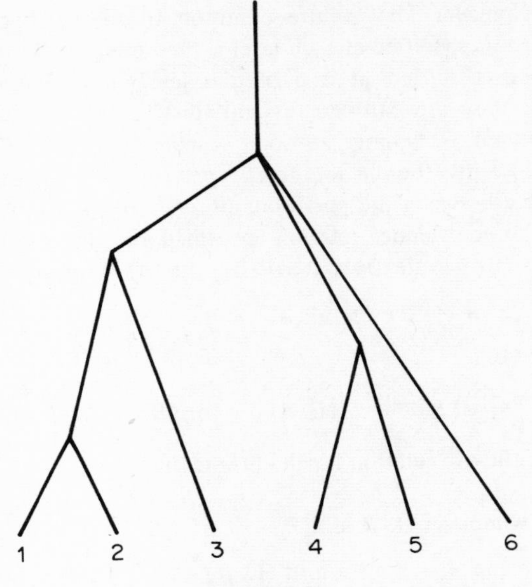

Figure 7.2.

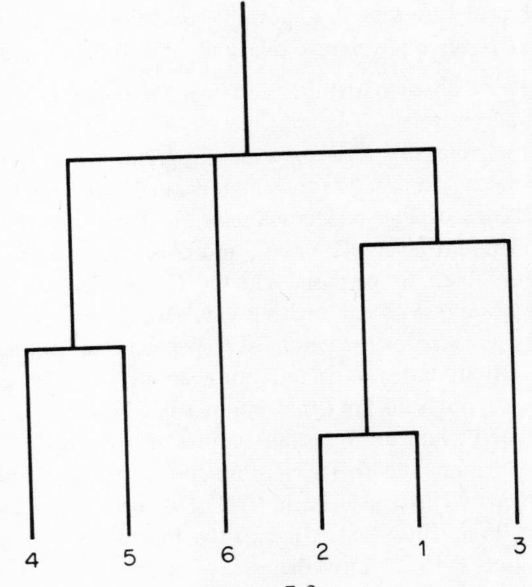

Figure 7.3.

base-line is changed. The feature common to all the diagrams is that the system of clusters specified at each level is the same.

The clusters specified at a particular level in a dendrogram have the property that they are pairwise disjoint—distinct clusters do not meet—and every element of P belongs to some cluster, possibly consisting of that element alone. Thus, the clusters form a *partition* of P. There is a natural 1-1 correspondence between the partitions of a set and the equivalence relations on it, where an equivalence relation is defined as follows. A *relation* on P is a subset r of $P \times P$. r is *reflexive* if it satisfies the following condition:

REF $\quad (A, A) \in r$ for all $A \in P$

and *symmetric* if

SYM $\quad (A, B) \in r \Rightarrow (B, A) \in r$ for all $A, B \in P$

A symmetric reflexive relation is called *transitive* if

T $\quad$ whenever $A, B, C \in P$

$$[(A, C) \in r \text{ and } (C, B) \in r] \Rightarrow (A, B) \in r$$

A symmetric reflexive transitive relation is called an *equivalence relation.* If r is an equivalence relation then the sets $C_A = \{B:(A, B) \in r\}$ form a partition of P. Conversely, if the sets $\{C_\alpha\}$ form a partition of P, then the relation $r = \bigcup_\alpha C_\alpha \times C_\alpha$ is an equivalence relation, and these transformations from equivalence relations to partitions and from partitions to equivalence relations are mutually inverse.

So in formalizing the idea of a dendrogram, we can make use of an equivalence relation to describe the clusters at each level, which appear as the equivalence classes of the associated relation. Next we must consider what happens for different levels. If $h \leqslant h'$, and C is a cluster at level h, then C is completely contained in just one cluster C' at level h': we say that the partitions are *nested.* At some high enough level, every element of P lies in a single cluster consisting of the whole of P. We can build these two restrictions into the definition, but this is not quite enough to define a dendrogram, because we have not said anything about what happens at levels where the system of clusters changes. We shall require this change to take place in a well-behaved manner, and this can most conveniently be done by requiring that the system of clusters at each level shall be the same as that at some slightly higher level. Thus we are led to the following formal definition of a dendrogram. Let E(P) denote the set of equivalence relations on P. A *dendrogram* is a function $c: [0, \infty) \rightarrow \mathrm{E}(P)$ satisfying the following conditions:

D1 $\quad 0 \leqslant h \leqslant h' \Rightarrow c(h) \subseteq c(h')$

D2 $\quad c(h) = P \times P$ for large enough h (we say:

$c(h)$ is *eventually* $P \times P$)

D3 $\quad$ Given h, there exists $\delta > 0$ such that

$c(h + \delta) = c(h)$

Note that we do not, at least for the time being, require that all elements of P should be distinct at level zero. Dendrograms with this property are said to be *definite,* and are those satisfying the following condition:

DD $\quad c(0) = \Delta P$, where $\Delta P = \{(A, A) : A \in P\}$

The input of a cluster method is a DC; the output is a dendrogram. Now that we have given formal definitions of both of these, it is possible to view a cluster method as some kind of transformation from DC's to dendrograms. We shall in fact want it to be a function, that is, we shall want just one dendrogram to be obtained from each DC. This remark may seem to be fatuous, in that cluster methods whether of the hierarchic numerically stratified type discussed here or of the more general type discussed in Chapter 8, do in fact produce just a single dendrogram (or a single one of whatever type of output they do produce). The point is that from given data the *same* dendrogram must always be produced; we shall see that this condition is by no means always satisfied. In the taxonomic context it is essential that we should have repeatability—this is discussed in detail in Chapter 9—and so we shall require that every hierarchic numerically stratified cluster method should be representable as a function from the set of DC's on P to the set of dendrograms on P.

7.2. THE ULTRAMETRIC INEQUALITY

The definition of a dendrogram as a function $c: [0, \infty) \to \mathrm{E}(P,)$ satisfying conditions D1-D3 certainly captures what is meant by a (numerically stratified) dendrogram, but in practice it is a somewhat unwieldy definition. Suppose that c is a dendrogram on P. Define

$$(Uc)(A, B) = \inf \{h : (A, B) \in c(h)\}$$

Uc is a DC, so $c \to Uc$ transforms dendrograms to DC's. If c is a dendrogram, and Uc is the associated DC, then

$$c(h) = \{(A, B) : (Uc)(A, B) \leqslant h\}$$

Thus U is actually 1-1 from the set of dendrograms on P to some subset of the set of DC's on P; condition DC3 guarantees 1-1-ness. It might be thought that if d is an arbitrary DC, and if T is defined by

$$(Td)(h) = \{(A, B) : d(A, B) \leqslant h\}$$

then Td would be a dendrogram, but this is not the case. $(Td)(h)$ is certainly a symmetric reflexive relation for all $h \geqslant 0$, but there is in general no reason why it should be an equivalence relation. Td is a dendrogram if and only if $(Td)(h)$ is an equivalence relation for all $h \geqslant 0$; that is, if and only if

$$(A, C) \in (Td)(h) \text{ and } (C, B) \in (Td)(h) \Rightarrow (A, B) \in (Td)(h) \text{ for all } h \geqslant 0.$$

This condition may be written as

$$d(A, C) \leqslant h \text{ and } d(C, B) \leqslant h \Rightarrow d(A, B) \leqslant h \text{ for all } h \geqslant 0$$

which in turn is the same as

DCU For all $A, B, C \in P$

$$d(A, B) \leqslant \max\{d(A, C), d(C, B)\}$$

This condition is known as the *ultrametric inequality,* and DC's satisfying it are called *ultrametric.* Note that ultrametric DC's are not quite the same as ultrametrics in the usual sense, because DC's need not necessarily satisfy

DCD $d(A, B) = 0 \Rightarrow A = B$

This is the *definiteness* condition. Definite ultrametric DC's are ultrametrics in the usual sense. The set of ultrametric DC's on P will be denoted by $\mathbf{U}(P)$, and the set of all DC's on P by $\mathbf{C}(P)$. So T is a 1-1 onto function from $\mathbf{U}(P)$ to the set of dendrograms on P, and U is its inverse. In terms of partitions, we may describe U by saying that $(Uc)(A, B)$ is the lowest level at which A, B lie in the same cluster for c.

Since T and U define a natural 1-1 correspondence, it is possible to identify the set of dendrograms on P with the set of ultrametric DC's on P, and for the purposes of the mathematical theory we shall do this, and will thus regard a cluster method of the hierarchic numerically stratified type (a *type A* method) as a function $D:\mathbf{C}(P) \rightarrow \mathbf{U}(P)$.* This means that we would expect to obtain as the output of a cluster method an ultrametric DC on the elements of P, rather than a list of the clusters at each level, it is the clusters which are actually wanted, so in practical terms the ultrametric DC is the wrong thing to obtain. However, it is a trivial task computationally to recover

---

* A similar characterization of a hierarchic dendrogram was given by Johnson (1967) and Hartigan (1967).

the clusters at each level from an ultrametric DC, so this aspect of the situation will cause us no concern. In Chapter 8 we shall see that problems arise in the non-hierarchic case, but fortunately these are soluble. The advantages to be gained from dealing with **U**(P) rather than the set of dendrograms on P are substantial, and arise from the fact that elements of **U**(P) are themselves DC's—we have $\mathbf{U}(P) \subset \mathbf{C}(P)$—and so the transformation D maps **C**(P) to a subset of itself. This is exploited extensively in Chapter 9. It is also very convenient conceptually to be able to consider a cluster method as a process of modifying a DC to a special kind of DC.

7.3. TYPE A CLUSTER METHODS

We now come to the description of some type A cluster methods. We shall find that, with one exception, they do not lend themselves readily to description in terms of a function $D:\mathbf{C}(P) \to \mathbf{U}(P)$, and we shall show why this is so. This remark does not contradict what has been said in previous paragraphs. It is certainly possible to regard each of the methods described as being given by such a function, and when later they are analysed mathematically we shall use this fact; but it is not necessarily convenient to write down an *explicit* function D in the general case.

The first type A method which we consider is the single-link method A_S: $\mathbf{C}(P) \to \mathbf{U}(P)$. This very simple method was apparently first introduced as a method of automatic classification by Florek and coworkers (1951a, 1951b) under the title 'dendritic method'. McQuitty (1957) and Sneath (1957) independently introduced slightly different versions of it. Its operation may conveniently be described in terms of graphs. Suppose d is a DC. Then $(Td)(h)$ defined by

$$(Td)(h) = \{(A, B) : d(A, B) \leqslant h\}$$

is a symmetric reflexive relation. It may be represented by a graph whose vertices are elements of P and whose edges link just those pairs of elements of P which lie in the relation $(Td)(h)$. The sets which we pick out to form the clusters at level h in the dendrogram obtained by the single-link method are simply the connected components of this graph. It is sometimes convenient to think rather of transforming the graph by adding edges until disjoint maximal complete subgraphs are obtained, and then taking the maximal complete subgraphs to be the clusters; this is, of course, an equivalent approach. Another way of thinking about the single-link method is to describe it by noting that $[TA_S(d)](h)$ is the smallest equivalence relation containing $(Td)(h)$. A description in which the function A_S is more explicitly given is as follows:

If d, d' are DC's, define $d \geqslant d'$ (*d dominates d'*) to mean

$$d(A, B) \geqslant d'(A, B) \text{ for all } A, B \in P$$

Then $A_S(d)$ is the largest element d' of $\mathbf{U}(P)$ such that $d' \leqslant d$. This form of description will be of great importance in Chapter 8 in the definition of non-hierarchic methods.

Another way of describing the operation of the single-link method is in terms of an agglomerative algorithm. Search the dissimilarity coefficient for the smallest value between distinct elements of P. Unite the pair A, B of elements of P separated by this smallest dissimilarity into a single group. Remove the two elements from the list and replace them with a single new element C whose distance from each of the remaining old elements X is the minimum of $d(A, X), d(B, X)$. Repeat this with $(P \backslash \{A, B\}) \cup \{C\}$ and the new DC, and continue until only one element is left. The groups formed by this process are the clusters in the single-link dendrogram, and the levels are the minimal DC values selected at each stage. It is easily seen that if it should happen that the smallest DC value is not unique, then there are two ways of dealing with this which lead to the same result. One is to select an arbitrary pair on which the smallest value occurs; the other is to apply the graphical process described above to the graph of $(Td)(h)$ where h is the minimal value taken by d on pairs of distinct elements, unite the vertex sets of components into groups, and calculate inter-group dissimilarities as minima of inter-element dissimilarities, and proceed as before. It is evident that either of these procedures will lead to the same result, because at each stage minimum distances are being considered. If equal DC values arise, we have to be a little careful about identifying groups with clusters in the dendrogram, because it will take several steps to unite a number of groups together if we adopt the first alternative. However, the difficulties are trivial.

This kind of process is known as an *agglomerative algorithm,* in contrast to a *divisive algorithm,* which operates by successive sub-division of P rather than by grouping elements of P together. Some authors have tried to make a distinction between agglomerative and divisive *methods* on this basis, but this, as pointed out in Chapter 6.2, is spurious, because a given method may very well be realized by numerous different algorithms, some agglomerative, some divisive, and some not really either. Failure to distinguish between a cluster method as such, and an algorithm for carrying it out, is very common. It is of course true that every well-defined algorithm defines a method, but the converse is not true, and indeed it can be quite difficult to find reasonable algorithms for many non-hierarchic methods.

Lance and Williams (1967) have developed the idea of an agglomerative algorithm in a general form, but, as pointed out by Sibson (1971) their approach has some deficiencies connected with what happens in the case of equal smallest dissimilarities. The Lance and Williams' general framework does not explicitly state how this situation is to be dealt with, and Sibson has shown that the arbitrary choice of a pair leads to failure of the algorithm to produce a well-defined cluster method in the majority of cases. Lance and Williams' formula for the recalculation of dissimilarities cannot immediately be extended to deal with situations more complicated than uniting groups two at a time. Nevertheless, their approach does suggest a useful form of algorithm, of which we shall give four special cases. The general form is as follows.

The smallest distance between distinct groups is found, and taken as the current level; all pairs between which this distance occurs are listed. The resultant graph, with groups as vertices and smallest distances corresponding to links, is divided into its connected components, and groups lying in the same connected component are united to form a smaller number of larger groups. New intergroup dissimilarities are calculated *in some way*, and the process is repeated. The starting point is the set of one-object groups, some of which may immediately be united at level 0; and the process finishes when the whole of P forms a single group. The element of generality comes in the choice of method of calculating new intergroup dissimilarities. Suppose that at some stage groups $i_1, \ldots, i_l \ldots, i_r$ coalesce to form i and $j_1, \ldots, j_m, \ldots, j_s$ coalesce to form j, and suppose that d is an intergroup dissimilarity coefficient. The four formulae for recalculating d which we shall consider are:

$$\text{S} \qquad d(i,j) = \min\, d(i_l, j_m)$$

$$\text{W} \qquad d(i,j) = |i|^{-1}\,|j|^{-1} \sum_l \sum_m |i_l|\,|j_m|\, d(i_l, j_m)$$

$$\text{U} \qquad d(i,j) = r^{-1}\, s^{-1} \sum_l \sum_m d(i_l, j_m)$$

$$\text{C} \qquad d(i,j) = \max\, d(i_l, j_m)$$

The resultant cluster methods A_S, A_W, A_U, A_C are *single-link, weighted-average-link, unweighted-average-link,* and *complete-link* respectively; the single-link method A_S is of course single-link as described earlier in this chapter, but the other methods are introduced here for the first time in this book. They are the four most commonly used type A cluster methods, although it is very probable that many users of the last three do not succeed in implementing them properly if equal smallest dissimilarities have to be considered.

7.4. PROPERTIES OF A_S, A_W, A_U, A_C

The single-link method is conceptually and computationally very simple, and, as will be shown in later chapters, it has a large number of satisfactory mathematical properties. It does, however, suffer from one substantial defect, commonly known as *chaining.* The meaning of this is easily seen by considering the effect of applying A_S to the DC specified by the following diagram, where all unmarked links are taken to be large.

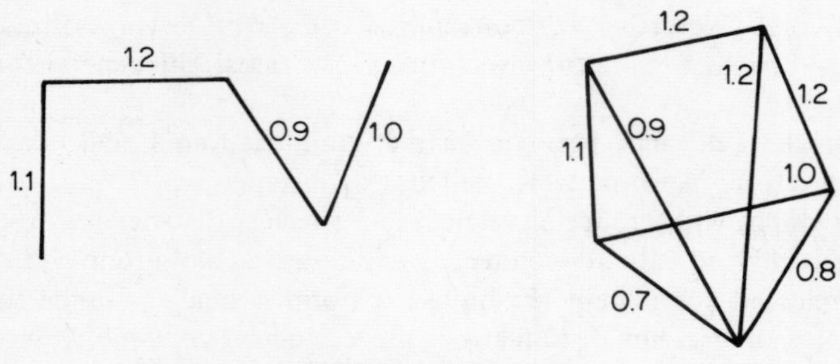

Figure 7.4.

Both clusters form at level 1.2, but there is clearly a sense in which the left-hand cluster is much less satisfactory than the right-hand cluster–it is in fact a 'chain'. There are grounds for supposing that in the biological context the production of chains by a cluster method is less of a defect than it might seem at first sight, because the existence of intermediates is often regarded as providing grounds for classifying together seemingly disparate OTU's at a lower level than might otherwise be chosen. However, in many applications it is a genuine defect of a cluster method, and much of the history of cluster analysis consists of attempts to find cluster methods which in some sense suffer less from chaining than does single-link. The original forms of the complete-link method, due to Sørensen (1948), and the various average-link methods, one of the first of which was suggested by Sokal and Michener (1958), can be regarded as attempts to find such methods. The well-defined forms of these methods are the methods A_W, A_U, A_C given above; it has already been pointed out that methods which are not well-defined have no place in any field of application where the opportunity to compare results is wanted, and we argue in Chapter 14 that this is generally the case in biological taxonomy. In making the methods well-defined we have had to unite groups in the same way as the single-link method; the difference lies in the different methods of recalculating dissimilarities. So 'genuine' chains, where all links are exactly equal, will be picked out by each method; the last three methods may however be able to break up chains where links are not all exactly equal, and this is in fact what happens. Consider the DC specified by the following table

	A	B	C
A			
B	$1+\epsilon$		
C	2	$1-\epsilon$	

ϵ small, either + ve or – ve or 0

The dendrograms obtained from this DC by the single-link method, the weighted- or unweighted-average-link method, and the complete-link method, are shown below in Figures 7.5-7.7 respectively. Each figure is preceded by a table showing the transformed DC values.

It can be seen that chains with slightly unequal links are indeed broken up by A_W, A_U, and A_C, but it is immediately apparent from the example that this is because the output DC values are not continuous functions of the

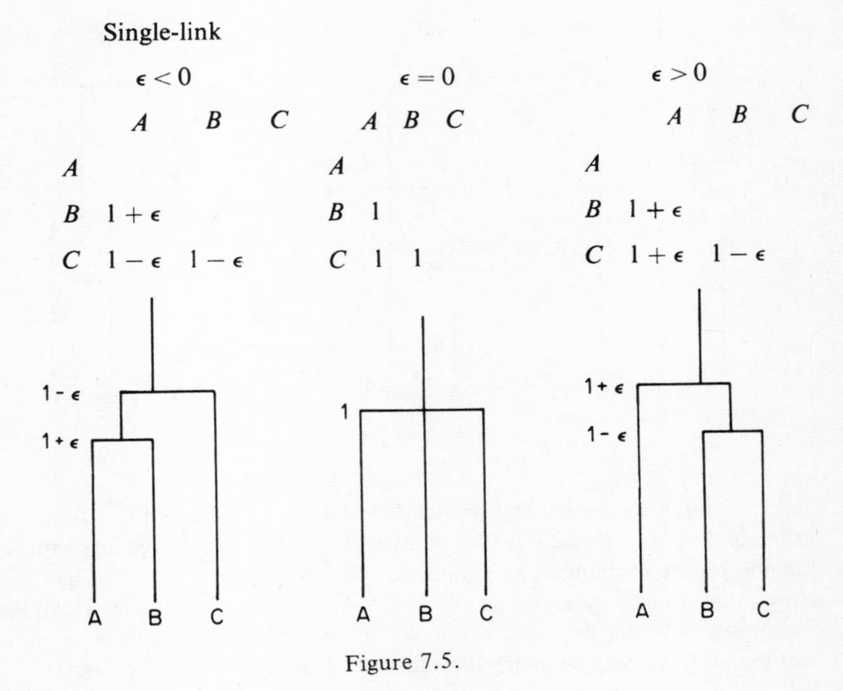

Figure 7.5.

input in these cases; they are continuous in the single-link case in this example, and it is shown in Chapter 9 that this is always so for the single-link method. Discontinuity is as crippling as ill-definedness if repeatability and comparability are needed, because the effects of computational rounding errors and of statistical sampling errors are not predictable in practice and can

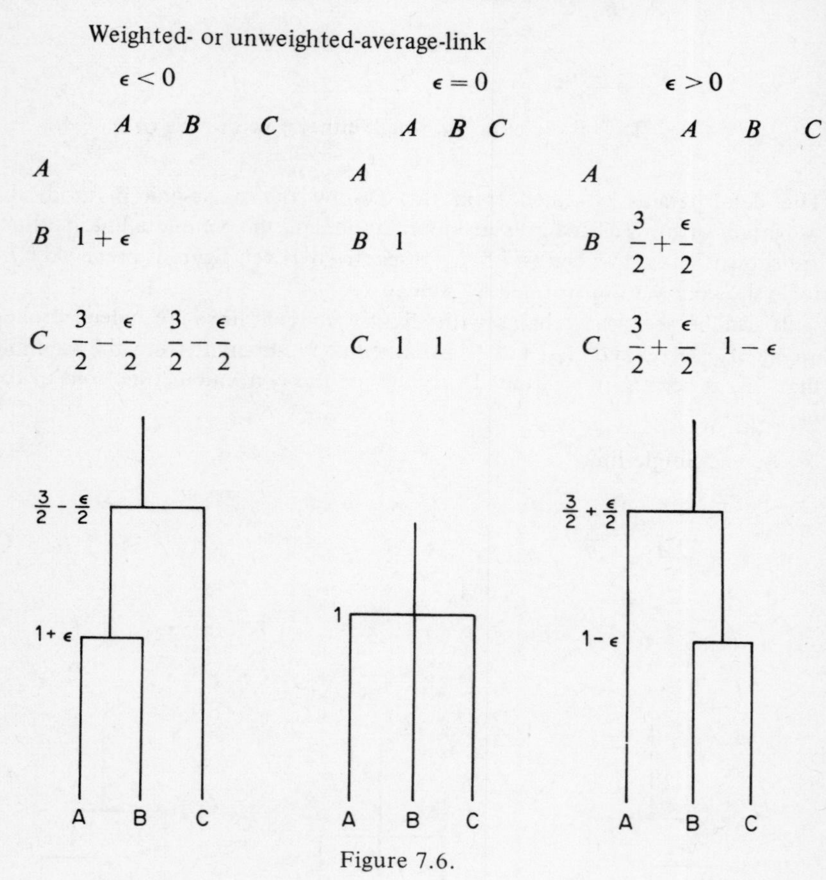

Figure 7.6.

lead to completely misleading results. We are thus led to reject the use of the methods A_W, A_U, and A_C in the biological context, and we are left with A_S and the resultant chains. The axiomatic framework established in Chapter 9 shows that it is a valid point of view to regard chaining as an inevitable concomitant of the use of a hierarchic method, and in Chapter 8 we show that the problem may be avoided by the use of non-hierarchic methods.

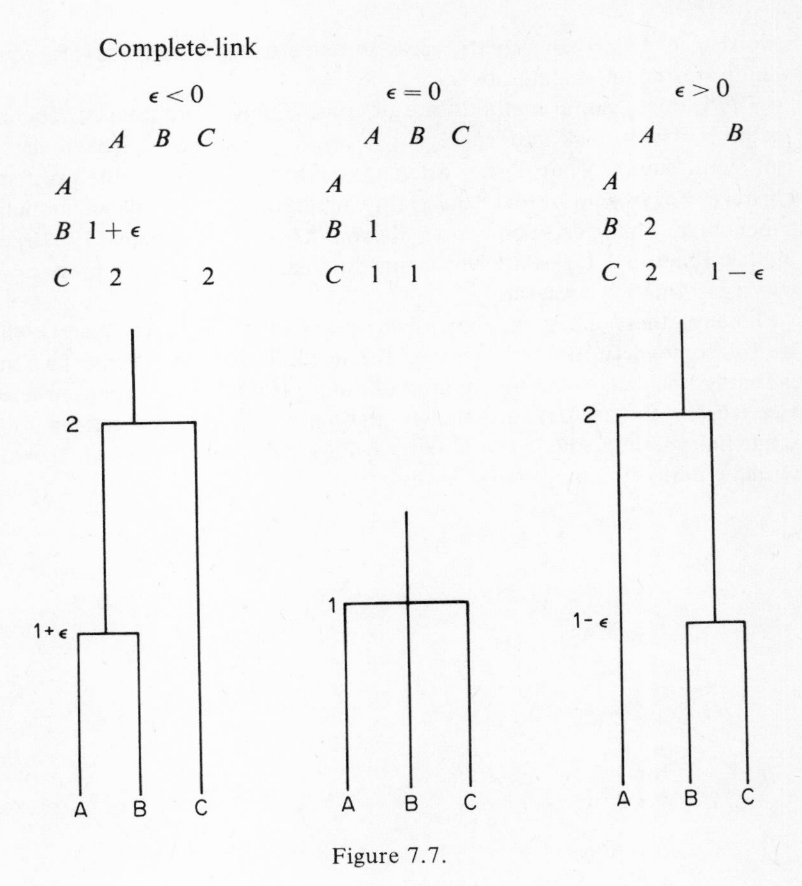

Figure 7.7.

7.5. OTHER METHODS

We mention finally some methods other than the four given above. These other methods are not numerically stratified, and so do not strictly fall within the scope of this chapter; a full discussion is deferred until Chapter 12.

There are a number of methods which pick out clusters in some nested way by means of a divisive algorithm. Usually only division of groups into two parts is permitted, and this gives rise to a difficulty similar to–perhaps 'dual to' is more accurate–that arising from uniting just two groups at a time in an agglomerative algorithm. Methods of this type have been suggested by Edwards and Cavalli-Sforza (1965), Macnaughton-Smith, Williams, Dale, and Mockett (1964), and Rescigno and Maccacaro (1961). The arbitrariness involved in binary division of groups renders these methods of dubious validity for taxonomic purposes.

The various 'clump methods' are of considerably more interest. These are methods which pick out subsets of P with properties of coherence and isolation ('clumps') but do not attempt to fit them together into a system of clusters. For the most part the clump methods are valuable in disciplines other than biological taxonomy: the methods of Needham (1961b) and Spärck Jones and Jackson (1967), for example, were developed primarily for use in information retrieval.

Finally, there are a number of methods which seek all clusters which satisfy some definition in terms of the dissimilarity coefficient. Two such methods, the ball-clustering method (Jardine, 1969b), and a method due to van Rijsbergen (1970a), are shown in Chapter 12.4 to be related to the single-link method, the clusters found being single-link clusters which satisfy strong criteria of homogeneity.

CHAPTER 8

Non-Hierarchic (Overlapping) Stratified Cluster Methods

8.1 THE BASIS OF THE GENERALIZATION

In the last chapter we were able to give convenient descriptions of four hierarchic cluster methods by using an approach suggested by the general form of agglomerative algorithm proposed by Lance and Williams (1967). This approach has the advantage that the methods are described in terms of algorithms which could be programmed as they stand. Although this is not always an efficient process, it shows that the methods are computable in finitely many steps, which is a logical property of some interest. This chapter is concerned with the introduction of non-hierarchic (overlapping) numerically stratified cluster methods. The kind of algorithm employed in the last chapter loses all record of the internal structure of the groups formed at each stage, treating the set of groups simply as a new set of objects. Because of this, it is necessary to find a different kind of approach to make possible the description of non-hierarchic cluster methods, which depend on the retention of a certain amount of information about the internal structure of groups.

The application of a hierarchic numerically stratified (type A) cluster method to a DC d produces a dendrogram. Because there is a natural 1-1 correspondence between the partitions of a set and the equivalence relations on that set, it is possible to formalize the idea of a dendrogram in rather a simple way in terms of equivalence relations. The idea of numerically stratified non-hierarchic cluster methods is to allow clusters at each level to overlap. For a general system of subsets of a set we no longer have a convenient descriptive tool such as the equivalence relations, if we were to ask that non-hierarchic cluster methods should be completely general, in the sense that the clusters produced at each level should be an arbitrary system of subsets covering the set P of OTU's, then there would be little hope of obtaining any reasonably manageable mathematical framework. There are, however, certain families of subsets which constitute a very natural generalization of the partitions. Suppose that r is a symmetric reflexive relation of P. A *maximal linked set* for r (*ML-set*) is a subset $S \subseteq P$ satisfying the following conditions.

ML1 $\quad S \times S \subseteq r$

ML2 $\quad A \notin S \Rightarrow$ there exists $B \in S$ such that $(A, B) \notin r$

An ML-set for r is thus a maximal subset of P on which the relation r is universal. It is often convenient to think of this in graphical terms. If we draw

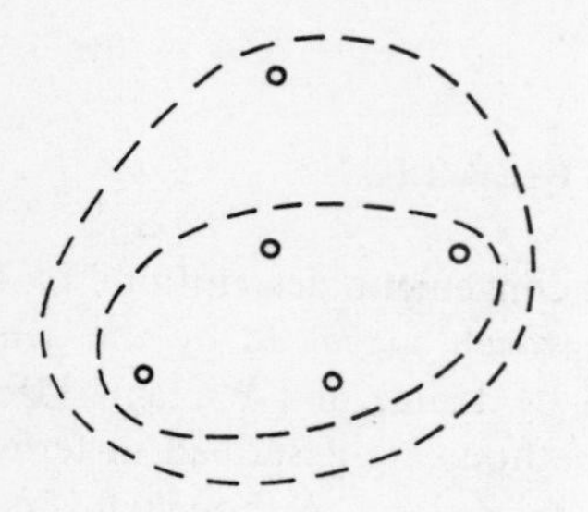

Figure 8.1.

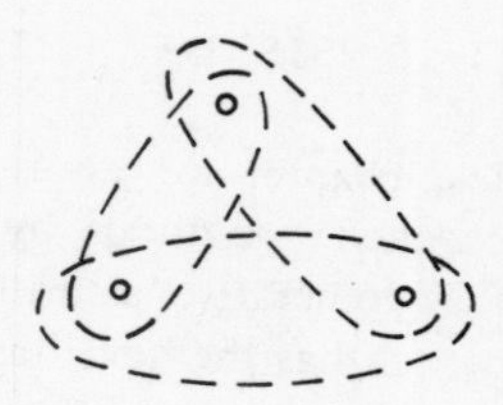

Figure 8.2.

a graph whose vertices are the elements of P and whose edges link just those pairs of elements of P contained in r, then the ML-sets are the vertex sets of maximal complete subgraphs. ML-sets have appeared in the literature under various other names: Needham (1961a) has called them *Kuhns clumps*, Harary and Ross (1957) have called them *cliques*, and many psychologists and sociologists have used the terminology *cliques* or *maximal cliques*; they have also often been called *maximal complete subgraphs*, a mild abuse of terminology. For various reasons, none of these names has seemed to us to be entirely satisfactory, and in consequence we introduce the name 'ML-set' which we shall use consistently throughout. If r is a symmetric reflexive relation, the system of ML-sets for r will be denoted by $ML(r)$. The set of symmetric reflexive relations on P will be denoted by $\Sigma(P)$. In the hierarchic case, we require that the system of clusters at each level shall be of the form $ML(r)$, where $r \in E(P)$, the set of symmetric reflexive *transitive* relations (equivalence relations) on P. To obtain a manageable generalization to non-hierarchic methods, we relax this to the condition that the system of clusters at each level shall be of the form $ML(r)$, where $r \in \Sigma(P)$. Thus we shall only be interested in systems of clusters which may be regarded as the set of all ML-sets of some symmetric reflexive relation. Systems of clusters such as those shown in Figures 8.1 and 8.2 will not be admissible within this framework.

This, of course, refers to clusters *at a single level.* Clusters at different levels may well be contained one within another, indeed we shall require a nesting of clusters at different levels in the non-hierarchic case just as we do in the hierarchic case.

8.2 NUMERICALLY STRATIFIED CLUSTERINGS

Having settled on systems of ML-sets as possible systems of clusters, we are now able to produce a generalization of the dendrogram in a suitable form. The generalization is the *numerically stratified clustering* (*NSC*), which we define to be a function $c : [0,\infty) \to \Sigma(P)$ satisfying the following conditions

NSC1 $\quad 0 \leqslant h \leqslant h' < \infty \Rightarrow c(h) \subseteq c(h')$

NSC2 $\quad c(h)$ is eventually $P \times P$

NSC3 $\quad$ Given $h \geqslant 0$, there exists $\delta > 0$ such that $c(h + \delta) = c(h)$

The conditions are in fact the same as conditions D1 – D3 in the definition of a dendrogram. The difference is that the range of the map c is extended from E(P) to $\Sigma(P)$. The *clusters* for c at level h are the elements of $ML(c(h))$, but for reasons which will become apparent shortly, we shall be reluctant to use the term 'clusters', preferring to speak of ML-sets for c at level h.

In Chapter 7 we established a 1-1 correspondence between the dendrograms on P and the ultrametric DC's on P. This correspondence extends to a natural 1-1 correspondence between the set of NSC's on P and the set $\mathbf{C}(P)$ of all DC's on P. Suppose $d \in \mathbf{C}(P)$. Define Td, an NSC, by

$$(Td)(h) = \{(A, B) : d(A, B) \leqslant h\}$$

If c is an NSC, define $T^{-1}c$, a DC, by

$$(T^{-1} c)(A, B) = \inf \{h : A, B) \in c(h)\}$$

The technical condition NSC3 ensures that this does in fact give a 1-1 correspondence. It is natural because no arbitrary choices are involved.

Condition NSC2, which ensures that for large enough h, $c(h)$ is universal and consequently has a single ML-set P, corresponds to the property that since P is finite, $d(A, B)$ is bounded. The possibility of identifying the NSC's on P with the DC's on P is fundamental to all that follows. It means that if we select some subset of the set of NSC's as the kind of output we would like a cluster method to have, then this set will correspond to some subset $\mathbf{Z} \subseteq \mathbf{C}(P)$. Hence the cluster method may be regarded as a function $D : \mathbf{C}(P) \to \mathbf{Z}$, that is as a function from $\mathbf{C}(P)$ to some subset of itself. From the mathematical point of view this is an agreeable situation, attention can be confined to objects of one kind, and the constraints imposed on cluster methods in Chapter 9 can be formulated neatly and shown to have real power. In practice, it is not a complete account of the situation. What we want as the output of a cluster method is not an element of $\mathbf{Z}$, which would be specified by a half-matrix of dissimilarities, but rather lists of the elements of $ML(c(h))$ for each splitting level h of c, together with the splitting levels themselves. A *splitting level* is a level h_0 such that $c(h) \neq c(h_0)$ if $h < h_0$, that is, a level at which c changes its value. Zero is a splitting level if and only if $c(0)$ is not the equality relation. Since this change of viewpoint represents a 1-1 correspondence, no further mathematical analysis of the process need be carried out, but it is certainly a non-trivial problem computationally except in a few simple cases, for example $\mathbf{Z} = \mathbf{U}(P)$, and it is an essential part of the process of cluster analysis. Methods for writing down the splitting levels and lists of ML-sets for each splitting level of c will be called *cluster listing methods.* The most efficient cluster listing method known to us is that developed by Moody and Hollis on the basis of a suggestion by Needham (1961a); this is described in Appendix 5.

Since the DC's and NSC's on P are identified under the correspondence T, it is even possible to regard the original DC d, the data for cluster analysis, as corresponding to an NSC. Using the correspondence, anything we can say about an NSC can also be said about a DC so for an arbitrary DC d we can define the ML-sets at level h to be the ML-sets for Td at that level. This means that we can recognize ML-sets *in the data.* These may not be the clusters we want to recognize eventually; in general there will be too many of them, since they contain complete information about d and the object of a cluster method may reasonably be regarded as the simplification of d by controlled discarding of information.

We may often want the eventually recognized clusters to be based on ML-sets in the data in some way; the effects of relating ML-sets in the data to those in the output are explored in Chapter 9: but it is possible that we might want to recognize clusters in a different way, and because of this we are being rather careful in our use of the word 'cluster'. The clusters we obtain will certainly be the ML-sets of the *output* DC at each level, because we have settled on the ML-sets of a DC, that is, of the corresponding NSC, as the kind of system of clusters which we are going to allow. What we are anxious not to prejudge is their relation to the ML-sets of the *data* DC.

8.3. SUBDOMINANT METHODS

The identification of the NSC's and DC's on P makes it possible to regard any function $D : \mathbf{C}(P) \to \mathbf{Z}$, where $\mathbf{Z} \subseteq \mathbf{C}(P)$, as being in some sense a cluster method. If $\mathbf{Z}$ is chosen misguidedly, then functions $D : \mathbf{C}(P) \to \mathbf{Z}$ are unlikely to be of much use as cluster methods. For example, there is no particular reason to expect that taking $\mathbf{Z} = \mathbf{M}(P)$, the set of metric DC's on P, will yield anything which we would normally want to regard as a cluster method unless we have chosen D to map in fact to some smaller subset of $\mathbf{M}(P)$ such as $\mathbf{U}(P)$. Thus it is necessary to find some way of choosing useful sets $\mathbf{Z}$, and then, given $\mathbf{Z}$, of choosing associated cluster methods $D : \mathbf{C}(P) \to \mathbf{Z}$. The four hierarchic cluster methods

$$A_{\mathrm{S}}: \mathbf{C}(P) \to \mathbf{U}(P) \quad \text{single-link}$$

$$A_{\mathrm{W}}: \mathbf{C}(P) \to \mathbf{U}(P) \quad \text{weighted-average-link}$$

$$A_{\mathrm{U}}: \mathbf{C}(P) \to \mathbf{U}(P) \quad \text{unweighted-average-link}$$

$$A_{\mathrm{C}}: \mathbf{C}(P) \to \mathbf{U}(P) \quad \text{complete-link}$$

were described in terms of algorithms of a kind which, as pointed out in Chapter 8.1, cannot be expected to extend to the non-hierarchic case. The only one of these type A cluster methods for which the function $\mathbf{C}(P) \to \mathbf{U}(P)$ admits of convenient description is the single-link method A_{S}. We have defined $d \geqslant d'$ (*d dominates* d') to mean

$$d(A, B) \geqslant d'(A, B) \text{ for all } A, B \in P$$

Then $A_{\mathrm{S}}(d)$ may be characterized as the element of $\mathbf{U}(P)$ which is maximal subject to the condition that it is dominated by d, we call $A_{\mathrm{S}}(d)$ the **U**-*subdominant* of d. All the non-hierarchic cluster methods introduced here will be defined in this way. With each chosen $\mathbf{Z}$ one method $D : \mathbf{C}(P) \to \mathbf{Z}$ will be associated, namely the method for which $D(d)$ is the maximal element of $\mathbf{Z}$ dominated by d–the **Z**-*subdominant* of d. Such methods will be called *subdominant methods.* They are of particular importance because they relate the ML-sets of the data to those of the output; this is discussed in detail in Chapter 9. In general a subset $\mathbf{Z} \subseteq \mathbf{C}(P)$ need not define a subdominant method; there might be no maximal dominated element, or no unique one. A condition which ensures that **Z** does define a subdominant method is the following.

Define a set **Y** of DC's to be *bounded* if there is a DC d_0 such that $d \in \mathbf{Y} \Rightarrow d \leqslant d_0$. If **Y** is a bounded set of DC's, define sup **Y** by

$$(\sup \mathbf{Y})(A, B) = \sup\{d(A, B) : d \in \mathbf{Y}\}$$

Then the condition is

S5 If **Y** is a bounded subset of **Z**, then $\sup \mathbf{Y} \in \mathbf{Z}$

It is very easily seen that this ensures that **Z** defines a subdominant method. If $d \in \mathbf{C}(P)$, take $D(d) = \sup\{d' : d' \in \mathbf{Z} \text{ and } d' \leqslant d\}$. Then clearly $D(d)$ is the **Z**-subdominant of d, and is unique. A set $\mathbf{Z} \subseteq \mathbf{C}(P)$ satisfying S5 will be called *sup-closed.* The rest of this chapter is devoted to the construction of suitable sup-closed sets $\mathbf{Z} \subseteq \mathbf{C}(P)$ and the investigation of the resultant subdominant cluster methods.

8.4. CONSTRAINTS ON NSC's

So far only two sup-closed subsets **Z** of $\mathbf{C}(P)$ have been introduced. These are the set $\mathbf{U}(P)$ of ultrametric DC's on P, for which the associated subdominant cluster method is the single-link method $A_S : \mathbf{C}(P) \rightarrow \mathbf{U}(P)$, and the set $\mathbf{C}(P)$ of all DC's on P, for which the subdominant method is the identity map $I : \mathbf{C}(P) \rightarrow \mathbf{C}(P)$; it is clear that if $D : \mathbf{C}(P) \rightarrow \mathbf{Z}$ is a subdominant method, then $D|\mathbf{Z}$ is the identity on **Z**. The case $\mathbf{Z} = \mathbf{C}(P)$ must not be dismissed immediately. Although it is mathematically trivial, it is important to remember that our intepretation of the input and output DC's differs; the use of the cluster method I corresponds to the statement that the clusters we want are precisely the ML-sets at each splitting level in the original DC. Usually this is not what is wanted, because the point of using a cluster method is to effect simplification of the data; equally, the method A_S may well effect too much simplification: the value of non-hierarchic cluster methods is that they provide an intermediate situation, and by choosing carefully we can simplify but avoid oversimplifying.

The ML-sets for an arbitrary element of $\Sigma(P)$ form a very special kind of system of subsets of P, but even this can be extremely complicated. What we shall do is to try to find some constraints on the system of ML-sets which will ensure that this complication is reduced far enough to make the clusters give a comprehensible description of structure in the original data DC. This can be done by imposing constraints which ensure that the ML-sets fit together in a reasonably simple way. The simplest such constraint is the transitivity condition, as follows:

If r is a symmetric reflexive relation on P ($r \in \Sigma(P)$) and $A, B, C \in P$, then

$$[(A, C) \in r \text{ and } (C, B) \in r] \Rightarrow (A, B) \in r$$

This condition ensures that the symmetric reflexive relation is an equivalence relation, and thus defines a partition. That is, distinct ML-sets are disjoint. We obtain non-hierarchic cluster methods by generalizing this condition so that the overlap between distinct ML-sets is restricted in size, but not actually required to be empty.

Restrictions on the overlap between distinct ML-sets fall roughly into three types, namely absolute, internal, and external. Absolute restrictions are given in terms which are self-contained; a restriction on the number of elements in the overlap is an example of an absolute restriction. Internal restrictions require reference back to something outside the relation $(Td)(h)$, for example to the level h or the DC d, but make no reference outside this domain; a restriction on the diameter of the overlap in terms of the current level or of some function of the DC values is an example of an internal restriction. External restrictions call on some function which assigns a size to each subset of P and require that no overlap should exceed a certain size, if the diameter of each set were measured in terms of a fixed DC d_0, and each overlap of distinct ML-sets of d at each fixed level were required to be of at most a certain d_0-diameter, then this would constitute an external restriction on d. We shall be concerned only with absolute or internal restrictions. Cluster methods defined as subdominant methods by sets **Z** characterized by absolute restrictions will be called *type B* cluster methods; and those defined similarly by sets **Z** characterized by internal restrictions will be called *type C* methods.

8.5. THE METHODS B_k

The methods B_k are type B cluster methods for $k = 1, 2, \ldots$ They are based on an absolute restriction, namely that the overlap between distinct ML-sets at the same level shall not contain more than $k-1$ elements of P. Thus B_1 is the single-link method A_S, and as k increases the methods B_k allow progressively more overlap between the ML-sets until when $k = p-1$ the overlap restriction becomes vacuous and $B_k = I$ for $k \geqslant p-1$. The methods B_k are the simplest and most obvious generalizations of the single-link method; B_k is called *(fine) k-clustering.* This family of methods was introduced by Jardine and Sibson (1968a, 1968b).

To establish the methods B_k, it will be sufficient, in the light of Chapter 8.1–Chapter 8.4, to construct sup-closed sets $\mathbf{C}_k(P)$ for $k = 1, 2, \ldots$ such that the DC's in $\mathbf{C}_k(P)$ correspond under T to the NSC's characterized by the property that the overlap between distinct ML-sets at each level contains at most $k - 1$ elements of P. The simplest way to do this is to start with a generalization of the transitivity condition which expresses this requirement as a constraint on symmetric reflexive relations. The condition is the (weak) k-transitivity condition, as follows.

T_k Let $r \in \Sigma(P)$. r is (*weakly*) *k-transitive* if whenever

$S \subseteq P$, $|S| = k$, A, $B \in P$, then

$$[\{A\} \times S \cup S \times S \cup S \times \{B\}] \subseteq r \Rightarrow (A, B) \in r$$

This condition is obtained from the ordinary transitivity condition T by replacing the single element C by the k-element set S, which must be completely linked. On the left-hand side of the implication the condition '$\{A\} \times S \subseteq r$' corresponds to '$(A, C) \in r$' and '$S \times \{B\} \subseteq r$' to '$(C, B) \in r$'. The effect of the condition is easy to think of in terms of ML-sets. Suppose that S_1, S_2 are distinct ML-sets for r. Then both $S_1 - S_2$ and $S_2 - S_1$ are non-empty. There exist $A \in S_1 - S_2, B \in S_2 - S_1$ such that $(A, B) \notin r$, because if this were not so, $S_1 \cup S_2$ would be an ML-set. It follows that $|S_1 \cap S_2| < k$ if r satisfies T_k; that is, the overlap between distinct ML-sets contains at most $k - 1$ elements. Conversely, suppose that r satisfies the condition that distinct ML-sets overlap in sets with at most $k-1$ elements. Then the left-hand side of the implication in T_k is true if and only if $\{A\} \times S$ and $S \times \{B\}$ lie wholly inside ML-sets for r. If they lie in the same ML-set, then $(A, B) \in r$ and the condition is satisfied. If they lie in distinct ML-sets then S must lie in the intersection of these two ML-sets and so must contain at most $k - 1$ elements, and again the condition is satisfied. If the left-hand side of the implication is false, then the condition is satisfied anyhow. Thus the k-transitivity condition is precisely the condition to apply to a symmetric reflexive relation to ensure that the overlap between distinct ML-sets contains at most $k - 1$ elements.

The associated set of NSC's consists of those NSC's for which $c(h)$ is k-transitive for all h. The set of k-transitive symmetric reflexive relations on P will be denoted by $J_k(P)$. An NSC such that $c[0, \infty) \subseteq J_k(P)$ will be called a (*fine*) *k-dendrogram*. $\mathbf{C}_k(P)$ is the set of DC's identified with the set of k-dendrograms under the correspondence T, but this is a very inconvenient way of characterizing it; a condition analogous to the ultrametric inequality is wanted if possible, and we can in fact find such a condition. It is the (weak) *k-ultrametric* inequality, as follows.

DCU_k Let $d \in \mathbf{C}(P)$. d is (*weakly*) *k-ultrametric* if whenever

$S \subseteq P, |S| = k, A, B \in P$, then

$d(A, B) \leqslant \max\{d(X, Y): X \in S \cup \{A, B\}, Y \in S\}$

If whenever $R \subseteq P$ we write $\text{diam}(d, R) = \max\{d(X, Y) : X, Y \in R\}$, then the k-ultrametric inequality can be written as

$$d(A, B) \leqslant \max\{\text{diam}(d, S \cup \{A\}), \text{diam}(d, S \cup \{B\})\}$$

It is easily seen that d is k-ultrametric if and only if $Td(h) \in J_k(P)$ for all $h \in [0, \infty)$, so $\mathbf{C}_k(P)$ is the set of k-ultrametric DC's.

There is a very simple condition which is equivalent to the k-ultrametric inequality, and which is of considerable utility. The condition is that on every $(k + 2)$-element subset of P, the largest dissimilarity value on the subset should occur more than once in the subset. Thus if $d(X, Y) = \text{diam}(d, R)$ where $X, Y \in R$, $|R| = k + 2$, there must be $X', Y' \in R$ with $d(X, Y) = d(X', Y')$ and $\{X,Y\} \neq \{X',Y'\}$. If this condition is taken to characterize $\mathbf{C}_k(P)$, then it is immediately obvious that $\mathbf{C}_k(P)$ is sup-closed, and consequently gives rise to a well-defined subdominant method $\mathbf{C}(P) \to \mathbf{C}_k(P)$: this is the method B_k. The implications

$$\text{T} = \text{T}_1 \Rightarrow \text{T}_2 \Rightarrow \ldots \Rightarrow \text{T}_{p-2} \Rightarrow \text{T}_{p-1} = \text{NN},$$

where NN is the vacuous condition, give rise to containment relations

$$E(P) = J_1(P) \subset J_1(P) \subset \ldots \subset J_{p-2}(P) \subset J_{p-1}(P) = \Sigma(P)$$

and so to

$$\mathbf{U}(P) = \mathbf{C}_1(P) \subset \mathbf{C}_2(P) \subset \ldots \subset \mathbf{C}_{p-2}(P) \subset \mathbf{C}_{p-1}(P) = \mathbf{C}(P)$$

It follows that $B_1 = A$s, $B_{p-1} = I$, and $B_k \circ B_j = B_k$ for $k \leqslant j$. As k increases, $B_k(d)$ may be thought of as giving progressively more and more information about d, until when $k \geqslant p-1$, $B_k(d) = d$ and complete information is recovered. The relation $B_k \circ B_j = B_k$ for $k \leqslant j$ is valuable for computation, since it means that in a process for finding $B_k(d)$ for $k = 1, 2, \ldots K$ we can work down the values of k and find $B_k(d)$ from $B_{k+1}(d)$—we do not need to go back to d itself.

$B_k(d)$ is calculable from d in finitely many steps. This is trivial, for the reason that $d(P \times P)$ is a finite set, and $[B_k(d)](P \times P) \subseteq d(P \times P)$. We do not at this stage give an efficient algorithm for calculating $B_k(d)$, because the best algorithm known to us, which is due to Moody and C. J. Jardine, is extremely complicated. It is discussed in Appendix 3. There is a very simple but inefficient algorithm based on the alternative characterization of k-ultrametric DC's; this works as follows. List all subsets of P having exactly $k + 2$ elements. Run through this list examining the maximum dissimilarity on each subset. If the maximum value is attained on a unique pair in the subset, reduce it to the next value, and go on to the next subset; if not, go on to the next subset without making any change. When the list can be run through completely without change then the original DC has been reduced to $B_k(d)$. We give this algorithm here simply to give the reader an idea of how B_k operates, and to show explicitly that a terminating algorithm exists.

The definition of B_k as the subdominant method associated with $\mathbf{C}_k(P)$ is by no means the simplest possible characterization of it, because, unlike some of the methods considered later in this chapter, B_k has special properties which derive essentially from its being related to an absolute restriction on overlap. For each $h \in [0, \infty)$, $[TB_k(d)](h) \in J_k(P)$, and, because B_k relates to an absolute restriction, $J_k(P)$ is fixed and independent of d, h. It is in fact possible to write

$$[TB_k(d)](h) = \gamma_k[(Td)(h)]$$

where γ_k is a function from $\Sigma(P)$ to $J_k(P)$ which maps each element of $\Sigma(P)$ to the smallest element of $J_k(P)$ containing it. If $D : \mathbf{C}(P) \to \mathbf{Z}$ is a cluster method, and if there exists a function $\gamma : \Sigma(P) \to \Sigma(P)$ such that

$$[TD(d)](h) = \gamma[(Td)(h)]$$

for all $d \in \mathbf{C}(P)$, $h \geqslant 0$, then D is called a *uniform cluster method.* We thus see that the methods B_k are uniform cluster methods, and that all we need to know is the function $\gamma_k : \Sigma(P) \to J_k(P)$. γ_k is most easily thought of in terms of operations on ML-sets. If r is a relation, then the sets in $ML(r)$ will in general have arbitrary overlap. Add pairs to r so that whenever S_1 and S_2 are in $ML(r)$ and $|S_1 \cap S_2| \geqslant k$, S_1 and S_2 are replaced by $S_1 \cup S_2$ in $ML(r^*)$. Continue until $ML(r^*)$ contains no pairs S_1, S_2 with $|S_1 \cap S_2| \geqslant k$. Then r^* is $\gamma_k(r)$.

This suggests a way of generating uniform cluster methods which are also subdominant methods. Let J be a subset of $\Sigma(P)$, $\gamma : \Sigma(P) \to J$ the function which associates with each element of $\Sigma(P)$ the smallest member of J containing it. Then if $\mathbf{Z}_J$ is the set of DC's corresponding to those NSC's for which $c[0, \infty) \subseteq J$, the subdominant method D defined by $\mathbf{Z}_J$ is given by

$$[TD(d)](h) = \gamma[(Td)(h)]$$

Some conditions on J are needed to make γ well-defined; the situation is investigated in full detail in Chapter 10, where in the case of subdominant cluster methods uniformity is shown to be implied by an important equivariance property.

8.6. THE METHODS B_k^c

The methods B_k^c—*coarse k-clustering*—are further type B cluster methods, and are quite closely related to the methods B_k. They are defined in terms of absolute restrictions on relations, but these restrictions do not have quite the simple interpretation in terms of overlap between ML-sets which was available

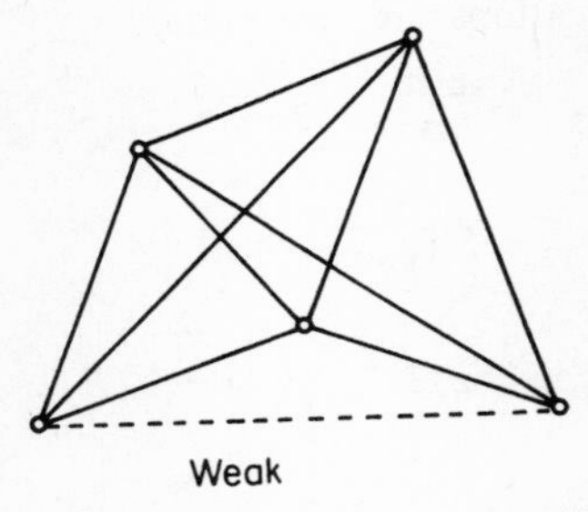

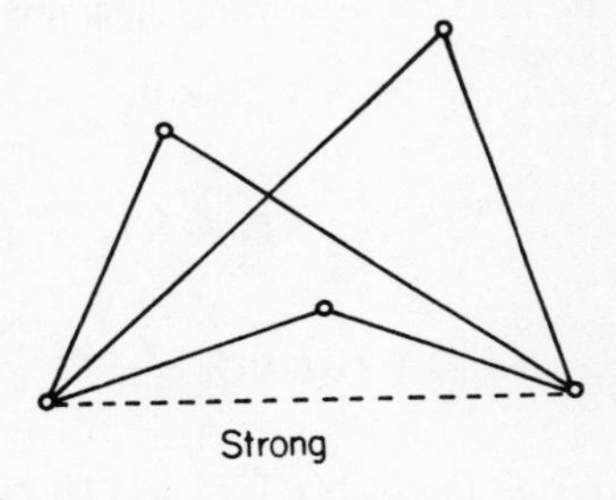

Figure 8.3.

in the case of fine k-clustering. The restriction on relations which gives rise to fine k-clustering is the weak k-transitivity condition T_k, which we repeat here.

T_k Let $r \in \Sigma(P)$. r is (*weakly*) *k-transitive* if whenever

$S \subseteq P, |S| = k, A, B \in P$, then

$[\{A\} \times S \cup S \times S \cup S \times \{B\}] \subseteq r \Rightarrow (A, B) \in r.$

The strong k-transitivity condition ST_k, on which coarse k-clustering is based, is similar to this, but the condition that S be completely linked is omitted.

ST_k Let $r \in \Sigma(P)$. r is *strongly k-transitive* if whenever

$S \cup P, |S| = k, A, B \in P$, then

$[\{A\} \times S \cup S \times \{B\}] \subseteq r \Rightarrow (A, B) \in r.$

The difference between strong and weak k-transitivity is illustrated in Figure 8.3, which is drawn to show what happens in the case $k = 3$; the 'deduced' link is dotted.

A *coarse k-dendrogram* is an NSC such that $c[0, \infty) \subseteq J_k^c(P)$, where $J_k^c(P)$ denotes the set of strongly k-transitive relations on P. The corresponding set of DC's is the set $\mathbf{C}_k^c(P)$ of DC's satisfying the strong k-ultrametric inequality, as follows:

DCSU$_k$ Let $d \in \mathbf{C}(P)$. d is *strongly k-ultrametric* if whenever

$S \subseteq P$, $|S| = k$, A, $B \in P$, then

$$d(A, B) \leqslant \max\{d(X, Y) : X \in \{A, B\}, Y \in S\}$$

$\mathbf{C}_k^c(P)$ is readily seen to be sup-closed; the resultant subdominant method is the *coarse k-clustering* method $B_k^c : \mathbf{C}(P) \rightarrow \mathbf{C}_k^c(P)$.

We have the following diagrams of implications and containments:

$$\begin{array}{cccccccc}
\mathrm{T} \Leftrightarrow & \mathrm{T}_1 & \Rightarrow \mathrm{T}_2 & \Rightarrow \ldots \Rightarrow & \mathrm{T}_{p-2} & \Rightarrow \mathrm{T}_{p-1} & \Leftrightarrow \mathrm{NN} \\
\Updownarrow & \Updownarrow & \Uparrow & & \Uparrow & \Updownarrow & \Updownarrow \\
\mathrm{T} \Leftrightarrow & \mathrm{ST}_1 & \Rightarrow \mathrm{ST}_2 & \Rightarrow \ldots \Rightarrow & \mathrm{ST}_{p-2} & \Rightarrow \mathrm{ST}_{p-1} & \Leftrightarrow \mathrm{NN}
\end{array}$$

$$\begin{array}{ccccccc}
E(P) = J_1(P) & \subset & J_2(P) \subset & \ldots \subset & J_{p-2}(P) \subset & J_{p-1}(P) & = \Sigma(P) \\
\| & & \cup & & \cup & \| & \| \\
E(P) = J_1^c(P) & \subset & J_2^c(P) \subset & \ldots \subset & J_{p-2}^c(P) \subset & J_{p-1}^c(P) & = \Sigma(P)
\end{array}$$

$$\begin{array}{ccccccc}
\mathbf{U}(P) = & \mathbf{C}_1(P) \subset & \mathbf{C}_2(P) \subset & \ldots \subset & \mathbf{C}_{p-2}(P) \subset & \mathbf{C}_{p-1}(P) & = \mathbf{C}(P) \\
\| & \| & \cup & & \cup & \| & \| \\
\mathbf{U}(P) = & \mathbf{C}_1^c(P) \subset & \mathbf{C}_2^c(P) \subset & \ldots \subset & \mathbf{C}_{p-2}^c(P) \subset & \mathbf{C}_{p-1}^c(P) & = \mathbf{C}(P)
\end{array}$$

It was possible to characterize B_k in the form

$$[TB_k(d)](h) = \gamma_k[(Td)(h)]$$

where γ_k is the function $\Sigma(P) \rightarrow J_k(P)$ which maps each relation in $\Sigma(P)$ to the smallest k-transitive relation containing it. Similarly,

$$[TB_k^c(d)](h) = \gamma_k^c[(Td)(h)]$$

where $\gamma_k^c : \Sigma(P) \rightarrow J_k^c(P)$ maps each relation in $\Sigma(P)$ to the smallest strongly k-transitive relation containing it. Thus B_k^c, like B_k, produces a nested sequence of uniform cluster methods, and B_k^c and B_k are related to one another.

Coarse k-clustering lacks the convenient interpretation in terms of overlap between ML-sets which was available for fine k-clustering. Certainly any strongly k-transitive relation is k-transitive, but the converse is not in general true unless $k = 1$ or $k \geqslant p - 1$. So the ML-sets for a strongly k-transitive relation have overlap containing at most $k - 1$ elements of P, but, unlike the k-transitive relations, they are not characterized by this property. Coarse k-clustering may be regarded as allowing overlaps of the same kind as fine k-clustering, but making less efficient use of them. In Chapter 10 it is shown that there are axioms which characterize fine k-clustering. For these reasons, fine k-clustering is much more important than coarse k-clustering, so that some justification is needed for introducing coarse k-clustering at all. There are three reasons why it is of interest. First, it will serve as a useful second example of a type B cluster method. Secondly, it is computationally much simpler, and there may be circumstances in which the computation of $B_k^c(d)$ is feasible but that of $B_k(d)$ is not. Thirdly, it is very easy in attempting to devise algorithms for fine k-clustering to produce algorithms which actually calculate $B_k^c(d)$ rather than $B_k(d)$, and to avoid confusion it is desirable to have the two families of methods clearly distinguished.

8.7. THE METHODS C_u

We now turn to the consideration of the first family of type C cluster methods, which are based on the internal restriction of controlling overlap in terms of overlap diameter in relation to the current level. These methods are due to Sibson.

We shall see in Chapter 9 that one of the most fundamental conditions imposed on a cluster method is the condition that the function $D : \mathbf{C}(P) \to \mathbf{Z}$ shall commute with scalar multiplication of dissimilarity values. Formally, the condition is as follows:

M3 If $\alpha > 0$, then $D(\alpha d) = \alpha D(d)$ where αd is defined by

$$(\alpha d)(A, B) = \alpha . d(A, B)$$

It is clear from this that if a condition making use of the diameter of the overlap is employed, then in order for the resultant cluster method to satisfy M3, the restriction must be on the ratio between the overlap diameter and some other quantity such as current level or average dissimilarity; we cannot simply lay down that the diameter of the overlap shall be at most some constant without violating M3.

The methods C_u (*u-diametric clustering*) are the subdominant methods associated with the restriction that the diameter of the overlap of distinct ML-sets at level h shall be at most uh, where u is a constant. It is clear that

$\phi(h) = uh$ is the only possible type of function of h which will lead to methods satisfying M3. We construct the set $\mathbf{E}_u(P)$ of DC's corresponding to NSC's which have this property; they are the DC's satisfying the *u-diametric inequality,* as follows:

DCDi$_u$ Let $d \in \mathbf{C}(P)$. d is *u-diametric* if whenever

$\varnothing \neq S \subseteq P$, A, $B \in P$, then, writing

$$l = \max \{ \text{diam } (d, S \cup \{A\}), \text{diam } (d, S \cup \{B\})\},$$

we have

$$\text{diam } (d, S) > ul \Rightarrow d(A, B) \leqslant l$$

We shall assume $u \geqslant 0$; the condition will not be used for $u < 0$. It is quite easy to check that if d is u-diametric, then the NSC Td has the property that if S_1, S_2 are distinct ML-sets at level h, then diam$(d, S_1 \cap S_2) \leqslant uh$; it follows immediately from this that $\mathbf{E}_u(P)$ is sup-closed. So there is a subdominant method $C_u : \mathbf{C}(P) \rightarrow \mathbf{E}_u(P)$; this method is called *u-diametric clustering.* If $u \geqslant 1$, the condition 'diam$(d, S) > ul$' cannot be satisfied, and DCDi$_u$ is vacuous, so for $u \geqslant 1$, $\mathbf{E}_u(P) = \mathbf{C}(P)$ and $C_u = I$.

We define a DC d to be *definite* if it satisfies the following condition:

DCD $\quad d(A, B) = 0 \Rightarrow A = B$ for all A, $B \in P$.

If d is definite, then for every $S \subseteq P$ we have $|S| > 1 \Leftrightarrow \text{diam}(d, S) > 0$. Thus if $u = 0$, the condition DCDi_0 can be written in the form

$$|S| > 1 \Rightarrow d(A, B) \leqslant \max \{ \text{diam } (d, S \cup \{A\}), \text{diam } (d, S \cup \{B\})\}$$

which is the same as condition DCU$_2$; we have [DCDi$_0$ and DCD] $\Leftrightarrow$ [DCU$_2$ and DCD]. It follows from this that on the set of definite DC's the methods C_0 and B_2 are the same. However, they do not behave in the same way on non-definite DC's, and the way that type B methods deal with non-definite DC's is one of the reasons for considering type C methods. The difficulties are connected with the confidence with which the initial OTU's are selected. If we encounter two OTU's of small or zero dissimilarity, then we are inclined to suppose that we may have taken duplicates of what should have been a single OTU. This will upset the operation of type B methods which count OTU's, but not the operation of type C methods based on diameter, so in this respect type C methods are valuable because they do provide this safeguard against errors in the selection of OTU's. However, there is a price to pay for this advantage. Type C methods are not uniform, and it is shown in Chapter 10 that this implies that they satisfy much weaker invariance (strictly, equivariance) properties than the type B methods. Thus the use of type C methods is only valid if the dissimilarities are regarded as having more than merely ordinal significance. It appears that if we are to use non-hierarchic methods, then we are faced with this choice of either assuming that OTU's are well chosen and using type B methods, or assuming that ratios of dissimilarities are meaningful and using type C methods. This is discussed further in Chapter 14. We may summarize by saying that type B and type C methods are robust with respect to the failure of two different types of assumption about the data.

The u-diametric inequality may be written in a much simpler form, involving only quadruples of elements of P, as follows:

DCDi$_u$ Let $d \in \mathbf{C}(P)$. d is *u-diametric* if whenever

$A, B, C_1, C_2 \in P$, then, writing

$$l = \max \{d(A, C_1), d(A, C_2), d(B, C_1), d(B, C_2), d(C_1, C_2)\}$$

we have

$$d(C_1, C_2) > ul \Rightarrow d(A, B) \leqslant l$$

This suggests the following algorithm for obtaining $C_u(d)$ from d. List all quadruples of elements of P. Run through this list. For each quadruple, if there is a unique largest dissimilarity $d^*(A, B)$, and if the other two elements of the quadruple are C_1 and C_2, then reduce $d^*(A, B)$ to the next largest dissimilarity value l if $d^*(C_1, C_2) > ul$; if $d^*(C_1, C_2) \leqslant ul$, or if the largest dissimilarity is not unique, go on to the next quadruple without making any change. If the complete list can be run through without any change being made, d^* is u-diametric, and the process stops. If initially $d^* = d$, then when the process stops, which it does after finitely many steps, we have $C_u(d) = d^*$. Thus $C_u(d)$ is computable from d in finitely many steps.

Just as B_k is nested, so C_u is nested for values of u which may be taken to be in the range $[0, 1]$, and $C_u \circ C_v = C_u$ if $u \leqslant v$. This is computationally convenient for precisely the same reasons that the relation $B_k \circ B_j = B_k$ for $k \leqslant j$ is convenient, since it means that when $C_u(d)$ is to be found for a number of values of u, this can be done by applying C_u to $C_v(d)$ rather than to d itself if $u \leqslant v$. $C_u = I$ when $u = 1$, so complete information about the initial DC is recovered in this case; but note that C_u is not the single-link method for any value of u.

8.8. OTHER TYPE C METHODS

u-diametric clustering relates permitted overlap diameter to current level. The alternative to this is to relate overlap diameter to some function of the DC values; we require that, for some function $\phi : \mathbf{C}(P) \to \mathscr{R}$, the overlap between distinct ML-sets at the same level should have diameter at most $\phi(d)$. Possible functions ϕ include

$$S_0(d) = \max\{d(A, B) : A, B \in P\}$$

$$S_1(d) = \Sigma\{d(A, B) : \{A, B\} \text{ a 2-element subset of } P\}$$

and the functions νS_0, νS_1, where $\nu \geqslant 0$.

The choice of the function ϕ is restricted by the requirement that the resultant subdominant method should satisfy condition M3. We say that a DC is ϕ-diametric if it satisfies the following condition.

DCDi$_\phi$ If $d \in \mathbf{C}(P)$, then d is ϕ-diametric if whenever $\emptyset \neq S \subseteq P$, A, $B \in P$, then, writing

$$l = \max\{\text{diam}(d, S \cup \{A\}), \text{diam}(d, S \cup \{B\})\},$$

we have

$$\text{diam}(d, S) > \phi(d) \Rightarrow d(A, B) \leqslant l.$$

The set of ϕ-diametric DC's is denoted by $\mathbf{E}_\phi(P)$, and the subdominant method associated with it by $C_\phi : \mathbf{C}(P) \to \mathbf{E}_\phi(P)$. If $\phi = \nu S_0$, νS_1, we write this as $C_\nu^0 : \mathbf{C}(P) \to \mathbf{E}_\nu^0(P)$, $C_\nu^1 : \mathbf{C}(P) \to \mathbf{E}_\nu^1(P)$. We shall make no use of these methods here, and we do not discuss them in detail: they are given simply as further examples of type C methods, and we have of course by no means exhausted the possibilities.

8.9. OTHER NON-HIERARCHIC METHODS

We now mention very briefly some ways in which external restrictions on overlap can arise. It is not our intention to explore the resultant cluster methods here, and we confine ourselves to observing that they are likely in many cases to be computationally unmanageable. We shall simply point out three lines of approach; the details can be filled in by analogy with the development of the types B and C methods considered already.

Probably the most obvious external constraint is provided by measuring overlap diameter in terms of the original DC rather than the transformed DC. Again, this will have to be related either to current level as in u-diametric clustering or to some function of DC values. This should lead to methods which will be reasonably acceptable from the computational point of view.

In Part I we constructed K-dissimilarity as a method of measuring dissimilarity over sets of OTU's, and although we were specifically interested in using it for two OTU's at a time, there is no reason in theory why it should not be used to provide a general measure of the size of a set of OTU's, and hence to measure overlap size and generate a cluster method. Almost certainly the computational problems involved would render such a method unrealizable at present, but in any case it seems probable that the use of K-dissimilarity on sets of more than two OTU's may involve developing a new approach in which the DC is only part of a more general structure. This is an unexplored possibility; all that can be said with confidence is that any such approach will lead to formidable programming difficulties.

The third possibility is the generalization of B_k to deal with a weight function on the OTU's. Each OTU would be given a weight, and the total weight of each overlap would be restricted in some way. Such an approach might provide an alternative to the use of some type C methods as a technique for increasing robustness with respect to duplication of OTU's. Methods of this type would be on the borderline of computational feasibility at present.

Another way in which other types of non-hierarchic method could arise is by the consideration of methods which are not subdominant methods. No examples of methods of this type are known, and it seems likely that they would be difficult to work with.

8.10. SUMMARY

At this point it is convenient to summarize what we have done in Chapters 7 and 8. In Chapter 7 we constructed four type A cluster methods, A_S, A_U, A_W, A_C. One of these, A_S, is the subdominant method associated with the set $\mathbf{U}(P)$ of ultrametric DC's. Taking this as a basis for generalization, in this chapter we constructed two families of type B cluster methods, namely B_k and B_k^c for $k \geqslant 1$; we have $B_1 = B_1^c = A_S$, and if $|P| = p$, $B_{p-1} = B_{p-1}^c = I$. Then we constructed a family of type C methods, C_u for $u \in [0, 1]$, based on relating diameter of overlap to level, and two families C_v^0, C_v^1, for $v \geqslant 0$, based on relating overlap to DC values; we observed that $C_1 = I$, and $C_0 = B_2$ on definite DC's.

The methods which we shall later make practical use of are B_k (fine k-clustering) and C_u (u-diametric clustering). Note that since $B_1 = A_S$ the single-link method is included in this system. We shall show that A_U, A_W, and A_C are unsatisfactory; we have already pointed out that B_k^c is less satisfactory than B_k. We shall not explore the possibilities of C_v^0, C_v^1, and other similar methods.

CHAPTER 9

An Axiomatic Approach to Cluster Analysis

9.1. DEFINITIONS AND CONSTRUCTIONS

Having in Chapters 7 and 8 prepared the ground by the introduction of examples of both hierarchic and non-hierarchic cluster methods, we are now in a position to advance the discussion of their properties from the comparatively informal analysis carried out in those chapters to a detailed investigation from an axiomatic point of view.

First, we recapitulate some old definitions and introduce some new ones. P will always denote a finite non-empty set with p elements called OTU's. A *dissimilarity coefficient* (*DC*) on P is a function $d : P \times P \to \mathscr{R}$, the real numbers, such that the following conditions are satisfied.

DC1	$d(A, B) \geqslant 0$	for all $A, B \in P$
DC2	$d(A, A) = 0$	for all $A \in P$
DC3	$d(A, B) = d(B, A)$	for all $A, B \in P$

This is a very general definition, and it is possible to impose a number of further conditions, of which four will be of interest. As it stands the definition of a DC allows distinct OTU's to differ by amount zero. In general this is not disturbing; it might be that some OTU had been counted twice. What would be rather disturbing would be for OTU's with zero dissimilarity to have different dissimilarities from other OTU's. Generally the DC's encountered in practice will satisfy a condition preventing this possibility, as follows.

$$\text{DCE} \qquad d(A_1, A_2) = 0 \Rightarrow d(A_1, B) = d(A_2, B) \text{ for all } A_1, A_2, B \in P$$

This is called the *evenness* condition, and DC's satisfying it are called *even.* It does not appear to be a condition of any great mathematical importance, at least in what follows here. Sometimes it is actually desirable to eliminate the possibility of distinct OTU's differing by zero amount. This can be ensured by imposing the *definiteness* condition, as follows.

$$\text{DCD} \qquad d(A, B) = 0 \Rightarrow A = B \text{ for all } A, B \in P$$

DC's satisfying this condition are called *definite.* The definiteness condition is of considerable importance in the theorems which follow, and frequently enters into the statements of the theorems as a condition for the truth of the results, or is made use of in the course of the proofs. Sometimes DC's calculated in a certain way may satisfy the well-known *metric (triangle) inequality*, as follows.

$$\text{DCM} \qquad d(A_1, A_2) + d(A_2, A_3) \geqslant d(A_1, A_3) \text{ for all } A_1, A_2, A_3 \in P$$

Such DC's are said to be *metric.* Definite metric DC's are metrics on P in the usual sense. A stronger inequality which is of considerably more importance in this context–although not, of course, in general–is the *ultrametric inequality*, which is the following.

$$\text{DCU} \qquad \max\{d(A_1, A_2), d(A_2, A_3)\} \geqslant d(A_1, A_3) \qquad \text{for all } A_1, A_2, A_3 \in P$$

Such DC's are said to be *ultrametric.* Definite ultrametric DC's are ultrametrics on P in the usual sense. The following implications are obvious.

$$\text{DCU} \Rightarrow \text{DCM} \Rightarrow \text{DCE} \Leftarrow \text{DCD}$$

It will be useful to have a system of notation for the sets of DC's characterized by these various conditions; the following notation will be employed.

$\mathbf{C}(P)$	for the set of DC's on P.
$\mathbf{C}'(P)$	for the set of definite DC's on P.
$\mathbf{M}(P)$	for the set of metric DC's on P.
$\mathbf{U}(P)$	for the set of ultrametric DC's on P.
$\mathbf{M}'(P)$, $\mathbf{U}'(P)$	for $\mathbf{M}(P) \cap \mathbf{C}'(P)$, $\mathbf{U}(P) \cap \mathbf{C}'(P)$.

Thus we have the following diagram.

$$\begin{array}{ccccc} \mathbf{U}(P) & \subseteq & \mathbf{M}(P) & & \\ \cup\!| & & \cup\!| & & \\ \mathbf{U}'(P) & \subseteq & \mathbf{M}'(P) & \subseteq & \mathbf{C}'(P) \end{array}$$

There are p^2 elements in $P \times P$, but these represent only $\frac{1}{2}p(p-1)$ 2-element subsets, so a DC can be specified by $\frac{1}{2}p(p-1)$ real non-negative numbers, one for each 2-element subset of P. Thus $\mathbf{C}(P)$ is in natural 1-1 correspondence with the closed positive cone in euclidean space of dimension $\frac{1}{2}p(p-1)$ whose co-ordinate axes are labelled by the 2-element subsets of P. Euclidean space is a finite-dimensional real normed space, so there is a unique norm topology, namely the usual topology for euclidean space. $\mathbf{C}(P)$ inherits a structure as a topological space because it is a subspace of euclidean space, and every subspace of $\mathbf{C}(P)$ inherits a topology from the topology for $\mathbf{C}(P)$. It is, of course, this topology which was involved in the brief discussion of continuity of type A methods at the end of Chapter 7. Occasionally it is useful to incorporate also one of the norms to provide a metric on $\mathbf{C}(P)$; the only such metrics of interest here are the following.

$$\Delta_0(d_1, d_2) = \max \{|d_1(A,B) - d_2(A,B)| : A, B \in P\}$$

$$\Delta_{\frac{1}{2}}(d_1, d_2) = \{\Sigma[d_1(A,B) - d_2(A,B)]^2\}^{\frac{1}{2}}$$

$$\Delta_1(d_1, d_2) = \Sigma|d_1(A,B) - d_2(A,B)|$$

where d_1, d_2 are DC's on P, and thus elements of $\mathbf{C}(P)$, and where the summations are taken over all 2-element subsets of P. Methods of comparing DC's are discussed further in Chapter 11. The DC defined by

$$d(A,B) = 0 \quad \text{for all } A, B \in P$$

will be denoted by $\overline{0}$; it is represented by the origin in the euclidean space of which $\mathbf{C}(P)$ is a subspace. If d_1, d_2 are DC's then, as before, we write $d_1 \leqslant d_2$ (d_1 is *dominated* by d_2) to mean

$$d_1(A,B) \leqslant d_2(A,B) \quad \text{for all } A, B \in P$$

In Chapter 7 the end-product of a cluster method was taken to be a dendrogram, axiomatized in an appropriate way; in Chapter 8 this definition was extended to that of a numerically stratified clustering by extending the range of the function c from the set of equivalence relations on P to the set $\Sigma(P)$ of arbitrary symmetric reflexive relations on P, or some subset of this.

We repeat the definition here. A *numerically stratified clustering* (*NSC*) is a function $c : [0, \infty) \to \Sigma(P)$ such that the following conditions hold.

NSC1 $\quad 0 \leqslant h \leqslant h' < \infty \Rightarrow c(h) \subseteq c(h')$

NSC2 $\quad c(h)$ is eventually $P \times P$

NSC3 $\quad$ Given $h \geqslant 0$ there exists $\delta > 0$ such that $c(h + \delta) = c(h)$

We shall be interested in special kinds of NSC, for example the dendrograms. A general NSC is the same as a $(p-1)$-dendrogram if $|P| = p$. The special kinds of NSC can be obtained by restricting c to map to some subset of $\Sigma(P)$. In Chapter 10 we investigate exactly when this does happen.

An NSC is called *definite* if it satisfies the following condition.

NSCD $\quad c(0)$ is the equality relation on P

It is called *hierarchical* if it satisfies the following condition.

NSCH $\quad c(h) \in \mathrm{E}(P)$ for all $h \geqslant 0$

A definite hierarchical NSC is a dendrogram in the sense in which the word is commonly used in the literature, with all OTU's distinguished at level zero; the term 'dendrogram' is used slightly differently here, where it means simply a hierarchical NSC, the definiteness condition being omitted in the definition.

If c is an NSC then $c(h)$ is, for each value of $h \geqslant 0$, a symmetric reflexive relation. The ML-sets for c at level h have been defined to be those maximal subsets Q of P for which $c(h)|Q$ is the universal relation. This definition has also been given in terms of graphs, namely that if we take a graph whose vertices are elements of P and whose edges link just those elements of P corresponding to pairs in the relation $c(h)$, then the ML-sets are the vertex sets of maximal complete subgraphs.

We also constructed in Chapter 8 the 1-1 correspondence T from the set $\mathbf{C}(P)$ of DC's on P to the set of NSC's on P; this is given by

$$(Td)(h) = \{(A, B) : d(A, B) \leqslant h\}$$
$$(T^{-1} c)(A, B) = \inf \{h : (A, B) \in c(h)\}$$

where T^{-1} is the inverse of T, d is a DC, and c is an NSC. T induces by restriction 1-1 correspondences between $\mathbf{C}'(P)$ and the definite NSC's, and between $\mathbf{U}(P)$ and the hierarchical NSC's. Because of the existence of this 1-1 correspondence in explicit form, it is possible at any stage of the exposition to consider either DC's or NSC's, and use will be made of this fact. Thus, we have seen that ML-sets may be recognized in a DC at level h as ML-sets at level

h for the corresponding NSC; it follows from this that a DC is naturally specified by a system of ML-sets at each level, these ML-sets possibly overlapping to any extent.

9.2. THE NATURE OF CLUSTER METHODS

We have already pointed out in Chapter 8 that the possibility of identifying NSC's and DC's, although of great utility, can be misleading. In particular, it is necessary to realize that, although ML-sets may be recognized in the data DC, it is possible that these may not be related to the type of cluster which is sought and which will appear as the ML-sets associated with the output DC. Even if the ML-sets of the data DC are required to be related to the ML-sets of the output DC in some way, an arbitrary DC generally has far too complicated a system of ML-sets for it to be possible readily to extract useful information directly from them. We shall thus select a particular type of NSC as output, the selection being based on the need to be able to appreciate what the system of clusters at each level looks like. Not all DC's under consideration will in general lie in the subset of $\mathbf{C}(P)$ corresponding to the chosen type of NSC, which might, for example, be hierarchic NSC's. Those which do not will have to be modified before a suitable NSC can be obtained. Let $\mathbf{Z}$ denote the subset of $\mathbf{C}(P)$ corresponding to the type of NSC sought as the result, and $\mathbf{A}$ the subset corresponding to the type of data arising from the problem. Thus, $\mathbf{A}$ might be $\mathbf{M}(P)$; very commonly it is conveniently taken to be the whole of $\mathbf{C}(P)$, as has been done in Chapters 7 and 8. We shall see that in many cases taking $\mathbf{A} = \mathbf{C}(P)$ in fact imposes no real restriction. $\mathbf{Z}$ might, for example, be $\mathbf{U}(P)$. A cluster method may then be thought of as a process which assigns to each element of $\mathbf{A}$ an element of $\mathbf{Z}$; that is, as a function $D : \mathbf{A} \to \mathbf{Z}$. $\mathbf{A}$ is called the *data set*; $\mathbf{Z}$ the *target set*. Before the constraints on D are examined, it is convenient to review the assumptions which have been made up to this stage.

Three substantial restrictions have already been placed on 'cluster method'. First, that the data is a DC; second, that the result is an NSC, and hence, under the correspondence T, is also a DC; and third, that the method can be represented by a function, in other words is one-valued. Each of these assumptions can be challenged, and indeed there are methods within the general heading of 'cluster analysis' for which they do not hold good. In making them we are not discarding the other methods but rather focussing attention on those which are of value in taxonomy, and the assumption will be examined in this light. The methods which are most commonly used either satisfy the assumptions or are clearly intended to do so.

There are some cluster methods which operate directly from attribute data without making any use of a DC as an intermediate stage, often these methods have some kind of information-theoretic background, and some of them lead not to a stratified clustering but to a simple partition or covering. Such methods are discussed in Chapter 12: they fall outside the scope of the treatment given in this chapter and the next.

Another form of data much closer to a DC is an ordinal-valued DC, in which only the ranking of the dissimilarities and not their actual values is of significance. There are two rather different ways in which such a DC can arise. In some cases a numerical valued DC may be reduced to ordinal form because of doubts about the significance of the actual numerical values; in other situations the DC may have been in ordinal form from the start. In the first case the correct procedure would appear to be the retention of the numerical values and the application of a method which is equivariant with respect to the action of all appropriate monotone transformations of the dissimilarity scale; methods of this type are investigated in detail in the next chapter. In the second case, since ordinals rather than reals are involved, it is difficult to see how to justify the use of any method which makes use of addition or multiplication operations, since it is inherent in an ordinal scale that no comparisons may be made which involve the sizes of the intervals between values. The details will not be explored here; it will be sufficient to observe that any of the methods described in the next chapter may be applied with confidence to genuinely ordinal data to yield an ordinally stratified result rather than a numerically stratified one. Some alternative methods, devised specifically for ordinal data, are mentioned in Chapter 12. The process of putting a numerically stratified DC into ordinal form is of dubious merit as a device for reducing the effects of errors if it is done without any accompanying account of the statistics of the situation, and it becomes unnecessary if methods equivariant under monotone transformations are employed.

The second assumption is that the result is in the form of an NSC. In this chapter and the next we are concerned only with numerically stratified cluster methods. It is possible to devise stratified systems of clusters where it is not in general possible to describe each level in terms of a symmetric reflexive relation, but such systems appear to be markedly inferior to NSC's as a means of handling data in the form of a DC, and they will not be discussed here. See also the remarks in Chapter 8.1.

The final point is the one-valuedness of the cluster method. Situations may arise in which the use to be made of the result of a cluster method renders it quite unimportant that the result should be unique, but very often the whole point of using a cluster method is to enable comparisons to be made, and in this case uniqueness is essential.

9.3. CONDITIONS ON CLUSTER METHODS

The above model for cluster methods which operate on a dissimilarity coefficient to produce a stratified clustering is quite general, although as we have pointed out, it does not cover certain cluster methods which have different starting points and different end-products. The model provides a peg on which further requirements about the properties and behaviour of cluster methods can be hung as desired; it is, for example, largely at this stage that a specification of what is meant by a cluster is built into the model. We emphasize that different cluster concepts and different requirements might lead to models within which different families of cluster methods proved acceptable.

The following mathematical conditions provide interpretations for the basic assumptions about cluster methods listed above, and for certain further requirements about the behaviour of cluster methods which are needed if they are to be appropriate for use in taxonomy. The plausibility of each requirement may be assessed by considering the consequences of use in taxonomy of a cluster method which fails to satisfy it. The usual pattern is that the constraints come in pairs, one a constraint on the function D, and the other a constraint on the sets **A**, **Z** to make the constraint on D a reasonable one. The axiomatic treatment which follows in this chapter and the next is based on Sibson (1970a).

Condition 1–Appropriateness

It would be unreasonable to try to represent the data by NSC's grossly ill-adapted to the purpose, and if the data shows structure of the type sought then it can be left unchanged. The mathematical form is as follows.

S1 $\quad \emptyset \neq \mathbf{Z}, \mathbf{Z} \subseteq \mathbf{A}$

M1 $\quad D|\mathbf{Z}$ is the identity on $\mathbf{Z}$

This embodies the assumption that when the data DC actually lies inside **Z**, then we may take the clusters to be the ML-sets of the data DC. It does not, however, imply any relationship between the ML-sets of d and the ML-sets–that is, the clusters–of $D(d)$ when d does not lie in **Z**.

Condition 2–Label freedom

The method used must be independent of any prior labelling of the objects. Otherwise said: D must commute with the action of the symmetric group on the objects. Suppose that ρ is a permutation of P. Then the conditions are as follows.

S2 $\quad d \in \mathbf{A} \Rightarrow d \circ (\rho \times \rho) \in \mathbf{A}$

$\quad d \in \mathbf{Z} \Rightarrow d \circ (\rho \times \rho) \in \mathbf{Z}$

M2 $\quad D(d) \circ (\rho \times \rho) = D[d \circ (\rho \times \rho)]$

This condition is related to well-definedness, in that it eliminates the possibility of rendering an ill-defined method well-defined by making arbitrary choices based on the order in which the OTU's are considered.

Condition 3–Scale freedom

The method must be independent of scale factors. Otherwise said: D must commute with multiplication by strictly positive scalars. Let $\alpha > 0$. Then the conditions are as follows.

S3 $\quad d \in \mathbf{A} \Rightarrow \alpha d \in \mathbf{A}$

$\quad d \in \mathbf{Z} \Rightarrow \alpha d \in \mathbf{Z}$

M3 $\quad D(\alpha d) = \alpha D(d)$

We have already seen in Chapter 8 that this condition cuts down the number of possible type C cluster methods. A good example of the need for such a condition is given by the use of K-dissimilarity as the DC–changing the base of the logarithms has the effect of introducing an overall scale factor.

For the above three conditions, the intuitive constraint involved in each case admits of only one reasonable interpretation. We now consider constraints such as cluster-preservation, where there may well be numerous possible interpretations leading to different mathematical conditions on cluster methods. We shall put forward one possible set of interpretations of the intuitive constraints of cluster-preservation, optimality, and stability, and investigate their consequences. The cluster-preservation and optimality constraints are interpreted by way of the ML-sets for the data DC. We have warned against thoughtlessly identifying these with, or relating them to, the clusters sought as output. We now argue that it is in fact reasonable to take the ML-sets in the initial DC as the basis for constructing clusters, provided that it is realized that this is not inevitable and may well be only one of a number of possible approaches. We suggest that this approach is much the simplest and most obvious one, and it also has a retrospective justification in that the methods to which it leads have various satisfactory mathematical properties. Needham (1961a) and Bonner (1964) have both suggested that the ML-sets at each level should form the basis from which clusters are constructed. Conditions 4 and 5 are responsible for relating ML-sets in the data to those in the output. Condition 4 is a cluster-preservation condition, and Condition 5 an optimality condition.

Condition 4–Preservation of clusters

Let S be an ML-set at level h in d. This ML-set may have other objects added to it at the same level in $D(d)$, but it must not be broken up. That is, if S is an ML-set at level h in d, then there must be an ML-set S' at level h in $D(d)$ such that $S \subseteq S'$; that is, S is contained in a cluster at level h. S' need not be unique, although it will be if $\mathbf{Z} = \mathbf{U}(P)$. In the case of this condition it is slightly less easy to see what the appropriate mathematical form is, but it may readily be checked that the following conditions have exactly the desired effect.

S4 If $d \in \mathbf{A}$, then there exists $d' \in \mathbf{Z}$ with $d' \leqslant d$

M4 $D(d) \leqslant d$

Condition 5–Optimality

Clustering objects together should not be done unnecessarily, but only in response to some observed structure in the data; that is, the resultant NSC, or DC in $\mathbf{Z}$, should in some sense, subject to the earlier conditions, be the best possible fit to the original data. Because of the form of S4 and M4, it is possible to take the following as the optimality condition.

M5 $[d' \in \mathbf{Z} \text{ and } D(d) \leqslant d' \leqslant d] \Rightarrow [d' = D(d)]$

This condition ensures that $D(d)$ is a local optimum; to obtain global optimality the following condition is imposed on $\mathbf{Z}$.

S5 If $\mathbf{Y}$ is a bounded subset of $\mathbf{Z}$, then the DC sup $\mathbf{Y}$ is an element of $\mathbf{Z}$

S5 has the important consequence that together with S3 it ensures that $\mathbf{Z} - \{\bar{0}\}$ is path-connected. This is established in the following lemma.

PATH-CONNECTEDNESS LEMMA

Conditions S3 *and* S5 *imply that* $\mathbf{Z} - \{\bar{0}\}$ *is path-connected.*

Proof

Suppose that d,d' are elements of **Z**. Consider the DC d_x defined by

$$d_x(A, B) = \max\{xd(A, B), (1 - x)d'(A, B)\}.$$

x is required to lie in the interval $[0,1]$. We have $d_0 = d$, $d_1 = d'$. The DC's xd, $(1 - x)d'$ are in **Z** for all $x \in (0, 1)$ by condition S3, and hence d_x is in **Z** for all $x \in (0, 1)$ by condition S5. Hence $d_x \in \mathbf{Z}$ for all $x \in [0, 1]$. If $d, d' \neq 0$ then $d_x \neq 0$ for all $d \in [0, 1]$, and since the function sending x to d_x is continuous, it is a path in $\mathbf{Z} - \{\bar{0}\}$ from d to d'.□

Since, by S1, we have $\mathbf{Z} \subseteq \mathbf{A}$, the set $\mathbf{Z} - \{\bar{0}\}$ lies in just one path-component of $\mathbf{A} - \{\bar{0}\}$, and hence, by the same kind of reasoning as that leading to the imposition of condition S1, it is unreasonable to represent elements of other path-components of $\mathbf{A} - \{\bar{0}\}$ by elements of $\mathbf{Z} - \{\bar{0}\}$. Hence it is desirable to impose the following condition.

S5b $\mathbf{A} - \{\bar{0}\}$ is path-connected.

Conditions 4 and 5 bring us back onto what is by now familiar ground, namely the subdominant methods; we define a subdominant method to be the method D specified by M4 and M5 when **A**, **Z** satisfy S1, S4, S5, S5b, and we observe immediately that every subdominant method satisfies M1. When dealing with subdominant methods we shall find it convenient to take $\mathbf{A} = \mathbf{C}(P)$, because if $D_{\mathbf{A}}$ is the subdominant method associated with **A**, **Z**, and D with **C**(P), **Z**, then $D|\mathbf{A} = D_{\mathbf{A}}$.

Condition 6–Stability

It is frequently stated that cluster methods should be *stable*, this being taken to mean that the overall effect of changes small by comparison with the scale of the system being investigated should itself be small. It is immediately possible to separate stability conditions into two types, namely those concerned with small changes in the DC, and those concerned with the introduction or removal of elements of P. These two types are really quite separate, and it would lead to needless confusion to consider them together. Conditions concerned with introduction or removal of elements of P will be called *fitting conditions*, and are considered later in this chapter. *Stability conditions* will be taken to be those concerned with small changes in the DC. Since any change in a DC can be accomplished by a sequence of small changes, it is not meaningful to ask that the output of a cluster method should be unaffected by small changes in the input DC; we should rather require that the response should be in some way well-behaved. The need for such a condition arises whenever there are sources of change in the input DC and we wish to make comparisons between the output DC's arising from them. There are at least three such sources of change for input DC's. The first is the occurrence of sampling errors in the measurement of dissimilarity. For example, we may estimate $K(\mu_1, \mu_2)$ (see Chapter 2 and Appendix 1) where μ_1, μ_2 are probability measures on a two-point space (i.e. μ_1, μ_2 are character states over a 2-state variable attribute) by the estimator

$$\overline{K} = K(\bar{\mu}_1, \bar{\mu}_2) \text{ where } \bar{\mu}_i = \left(\frac{r_i}{n}, \frac{n - r_i}{n}\right)$$

which has a small positive bias. We know in numerical terms how the sampling errors behave (see Appendix 1), but they cannot be eliminated, only made very small with high probability by taking large samples. Thus for comparability and repeatability to be possible, we need to consider cluster methods which do not magnify such errors unduly. Much the same remarks apply to the second source of change, namely the inevitable rounding errors introduced in computation. The third source of change is rather different in origin. In taxonomic studies it may happen that the dissimilarities between OTU's settle down in relation to one another as the number of attributes considered is increased. It is not the actual values which appear to approach limits, but their ratios, so we shall be considering small changes in ratios of dissimilarity values. If we are dealing with methods satisfying S3, M3 then it is possible to view this in terms of small changes in dissimilarity values. It is clearly going to be meaningless to ask for information about the details of the way in which such changes are transmitted by the cluster method unless we know that as the size of change tends to zero, so does its effect. Thus we shall impose the following condition.

M6 $D : \mathbf{A} \rightarrow \mathbf{Z}$ is continuous

It is best to regard this condition not as the condition for stability, but rather as a necessary condition for the possibility of investigating stability. In Chapter 7 an example was given to illustrate discontinuity in the cluster methods A_W, A_U, A_C, and it was stated that the single-link method A_S was continuous. We shall see that this is a special case of a general result about subdominant methods. Professor Benzécri (personal communication) has pointed out that since it is to be expected that any finite set will be given the discrete topology, any cluster method leading directly to a taxonomic hierarchy or to a simple clustering, will, since $\mathbf{A}$ is path-connected, be either constant or discontinuous; he goes on to suggest that in such circumstances either it is misguided to seek stability or it is necessary to find some different interpretation of it. When the more flexible kind of clustering described here is employed as the output of the method, stability appears to be a perfectly valid concept, and the imposition of continuity as a necessary condition for it certainly leads to no absurdities.

The various conditions on **A**, **Z** and D suggested above are by no means independent of one another. It is easily seen that if **A**, **Z** satisfy S4 and S5 and D satisfies M4 and M5, then D is determined by **A**, **Z**. If S4, S5, M4, M5 hold, then so do the following implications.

S1 $\Rightarrow$ M1
S2 $\Rightarrow$ M2
S3 $\Rightarrow$ M3

The dependence of M6 on the other conditions may be less apparent; the following lemma makes the situation clear.

FIRST CONTINUITY LEMMA

If conditions S3, S4, S5 *and* M4, M5 *hold, then* D *is continuous on* $\mathbf{A} \cap [\mathbf{C}'(P) \cup \{\bar{0}\}]$, *and this is the strongest general result.*

Proof

Continuity at $\bar{0}$, if it lies in **A**, is obvious. Suppose $d \in [\mathbf{A} \cap \mathbf{C}'(P)]$. Let l be the minimal value taken by d on a pair of distinct elements of P; $l > 0$ because $d \in \mathbf{C}'(P)$. Choose $\epsilon > 0$ such that $\epsilon < l$, and suppose that d' is such that $\Delta_0(d, d') \leqslant \epsilon$. Then

$$[(l - \epsilon)/l]\, d \leqslant d' \leqslant [(l + \epsilon)/l]\, d.$$

The conditions in the statement of the theorem imply M3 and that

$$d_1 \leqslant d_2 \Rightarrow D(d_1) \leqslant D(d_2)$$

so

$$[(l - \epsilon)/l]\, D(d) \leqslant D(d') \leqslant [(l + \epsilon)/l]\, D(d).$$

Now suppose that $m = \Delta_0(D(d), \bar{0})$. Then

$$\Delta_0(D(d), D(d')) \leqslant m\epsilon/l$$

and so D is continuous at d in terms of the metric Δ_0. But Δ_0 induces the usual topology on $\mathbf{C}(P)$, and d is any element of $\mathbf{A} \cap \mathbf{C}'(P)$, so it follows that D is continuous on $\mathbf{A} \cap \mathbf{C}'(P)$ and hence on $\mathbf{A} \cap [\mathbf{C}'(P) \cup \{\bar{0}\}]$. That this is the strongest result may be seen by considering $\mathbf{A} = \mathbf{C}(P)$, $\mathbf{Z} = [\mathbf{C}'(P) \cup \{\bar{0}\}]$; in this case D is continuous precisely on $[\mathbf{C}'(P) \cup \{\bar{0}\}]$.

This result shows that to take continuity as a necessary condition for stability imposes little further restriction on a subdominant method and it emphasizes the basic nature of the cluster-preservation and optimality conditions; this is a desirable but hardly surprising state of affairs. Because of the lemma, and of the uniqueness of D given **A** and **Z**, the investigation of cluster methods acceptable within the model, that is, subdominant methods, reduces to a search for pairs **A**, **Z** satisfying the S conditions and an investigation of the continuity of D on the set $\mathbf{A} - [\mathbf{C}'(P) \cup \{\bar{0}\}]$. This investigation is facilitated by the observation that if **A**, **Z** satisfy the S conditions, then so do $\mathbf{C}(P)$, $[\mathbf{Z} \cup \{\bar{0}\}]$, and the cluster map D for **A**, **Z** is the restriction to **A** of the map for $\mathbf{C}(P)$, $[\mathbf{Z} \cup \{\bar{0}\}]$. A typical example of a cluster method is that determined by $\mathbf{A} = \mathbf{C}(P)$, $\mathbf{Z} = \mathbf{U}(P)$, which is of course the single-link method A_S. This is continuous on the whole of $\mathbf{C}(P)$; moreover it is the only hierarchic method acceptable within the model. Note in particular that the methods described in Chapter 7 as weighted and unweighted average-link, and complete-link methods do not satisfy conditions M4, M5, and M6; thus even with any alternative versions of the cluster preservation and optimality conditions, these methods would be rejected on grounds of discontinuity.

9.4. SEQUENCES OF METHODS: FITTING CONDITIONS

It is usually of little value to have cluster methods which operate on just one set P; the situation which occurs in practice is that a method is defined for finite sets of all sizes in such a way that the method as applied over the individual sets obeys some kind of fitting-together condition. As was pointed out above, this is sometimes referred to as stability. The 'stability' discussed under Condition 6 and the fitting-together conditions discussed here are quite unrelated. To permit the incorporation of fitting-together conditions, the model so far established must be elaborated considerably. In an attempt to make the arguments as perspicuous as possible, the introduction of this elaboration has been delayed up to this point. The modification is made by stringing together a lot of models of the basic type. Instead of a single set P, it is necessary to consider a system of sets (P_i) which will be denoted by P_*. i ranges through integers not less than 1, and $|P_i| = i$. If $i \leqslant j$ let E_{ij} be the set of embeddings (1-1 functions) $\phi: P_i \to P_j$. Let $\mathbf{C}(P_*)$ denote the sequence $(\mathbf{C}(P_1), \mathbf{C}(P_2), \ldots)$. If $\phi \in E_{ij}$, then ϕ induces a continuous map $\phi^* : \mathbf{C}(P_j) \to \mathbf{C}(P_i)$ given by

$$(\phi^* d)(A, B) = d(\phi A, \phi B)$$

where $A, B \in P_i, d \in \mathbf{C}(P_j)$. If $\phi \in E_{ii'}, \psi \in E_{i'i''}$, then $\psi \circ \phi \in E_{ii''}$ and ϕ^*, ψ^* satisfy $(\psi \circ \phi)^* = \phi^* \circ \psi^*$. ϕ^* preserves each of the subsystems $\mathbf{E}, \mathbf{C}', \mathbf{M}, \mathbf{U}$. E_{ii} is simply the set of permutations of P_i, and if $\rho \in E_{ii}$ then $\rho^* d = d(\rho \times \rho)$. The fitting-together conditions are appropriate generalizations of the label-freedom conditions S2 and M2 suitable for use in connection with sub-dominant methods. In some circumstances it may be desirable to impose stronger or different conditions, although the point will not be explored further here. A *data series* is a sequence of sets $\mathbf{A}_* = (\mathbf{A}_1, \mathbf{A}_2, \ldots)$ with $\mathbf{A}_i \subseteq \mathbf{C}(P_i)$ and a *target series* is similarly a sequence $\mathbf{Z}_* = (\mathbf{Z}_1, \mathbf{Z}_2, \ldots)$. A *cluster series* is a sequence of functions $D_* = (D_1, D_2, \ldots)$ with $D_i : \mathbf{A}_i \to \mathbf{Z}_i$. The S and M conditions can be imposed termwise on these sequences, but of course they do not tie together what goes on for different values of i. This is accomplished by the fitting conditions, which are as follows.

SF If $\phi \in E_{ij}$, then ϕ^* maps $\mathbf{A}_j$ *onto* $\mathbf{A}_i$ and $\mathbf{Z}_j$ *onto* $\mathbf{Z}_i$

MF $D_i[\phi^*(d)] \geq \phi^*[D_j(d)]$

Condition SF requires that if d is a DC in $\mathbf{A}_j$ (or $\mathbf{Z}_j$), then its restriction to the subset ϕP_i of P_j should be in $\mathbf{A}_i$ (or $\mathbf{Z}_i$), and moreover that if d is defined only on ϕP_i and lies in $\mathbf{A}_i$ (or $\mathbf{Z}_i$), then it can be extended to a DC d' on the whole of P_j such that d' lies in $\mathbf{A}_j$ (or $\mathbf{Z}_j$). Condition MF is best interpreted by thinking of the effect of removing objects from P_j to leave ϕP_i; the resultant classification on ϕP_i must have no clusters which are not contained in clusters at the same level for the restriction to ϕP_i of the original classification on P_j, because removing elements of P can only break up clusters at any fixed level, never unite them.

The conditions SF and MF respectively imply S2 and M2 in each pair $\mathbf{A}_i$, $\mathbf{Z}_i$ and if S4, S5 and M4, M5 hold in each pair then the expected implication

$$\text{SF} \Rightarrow \text{MF}$$

also holds. A simple example of a cluster series is the single-link series. This is given by taking $\mathbf{A}_* = \mathbf{C}(P_*)$, $\mathbf{Z}_* = \mathbf{U}(P_*)$. The methods described in Chapter 8 are more general examples of cluster series. The series B_{k*} is given by $\mathbf{A}_{ki} = \mathbf{C}(P_i)$, $\mathbf{Z}_{ki} = \mathbf{C}_k(P_i)$, and the series B^c_{k*} is given by $\mathbf{A}_{ki} = \mathbf{C}(P_i)$, $\mathbf{Z}_{ki} = \mathbf{C}^c_k(P_i)$. $B_{1*} = B^c_{1*}$ is the single-link series, and since $\mathbf{C}_k(P_i) = \mathbf{C}^c_k(P_i) = \mathbf{C}(P_i)$ if $i \leq (k + 1)$, B_{ki}, B^c_{ki} are the identity if $i \leq (k + 1)$, so the families of series B_k, B^c_k have the property that for fixed i, B_{ki}, B^c_{ki} are the identity if k is large enough.

The sets P_i must be regarded as selected 'test' sets rather than as actual sets of OTU's. In dealing with an actual set of OTU's we would identify it with a set P_i of the appropriate size by means of a 1-1 correspondence.

This completes the construction of the basic model for cluster methods and series. In some ways it is a very powerful model. For example, the uniqueness of the series given $\mathbf{A}_*$ and $\mathbf{Z}_*$ means that if $\mathbf{A}_* = \mathbf{C}(P_*)$ and $\mathbf{Z}_* = \mathbf{U}(P_*)$ then there is just one series (single-link) which is acceptable, and that all other hierarchical series are ruled out immediately *within this model.* This is not to say that such series are of no value for doing other things; the point is that they have no place in this model, which is designed to incorporate a certain set of basic assumptions; to justify their use in other circumstances it would be necessary to construct a model appropriate to the particular situation. In other ways this model is extremely weak. For example, a perfectly acceptable cluster series within the model would be that obtained by taking $\mathbf{A}_* = \mathbf{C}(P_*)$, $\mathbf{Z}_* = \mathbf{M}(P_*)$ but this is, to say the least, not what is usually thought of as a classificatory method. In the next chapter the possibility of adding further axioms to strengthen the model is investigated, and we revert to working in terms of cluster methods rather than cluster series.

CHAPTER 10

Flat Cluster Methods

10.1. MONOTONE EQUIVARIANCE

It is not uncommon for the actual numerical values taken by a dissimilarity coefficient to have little significance in isolation, but only to be meaningful insofar as they are related to other values. The conditions

S3 If $\alpha > 0$ then $d \in \mathbf{A} \Rightarrow \alpha d \in \mathbf{A}$

$$d \in \mathbf{Z} \Rightarrow \alpha d \in \mathbf{Z}$$

M3 $D(\alpha d) = \alpha D(d)$

in the last chapter already incorporate the idea that scale factors are of no importance, and that the ratios between dissimilarities, rather than the actual values, provide the basis on which classification is to be carried out. Recall, for example, K-dissimilarity as defined in Part I. The value of this incorporates a scale factor arising from the choice, conventional in information theory, of 2 as the base for logarithms. Although there are reasons why the use of base 2 is convenient, the choice is really quite arbitrary, and changing it simply changes the dissimilarity values by a constant multiplier. Conditions S3 and M3 ensure that this leads to no anomalies. In some cases in which dissimilarities are used there may be no reason for using the dissimilarity as it stands rather than (say) its square root, or some other function of it which still yields a DC. In this situation it may still be possible to regard the order of the values, which is unchanged by such transformations, as containing relevant information. Usually more is significant than just the ordering, but it is rash to employ without careful justification methods such as C_u which assume this. Any method involving algebraic operations (for example, multiplication or addition) on the dissimilarity values rather than just order operations (for example, taking max or min) is liable to be subject to criticism on these grounds even if it is otherwise acceptable. What is needed is that the method should commute with all suitable monotone transformations of the dissimilarity scale rather than just with multiplication by constants. The appropriate type of transformation to consider in these circumstances is an invertible monotone transformation of $[0, \infty)$; it is clear that such a transformation and its inverse, which is a transformation of the same type, must both be continuous, and thus that the relevant transformations are the self-homeomorphisms of $[0, \infty)$. The point 0 is left fixed by all such transformations, which will be called *dissimilarity transformations.* The requisite conditions are as follows.

S3+ If θ is a dissimilarity transformation, then

$$d \in \mathbf{A} \Rightarrow \theta d \in \mathbf{A}$$

$$d \in \mathbf{Z} \Rightarrow \theta d \in \mathbf{Z}$$

M3+ $D(\theta d) = \theta D(d)$

Cluster methods $D:\mathbf{A} \to \mathbf{Z}$ such that S3+, M3+ are satisfied are called *monotone equivariant* (*ME*) cluster methods. If $\mathbf{D}:\mathbf{A} \to \mathbf{Z}$ is a subdominant method, then

$$\text{S3+} \Rightarrow \text{M3+}$$

so a subdominant method is ME if and only if $\mathbf{A}$, $\mathbf{Z}$ satisfy S3+. The methods C_u are not ME. If D is an ME method, then $[D(d)](P \times P) \subseteq d(P \times P)$, but this is not a sufficient condition to make D ME; a counter-example is given by C_u, which satisfies this condition but is not ME.

10.2. FLAT CLUSTER METHODS: THE FLATNESS AND REDUCTION THEOREMS

It will be convenient at this point to introduce two further technical conditions to eliminate a trivial case and to ensure that $\mathbf{Z}$ is well-behaved near the set of non-definite DC's. The conditions are the following.

S7 $\quad \mathbf{Z} \neq \{\bar{0}\}$

S8 $\quad \mathbf{Z}$ is closed in $\mathbf{C}(P)$

We shall now investigate the ME subdominant methods $D:\mathbf{A} \to \mathbf{Z}$ satisfying S2, S7, S8. If the pair **A**, **Z** satisfies these conditions then so does $\mathbf{C}(P)$, **Z** and moreover the map $D:\mathbf{A} \to \mathbf{Z}$ defined by **A**, **Z** is the restriction to **A** of the map $D:\mathbf{C}(P) \to \mathbf{Z}$ defined by $\mathbf{C}(P)$, **Z**. Thus there is no loss of generality in taking $\mathbf{A} = \mathbf{C}(P)$, and if this is done then S5b is automatically satisfied. Such a cluster method D is called a *flat cluster method,* and it is of interest to know how to recognize and describe such cluster methods. The required results are given by the theorems below. First, a definition. An *indicator family* is a subset J of the set $\Sigma(P)$ of symmetric reflexive relations on P satisfying the following conditions.

IF1	$P \times P \in J$
IF2	If $r \in J$ and ρ is a permutation of P, then $(\rho \times \rho)r \in J$
IF5	If $K \subseteq J$ then the relation $\cap K$ (the intersection of the elements of K) is in J
IF7	$J \neq \{P \times P\}$

The conditions are numbered to correspond with the S and M conditions with which they are associated. Note in particular that IF5 is related to S5. $\cap K$ is the relation r defined by

$$(A, B) \in r \Leftrightarrow [(A, B) \in r' \text{ for all } r' \in K].$$

If J is an indicator family, define a subset $\mathbf{Z}_J$ of $\mathbf{C}(P)$ by

$$\mathbf{Z}_J = \{d: (Td)(h) \in J \text{ for all } h \geqslant 0\}.$$

FLATNESS THEOREM

The pair $\mathbf{C}(P)$, $\mathbf{Z}$ *defines a flat cluster method if and only if* $\mathbf{Z} = \mathbf{Z}_J$ *for some indicator family* J.

Proof

It is quite easy to prove that if J is an indicator family then $\mathbf{C}(P)$, $\mathbf{Z}_J$ defines a flat cluster method. Certainly $\mathbf{Z}_J \subseteq \mathbf{C}(P)$, by definition, and $\bar{0} \in \mathbf{Z}_J$ since $P \times P \in J$ by IF1; hence S1 is satisfied, and also S4. S2, S5, S7 follow immediately from IF2, IF5, IF7.

If θ is a dissimilarity transformation, then we have

$$[T(\theta d)](\theta h) = (Td)(h)$$

so S3+ is satisfied by the pair $\mathbf{C}(P)$, $\mathbf{Z}_J$. Now let $d \in \mathbf{C}(P) - \mathbf{Z}_J$. There must exist some $h \geqslant 0$ for which $(Td)(h) \notin J$, and hence, because of condition NSC3, some $h' > 0$ and some $\delta > 0$ such that $\delta < h'$ and

$$(Td)(h'') = (Td)(h') \notin J \text{ for all } h'' \in [h' - \delta, h' + \delta].$$

Now if $\Delta_0(d, d') < \delta$ then

$$(Td')(h') = (Td)(h') \notin J,$$

so there is a neighbourhood of d containing no elements of $\mathbf{Z}_J$. It follows that $\mathbf{C}(P) - \mathbf{Z}_J$ is open in $\mathbf{C}(P)$, and hence that $\mathbf{Z}_J$ is closed in $\mathbf{C}(P)$, which is condition S8. Thus we conclude that $\mathbf{C}(P)$, $\mathbf{Z}_J$ defines a flat cluster method.

The proof of the converse is rather more complicated. The idea of the proof is to construct from some given $\mathbf{Z}$ an explicit subset J of $\Sigma(P)$ which is shown to be an indicator family such that $\mathbf{Z} \subseteq \mathbf{Z}_J$. This is quite simple; the harder part is showing that $\mathbf{Z}$ is in fact the whole of $\mathbf{Z}_J$. So suppose that $\mathbf{C}(P)$, $\mathbf{Z}$ defines a flat cluster method. Let J be the subset of $\Sigma(P)$ defined by

$$J = \{(Td)(h) : d \in \mathbf{Z}, h \geqslant 0\}.$$

If $r \in J$, then it is possible, by NSC3, to assume that $r = (Td)(h)$ for some $h > 0$. $\mathbf{Z}$ is non-empty, by S1, and so there exists at least one $d \in \mathbf{Z}$. $(Td)(h) = P \times P$ if h is large enough, so $P \times P \in J$, and J satisfies IF1. IF2, IF7 are immediate consequences of S2, S7 respectively. Now suppose $K \subseteq J$. K is finite, so has elements $r_1, \ldots, r_n$ with $r_i = (Td_i)(h_i)$ for some $d_i \in \mathbf{Z}, h_i > 0$. Let $d_i' = h_i^{-1} d_i$. $d_i' \in \mathbf{Z}$ by S3+ (or indeed by S3 – we do not yet need the full strength of S3+), and we have $(Td_i')(1) = (Td_i)(h_i) = r_i$. By S5 we have $\sup d_i' \in \mathbf{Z}$. But $[T(\sup d_i')](1) = \cap r_i = \cap K$, which must thus be in J, so J satisfies IF5 and is an indicator family. Clearly $\mathbf{Z} \subseteq \mathbf{Z}_J$, and it remains to prove that $\mathbf{Z} = \mathbf{Z}_J$; it is only in this part of the proof that the full strength of S3+ is made use of. First we show that

$$\mathbf{Z}_J \cap \mathbf{C}'(P) = \mathbf{Z} \cap \mathbf{C}'(P).$$

By S1, $\mathbf{Z} \neq \emptyset$, and by S7, $\mathbf{Z} \neq \{\bar{0}\}$, so there is some element $d \neq \bar{0}$ in $\mathbf{Z}$. Let

$$L = \max\{d(A, B) : A, B \in P\}.$$

If ρ is any permutation of P, then $d \circ (\rho \times \rho) \in \mathbf{Z}$ by S2, and

$$\{d(\rho \times \rho) : \rho \text{ is a permutation of } P\}$$

is thus a bounded subset of $\mathbf{Z}$. By S5, $\sup d \circ (\rho \times \rho) \in \mathbf{Z}$; but this is simply the DC δ^L given by

$$\begin{aligned} \delta^L(A, B) &= 0 \quad \text{if } A = B \\ &= L \quad \text{if } A \neq B. \end{aligned}$$

By S3, $\delta^l \in \mathbf{Z}$ for all $l > 0$, and by S8, $\vec{0} \in \mathbf{Z}$, so $\delta^l \in \mathbf{Z}$ for all $l \geqslant 0$. If $d \in \mathbf{Z}$, let d^ϵ denote the DC $\sup\{d, \delta^\epsilon\}$. This lies in $\mathbf{Z}$ for all $\epsilon \geqslant 0$ by S5, and if $\epsilon > 0$

it is in $\mathbf{Z} \cap \mathbf{C}'(P)$. Suppose $r \in J$. Then $r = (Td)(h)$ for some $d \in \mathbf{Z}$, $h > 0$. If $0 < \epsilon < h$, then $r = (Td^{\epsilon})(h)$, so if $r \in J$ there exists $d \in \mathbf{Z} \cap \mathbf{C}'(P)$ such that $r = (Td)(h)$ for some $h > 0$.

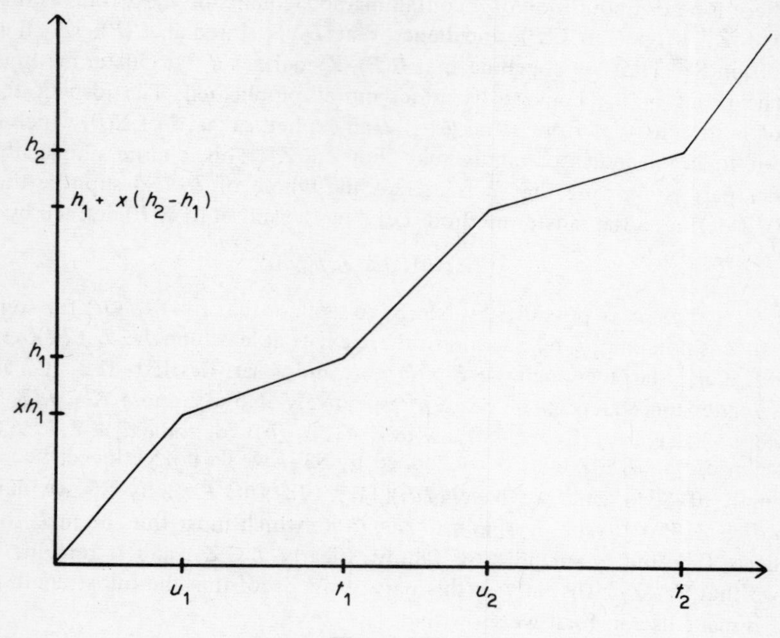

Figure 10.1.

Now let ΔP denote the equality relation on P, that is, the set $\{(A, A): A \in P)\}$. If there exists any $r \in J$ distinct from ΔP and $P \times P$, define $d[h_1, r, h_2] \in \mathbf{Z}_J$, where $0 < h_1 < h_2 < \infty$, by

$$
\begin{aligned}
\{T(d[h_1, r, h_2])\}(h) &= \Delta P \quad \text{if } 0 \leqslant h < h_1 \\
&= r \quad \text{if } h_1 \leqslant h < h_2 \\
&= P \times P \text{ if } h_2 \leqslant h < \infty.
\end{aligned}
$$

$r \in J$, so $r = (Td)(h)$ for some $d \in \mathbf{Z} \cap \mathbf{C}'(P)$, $h > 0$. Choose a particular such d, and let $t_1 = \inf\{h: (Td)(h) = r\}$, $t_2 = \inf\{h: (Td)(h) = P \times P\}$. Then $0 < t_1 < t_2 < \infty$. Choose u_1, u_2 such that $0 < u_1 < t_1 < u_2 < t_2 < \infty$ and $(Td)(u_1) = \Delta P$, $(Td)(u_2) = r$. Note that, by NSC3, $(Td)(t_1) = r$, $(Td)(t_2) = P \times P$.

Let θ_x denote the piecewise-linear dissimilarity transformation which is specified for $0<x<1$ by being linear on $[0, u_1]$, $[u_1, t_1]$, $[t_1, u_2]$, $[u_2, t_2]$, $[t_2, \infty]$ and by taking the values

$$\begin{aligned}
\theta_x(0) &= 0\\
\theta_x(u_1) &= xh_1\\
\theta_x(t_1) &= h_1\\
\theta_x(u_2) &= h_1 + x(h_2 - h_1)\\
\theta_x(h) &= h_2 + h - t_2 \text{ if } h \geqslant t_2.
\end{aligned}$$

The function is illustrated in Figure 10.1.
This is a dissimilarity transformation, and so, by S3+, we have $\theta_x d \in \mathbf{Z}$ for all $x \in (0, 1)$, and since $\theta_x d$ is bounded over $x \in (0, 1)$ it follows from S5 that $\sup\ \theta_x d \in \mathbf{Z}$. It is easy to check that $\sup\ \theta_x d$ is simply $d[h_1, r, h_2]$, and it follows that if there exists any element r in J which is neither of ΔP, $P \times P$ then the DC $d[h_1, r, h_2]$ lies not only in $\mathbf{Z}_J$ but in $\mathbf{Z}$.

We have already shown that all DC's of the form δ^l lie in $\mathbf{Z}$ for $l \geqslant 0$. Now let d be an arbitrary element of $\mathbf{Z}_J \cap \mathbf{C}'(P)$ which is not equal to δ^l for any l. We may suppose that Td is given by the following scheme.

$$\begin{aligned}
(Td)(h) &= \Delta P \quad \text{if } 0 \leqslant h < h_1\\
&= r_1 \in J \text{ if } h_1 \leqslant h < h_2\\
&= r_2 \in J \text{ if } h_2 \leqslant h < h_3\\
&\ \ \vdots\\
&= r_n \in J \text{ if } h_n \leqslant h < h_{n+1}\\
&= P \times P \text{ if } h_{n+1} \leqslant h < \infty
\end{aligned}$$

Since d is not a δ^l, we have $n \geqslant 1$. If $n = 1$, then d is simply $d[h_1, r_1, h_2]$ and we know already that this lies in $\mathbf{Z}$. Now put $d_i = d[h_1, r_i, h_{i+1}]$ for $i = 1, \ldots, n$. Then $d_i \in \mathbf{Z}$ for $i = 1, \ldots, n$, and $\{d_i\}$ is bounded, so by S5 we have $\sup\ d_i \in \mathbf{Z}$; but $\sup\ d_i$ is equal to d and we conclude that $\mathbf{Z}_J \cap \mathbf{C}'(P) = \mathbf{Z} \cap \mathbf{C}'(P)$.

Finally, let d be an arbitrary element of $\mathbf{Z}_J$. Any neighbourhood of d contains a d^ϵ for small enough $\epsilon > 0$, because $\Delta_0(d, d^\epsilon) \leqslant \epsilon$. Moreover $d^\epsilon \in \mathbf{Z}_J \cap \mathbf{C}'(P) = \mathbf{Z} \cap \mathbf{C}'(P)$, so d is a limit point of $\mathbf{Z} \cap \mathbf{C}'(P)$ and *a fortiori* of $\mathbf{Z}$. $\mathbf{Z}$ is closed, by S8, whence we conclude that $d \in \mathbf{Z}$, and hence that $\mathbf{Z}_J = \mathbf{Z}$.□

The flatness theorem establishes a natural 1-1 correspondence between the flat cluster methods and the indicator families. The next theorem shows how the map $D:\mathbf{C}(P) \to \mathbf{Z}$ is related to a map $\gamma:\Sigma(P) \to J$, where $\mathbf{Z} = \mathbf{Z}_J$, and completes the reduction of the problem of describing flat cluster methods to that of describing indicator families. Suppose $\mathbf{C}(P)$, $\mathbf{Z}$ specifies a flat cluster method D, and that $\mathbf{Z} = \mathbf{Z}_J$ where J is an indicator family. Let r be an arbitrary element of $\Sigma(P)$ and let K be the set of elements of J such that $r' \in K$ if and only if $r' \supseteq r$. K is non-empty, because $P \times P \in J$. $\cap K \in J$ by IF5, and $\cap K \supseteq r$; $\cap K$ is the unique smallest element of J containing r, and will be denoted by $\gamma(r)$. D and γ are related by the reduction theorem, as follows.

REDUCTION THEOREM

With the above notation

$$\{T[D(d)]\}(h) = \gamma[(Td)(h)]$$

for all $d \in \mathbf{C}(P)$, $h \geqslant 0$.

Proof

If $r_1 \subseteq r_2$, then $\gamma(r_1) \subseteq \gamma(r_2)$, so $h_1 \leqslant h_2$ implies that $\gamma[(Td)(h_1)] \subseteq \gamma[(Td)(h_2)]$. From this follows that $c(h) = \gamma[(Td)(h)]$ defines an NSC and that $T^{-1}c$ ($= D'(d)$, say) is an element of $\mathbf{Z}_J = \mathbf{Z}$ such that $D'(d) \leqslant d$, and hence $D'(d) \leqslant D(d)$. But if $D'(d) \neq D(d)$ there is some h for which $\{T[D(d)]\}(h)$ is strictly smaller than $\{T[D'(d)]\}(h)$, that is, than $\gamma[(Td)(h)]$, and this contradicts the definition of γ. Thus $D(d) = D'(d)$ and the formula holds.□

It is intuitively very plausible that if D is a flat cluster method then $T[D(d)]$ at level h should depend only on Td at the same level. The flatness and reduction theorems establish this result in a precise form, and effect the reduction of the analysis of flat cluster methods to that of indicator families. Thus, as promised in Chapter 8.5, we have shown that every flat cluster method–that is, every ME subdominant method satisfying certain 'tidying-up' conditions–is a uniform method. To justify the use of the methods B_k, B_k^c within this model it is necessary to do no more than to observe that J_k, J_k^c are indicator families. It is because of the relationships to indicator families that flat cluster methods are easy to work with in practice. γ can be thought of as operating on the graph at level h for d to produce the graph for $D(d)$ at the same level, and this makes it easy to visualize the way the methods work; for similar reasons flat cluster methods are a practical proposition from the computational point of view. By no means all of the methods acceptable within the model in Chapter 9 are flat. In particular, any method involving algebraic operations on the dissimilarities will not be flat; we have already pointed out that C_u is not ME, and thus not flat.

Condition M6 has not yet been discussed for flat cluster methods. The required result is the second continuity lemma.

SECOND CONTINUITY LEMMA

Let D: $\mathbf{C}(P) \rightarrow \mathbf{Z}$ be a flat cluster method. Then D is continuous on the whole of $\mathbf{C}(P)$.

Proof

Since D is flat it follows that $D(d^\epsilon) = D(d)^\epsilon$ for all $d \in \mathbf{C}(P)$, $\epsilon \geqslant 0$. This is a corollary of the reduction and flatness theorems. After the first continuity lemma it will be enough to show that D is continuous at each $d \in \mathbf{C}(P) - [\mathbf{C}'(P) \cup \{\bar{0}\}]$. Let l be the minimal value other than zero taken by d, and suppose that $\Delta_0(d, d') < \epsilon < l$. Then

$$[(l-\epsilon)/l]\, d \leqslant d' \leqslant \{[(l+\epsilon)/l]\, d\}^\epsilon$$

whence

$$[(l-\epsilon)/l]\, D(d) \leqslant D(d') \leqslant \{[(l+\epsilon)/l]\, D(d)\}^\epsilon$$

and so $\Delta_0(D(d), D(d')) \leqslant m\epsilon/l$ where $m = \Delta_0(D(d), \bar{0})$. The result follows as in the first continuity lemma.

Flatness is a termwise phenomenon as far as cluster series are concerned, so the generalization of it is rather trivial. The only point to note is the form of the fitting condition on indicator families. If $\phi \in E_{ij}$, then ϕ induces $\phi^! : \Sigma(P_j) \rightarrow \Sigma(P_i)$ defined by

$$(A, B) \in \phi^!\, r \Leftrightarrow (\phi A, \phi B) \in r$$

where $A, B \in P_i$, $r \in \Sigma(P_j)$. The condition is

IFF $\quad$ $\phi^!$ maps J_j *onto* J_i

J_i, J_j are the indicator families in $\Sigma(P_i)$, $\Sigma(P_j)$ respectively. An *indicator series* J_* is a sequence $(J_1, J_2, \ldots)$ of indicator families satisfying this condition. Further details will not be given, as no more points of interest arise.

10.3. MEASURES OF ASSOCIATION

In connection with the definition of B_k, we stated in Chapter 8 that a k-transitive relation is characterized by the property that the overlap between distinct ML-sets contains at most $k-1$ OTU's, and that it was because of this simple characterization that the methods B_k were of greater utility than the methods B_k^c, since the latter are based on strongly *k-transitive relations* which have no such simple characterization. Whilst the methods B_k are known to be of value in applications, it may seem rather arbitrary to make use of the sort of step-by-step process involved in increasing by 1 the integer value of k. Clearly the size of the overlap takes the role of a measure of association between ML-sets and the k-transitivity condition is obtained simply by restricting the degree of association between distinct ML-sets; as k is increased the permitted level of association between distinct ML-sets is raised until eventually no restriction is placed on the ML-sets at all. We have called this type of restriction an *absolute* restriction in Chapter 8. We now consider whether there are any other types of absolute restriction based on association between ML-sets. Because only flat cluster methods will be considered in this context, the flatness and reduction theorems allow us to work entirely in terms of symmetric reflexive relations.

Let $\theta(P)$ denote the set of unordered pairs of subsets U, V of P such that both $U-V$ and $V-U$ are non-empty. If $\{U, V\} \in \theta(P)$, define $R\{U, V\}$ to be the symmetric reflexive relation which is universal on U and on V and is the equality relation elsewhere. Thus

$$R\{U, V\} = U \times U \cup V \times V \cup \Delta P$$

The ML-sets for $R\{U, V\}$ are precisely U, V, and the 1-element subsets of $P-(U \cup V)$. An association measure on P will be defined as follows. Let ω be a function from $\theta(P)$ to the non-negative real numbers. ω will be called an *association measure* if it satisfies the following conditions.

AM1 $\quad \omega\{U, V\} = 0$ if and only if $U \cap V = \varnothing$.

AM2 $\quad$ If $r \in \Sigma(P)$, and $\omega'(r)$ is defined to be the maximum value of ω on distinct ML-sets of r then J_κ defined by $J_\kappa = \{r \in \Sigma(P) \colon \omega'(r) \leqslant \kappa\}$ is an indicator family for all $\kappa \geqslant 0$.

The association measure theorem now shows that in fact the apparent freedom of choice available by using a general association measure is illusory; we gain nothing at all.

ASSOCIATION MEASURE THEOREM

If ω is an association measure on P, then ω is a monotone increasing function of $|U \cap V|$, $|U - V|$, $|V - U|$.

Proof

$\omega\{U, V\} = \nu\omega'(R\{U, V\})$ so it follows from IF2 that if ρ is any permutation of P, then $\omega\{U, V\} = \omega\{\rho, U\rho V\}$, and hence that ω depends only on $|U \cap V|$,

$|U-V|$, $|V-U|$, $|P-(U \cup V)|$. Since these sum to $|P| = p$, for fixed P, ω is a function of the first three of these quantities. Now J_0 is the set of equivalence relations on P, by condition AM1, and $J_\kappa \subseteq J_{\kappa'}$ for $\kappa \leqslant \kappa'$, so all equivalence relations lie in J_κ for all $\kappa \geqslant 0$. Let r be a relation in $\Sigma(P)$, let $\omega'(r) = \kappa$, and suppose that U, V are distinct ML-sets for r for which $\omega\{U, V\} = \kappa$. Let W be a subset of P such that both $(U - V) \cap W$ and $(V-U) \cap W$ are non-empty; W may meet $U \cap V$ in any manner. Let r' be the equivalence relation whose equivalence classes are $W, P-W$. Then $r \cap r' \in J_\kappa$ by IF5. $U \cap W$, $V \cap W$ are distinct ML-sets for $r \cap r'$ so $\omega\{U \cap W, V \cap W\} \leqslant \kappa$, that is, ω is monotone increasing in each of $|U \cap V|$, $|U - V|$, $|V - U|$.□

The association measure theorem does not, as it stands, eliminate completely the possibility of choosing other interesting association measures on which to base the construction of indicator families. However, it is possible to reason further as follows. It is reasonable to expect to attach less significance to a small overlap between large clusters than to an overlap of the same size between small clusters; or, at least, one would not expect to attach *more* significance to a small overlap between large clusters than to an overlap of the same size between small clusters. Now if ω were taken to be strictly increasing in $|U - V|$, $|V - U|$–and it is, of course, symmetric in these quantities–this would have the effect of regarding large clusters with a given overlap as being more closely associated than small clusters of the same sized overlap. Thus the best which can be achieved is to make ω independent of $|U-V|$, $|V-U|$ and hence a monotone function of $|U \cap V|$ only. In this case the indicator family J_κ is the set of k-transitive relations on P for some value of k, and the methods B_k are recovered as the only acceptable flat cluster methods which are obtainable from a measure of association between clusters.

The theorems in this chapter show that the methods B_k occupy a rather special position amongst the flat cluster methods. In Chapter 8 it was pointed out that there was an interpretation for the operation of B_k in terms of cluster overlap which was lacking for B_k^c; what we have now shown is that the methods B_k are the only reasonable flat cluster methods based on a measure of association between clusters. This observation is valuable in that it isolates (B_k) as a family of methods with special properties, but it does not rule out the other methods altogether. B_k^c, for example, might be of value for other purposes than those we shall be interested in, although for our purposes what we have shown indicates the superiority of B_k. The results in this chapter apply strictly to flat methods, and are consequently of no relevance to such non-flat methods as C_u.

CHAPTER 11

Evaluation of Clusterings

11.1. THE NATURE OF THE PROBLEMS

We now turn from the discussion of criteria whereby cluster methods may be judged to that of the performance of cluster methods in particular cases, relating this to questions such as the validity of seeking clusters of a certain kind in a data DC, and to the comparison of DC's in general.

It is useful to think in terms of an analogy with estimation processes on the line. Suppose that we have a large sample from a distribution of unknown form, but possessing moments of all orders. We know that it is reasonable to estimate the mean of the parent distribution by the sample mean, but if in the sample we find that, say, the ratio between the fourth moment about the mean and the square of the second moment about the mean is very large by comparison with that expected for samples from familiar standard distributions, then we shall be inclined to investigate the possibility that the parent distribution is, for example, bimodal, in which case the mean is, for many purposes, not particularly useful. Thus one statistic of the original data can be used to estimate the mean; another will provide us with some guidance about whether this is in fact a useful parameter to estimate.

In cluster analysis it is only in very special cases that we have a precise statistical model available, but nevertheless we can profitably adopt an approach not too far removed from the statistical one. Just as equivariance principles illustrate the utility of the sample mean as an estimator for the true mean, so the axioms in Chapter 9 point to the single-link method as the appropriate method for hierarchic cluster analysis, and, more generally, pick out the subdominant methods as mathematically acceptable. Our first object in this chapter is to give some methods for the *post facto* evaluation of the extent to which some given data is suited to analysis by a chosen cluster method. The most obvious way of doing this is to compare the input DC d with the output DC $D(d)$; if they are not too different by some suitably chosen criterion, then we may regard this as evidence in favour of the use of D in this case. This is not a particularly subtle approach, and, except in cases where a statistical model is available or may reasonably be supposed to exist, it simply provides a rule-of-thumb technique to assist in the sensible application of the methods.

It is natural to turn from the problem of considering comparison of d with $D(d)$ to that of comparing arbitrary DC's d_1 and d_2, these may either be two different input DC's or two different output DC's, and we may often wish to make a comparison of this kind when checking the validity of a numerical taxonomic study. We might wish to show, for example, that provided a sufficient number of attributes were considered, the pattern of classification which resulted was insensitive to the exact choice made, or we might wish to compare classifications produced from morphological attributes on the one hand, and biochemical attributes on the other.

Sometimes we may wish to compare relations. For example, we may wish to know how well a partition of a set of OTU's suggested by a previous author matches the relation $(Td)(h)$ or $[T(D(d))](h)$ at a particular level h, or how well the preassigned partition can be matched by choice of h.

Finally, we may wish to comment on the validity of individual clusters—how well isolated they are from other clusters, or how tightly linked internally.

11.2. COMPARISON OF INPUT AND OUTPUT DC'S

This section is concerned with the problem of judging how well a cluster-method D has worked in a particular case. Both d and $D(d)$ are represented by points in euclidean space of dimension $\frac{1}{2}p(p-1)$, so that one obvious way of comparing them is simply to measure their distance apart in this space. We can use the usual euclidean metric for this, but since we have a preferred set of axes in the space, other metrics can also be used. For DC's d_1, d_2, we define

$$\Delta_0(d_1, d_2) = \max\{|d_1(A,B) - d_2(A,B)| : A, B \in P\}$$

$$\Delta_{\frac{1}{2}}(d_1, d_2) = \{\Sigma[d_1(A,B) - d_2(A,B)]^2\}^{\frac{1}{2}}$$

$$\Delta_1(d_1, d_2) = \Sigma|d_1(A,B) - d_2(A,B)|$$

where in the last two cases the summation is taken over the $\frac{1}{2}p(p-1)$ two-element subsets of P. Sometimes it is convenient to divide by a normalizing factor to remove the effect of dimensionality, and thus to consider

$$\bar{\Delta}_0 = \Delta_0$$
$$\bar{\Delta}_{\frac{1}{2}} = [\tfrac{1}{2}p(p-1)]^{-\frac{1}{2}}\Delta_{\frac{1}{2}}$$
$$\bar{\Delta}_1 = [\tfrac{1}{2}p(p-1)]^{-1}\Delta_1$$

The Δ's, or for that matter the $\bar{\Delta}$'s, are all metrics on $\mathbf{C}(P)$ yielding the usual topology for $\mathbf{C}(P)$. We cannot use them directly for comparison between d and $D(d)$ for the following reason. If $f(d, D(d))$ measures the goodness-of-fit between d and $D(d)$, then by condition M3 in Chapter 9 we have

$$f(\alpha d, D(\alpha d)) = f(\alpha d, \alpha D(d))$$

In the majority of cases in which we investigate goodness-of-fit, we want this to be independent of the constant multiplier α. If we define

$$S_0(d) = \Delta_0(d, \bar{0})$$
$$S_{\frac{1}{2}}(d) = \Delta_{\frac{1}{2}}(d, \bar{0})$$
$$S_1(d) = \Delta_1(d, \bar{0})$$

where, as above, $\bar{0}$ is the zero DC represented by the origin in the associated euclidean space, then we have

$$S(\alpha d) = \alpha S(d), \neq 0 \text{ if } d \neq \bar{0}$$

So, for $d \neq \bar{0}$, we can define

$$\hat{\Delta}(d;D) = \Delta(d, D(d))/S(d)$$

(with subscripts $0, \frac{1}{2}, 1$ throughout) and observe that for $\alpha > 0$ we have

$$\hat{\Delta}(\alpha d; D) = \hat{\Delta}(d; D)$$

It is immaterial whether we think of constructing the $\hat{\Delta}$'s by prior normalization of d to norm 1 or posterior normalization of $\Delta(d, D(d))$ by division by $S(d)$; the important point is that $\hat{\Delta}$ is scale-free, and can be used as a measure of goodness-of-fit in any case where $d \neq 0$. The quantities $\Delta(d, D(d))/S(d)$ are not symmetrical in d and $D(d)$, so they are not metrics; but they do provide a useful standardized method of comparison between d and $D(d)$. $\hat{\Delta}_0$ is comparatively insensitive, and $\hat{\Delta}_{\frac{1}{2}}$ involves attaching meaning to squaring operations on DC values, which may not be justifiable. For many purposes the most reliable guide to the performance of the cluster method D is given by $\hat{\Delta}_1$. Clearly $\hat{\Delta}(d; D) = 0$ if and only if $d \in \mathbf{Z}$, the target set for D—assuming that D satisfies M1. So it is reasonable to ask how large $\hat{\Delta}_\mu$ can be for $\mu = 0, \frac{1}{2}, 1$ and for different choices of $\mathbf{Z}$ and different values of $|P| = p$ before we regard the data as being inherently ill-suited for classification by the method D. Except in a few special cases we do not regard the deviation of d from the set $\mathbf{Z}$ as being due to precisely specifiable random effects. It is thus impossible to provide a statistical answer to the question of how well $D(d)$ and d should fit in general, but experience in the use of $\hat{\Delta}$ indicates that it does provide a good practical guide. In particular, the decreasing sequence

$$\hat{\Delta}(d; B_1), \hat{\Delta}(d; B_2), \ldots$$

gives considerable guidance about how much overlap between clusters may profitably be allowed. In cases where there is a precise statistical model for the generation of the DC d, a sampling distribution for $\Delta(d; D)$ can be obtained; although this is usually a laborious simulation process rather than a neat analytical argument, it can on occasion be useful.

Amongst the statistical models for generation of a DC which have been suggested are the distances between points randomly distributed in euclidean space (Sneath, 1966a), and the random placing of edges in a graph, where an edge represents a value of the DC less than some threshold (Ogilvie, 1968). Neither of these models seems appropriate for dissimilarities between OTU's used as a basis for the description of patterns of differentiation of organisms.

It is convenient at this point to consider briefly an alternative to conditions M4 and M5 which the existence of a goodness-of-fit measure suggests. Instead of the cluster preservation and optimality conditions we could introduce just a different optimality condition requiring that $D(d)$ should be that element of $\mathbf{Z}$ which best fits d. Unfortunately this very simple idea is rather unsound, both in practice and in theory. If we take even a very simple set $\mathbf{Z}$, such as $\mathbf{Z} = \mathbf{U}$, there is no known computationally feasible process for finding the $D(d) \in \mathbf{Z}$ such that, say, $\Delta_{\frac{1}{2}}(d, d')$ is minimized for $d' \in \mathbf{Z}$ by $d' = D(d)$. When we come to examine the theoretical aspects we find that it is easy to construct simple small examples in which $D(d)$ can be found, and is non-unique; so that although it would undoubtedly be of interest to be able to find minimizing elements of $\mathbf{Z}$ for some purposes, non-uniqueness renders their value in taxonomy very limited. Also there are reasons for supposing that at least in the case of Δ_1, the minimizing element or elements of $\mathbf{U}$ are by no means the appropriate classifications of the data. This is discussed further in Chapter 11.4.

If we impose condition M4, and thus consider minimization of $\Delta(d, d')$ subject to $d' \leqslant d$, then it is very easily seen that because the sign of each term

$(d(A, B) - d'(A, B))$ is positive, minimization of any $\Delta(d, d')$ is in fact equivalent to condition M5, and this observation stresses further the rôle of M5 as an optimality condition. It also indicates that the Δ's are more closely related to the operation of subdominant methods than might at first sight appear to be the case.

11.3. SCALE-FREE COMPARISONS OF DC'S

We may often wish to compare DC's calculated in different ways, or obtained from different sets of attributes, either before or after they are transformed by a cluster method. This comparison needs to be free of scale factors in *both* DC's; we are no longer in the situation of the last section, where a scale factor introduced in d appears also in $D(d)$. The reason for our usual unconcern with scale-factors is that no natural scale is available; we expect dissimilarities to go on increasing as more attributes are brought in, and it is the pattern of the values as expressed by their ratios to one another, rather than the individual values, which are the focus of attention.

We suggest three methods of making comparisons of this kind. The first can be used for the scale-free comparison of any two non-zero DC's d_1, d_2, and the comparison is effected simply by normalizing each of d_1, d_2 and then using the appropriate Δ to compare them. Thus we have

$$\Delta_0^*(d_1, d_2) = \Delta_0(d_1/S_0(d_1), d_2/S_0(d_2)),$$
$$\Delta_{\frac{1}{2}}^*(d_1, d_2) = \Delta_{\frac{1}{2}}(d_1/S_{\frac{1}{2}}(d_1), d_2/S_{\frac{1}{2}}(d_2)),$$
$$\Delta_1^*(d_1, d_2) = \Delta_1(d_1/S_1(d_1), d_2/S_1(d_2)).$$

The second method of comparison is a slight variant of this group of methods, and again can be used to compare any two non-zero DC's. It consists simply of taking the difference between d_1 and d_2 to be the angle between the rays $\bar{0}d_1$ and $\bar{0}d_2$. The formula is as follows:

$$\alpha(d_1, d_2) = \cos^{-1}\left\{\frac{[S_{\frac{1}{2}}(d_1)]^2 + [S_{\frac{1}{2}}(d_2)]^2 - [\Delta_{\frac{1}{2}}(d_1, d_2)]^2}{2S_{\frac{1}{2}}(d_1)\,S_{\frac{1}{2}}(d_2)}\right\}$$

That this is closely related to the first method of comparison may be seen from the alternative formula

$$\alpha(d_1, d_2) = 2\sin^{-1} \tfrac{1}{2}\Delta_{\frac{1}{2}}^*(d_1, d_2)$$

so α is just an alternative way of looking at $\Delta_{\frac{1}{2}}^*$.

The third method of comparison is rather different, and can only be used to compare definite DC's d_1, d_2. In this case we can use the fact that **C**(P) is the positive cone in $\frac{1}{2}p(p-1)$-dimensional euclidean space. If we have two rays not lying in the boundary of the cone then they define a plane meeting the boundary of the cone in two further rays, thus giving a planar pencil of four rays. The modulus of the logarithm of the cross-ratio of this pencil, taken in the appropriate order, is the Hilbert projective metric between the original two rays. A detailed treatment of projective metrics may be found in Busemann and Kelly (1953). All we shall need to know here is a formula to give the value of the projective metric $\pi(d_1, d_2)$ between two definite DC's d_1, d_2; it is easy to show that the value is given by

$$\pi(d_1, d_2) = \log m_1 m_2$$

where

$$m_1 = \max\{d_1(A, B)/d_2(A, B) : A, B \in P\}$$
$$m_2 = \max\{d_2(A, B)/d_1(A, B) : A, B \in P\}$$

This is simple to work out computationally, and provides an interesting comparison with the Δ^*'s.

Many of the observations we have already made about the use of the Δ's to compare d and $D(d)$ apply also to the use of the Δ^*'s and π to compare d_1 and d_2; the Δ^*'s and π are best regarded simply as useful *ad hoc* guides for the comparison of DC's, and we should hesitate to read too much into them. Nevertheless they can, if properly used, be of considerable value, and they present no computational difficulties.

Sokal and Rohlf (1962) proposed as a measure for the comparison of DC's what they called the 'cophenetic correlation coefficient'. This provides another *ad hoc* measure for making comparisons; justification for its use rests on its practical utility rather than on any resemblance to statistical correlation measures. It has been widely used for comparison of DC's.

11.4. COMPARISON OF RELATIONS

Sometimes we may wish to compare, not two DC's as a whole, but two (symmetric reflexive) relations or a relation and a DC. Undoubtedly the most convenient quantity for this purpose is the symmetric difference of the relations; its use was originally suggested in this context by Zahn (1964). Extensive investigations employing this measure have been carried out by

Lerman (1970). If r_1 and r_2 are the two symmetric relations, that is, symmetric subsets of $P \times P$ containing the diagonal ΔP, then we define

$$\delta(r_1, r_2) = \tfrac{1}{2}|[r_1 \cap (P \times P - r_2)] \cup [(P \times P - r_1) \cap r_2]|.$$

The factor $\frac{1}{2}$ is introduced to eliminate the redundancy which would otherwise arise because all relations considered are symmetric. δ is a metric on the set $\Sigma(P)$ of symmetric reflexive relations on P, and its maximum value is $\delta(\Delta P, P \times P)$ which is $\frac{1}{2}p(p-1)$. δ is closely related to Δ_1 by the formula

$$\Delta_1(d_1, d_2) = \int_0^\infty \delta[(Td_1)(h), (Td_2)(h)]\, dh.$$

δ, or the normalized form $\bar{\delta} = \delta / \frac{1}{2}p(p-1)$, provides a measure of comparison between relations which is of similar utility to the measures Δ_1, $\bar{\Delta}_1$ between DC's.

The formula relating Δ_1 and δ to one another provides a means whereby we may obtain a little more information about the absolute minimization processes discussed at the end of Chapter 11.2. Suppose that J is an indicator family, and that $\mathbf{Z}_J$ is the derived target set. For example J might be the set $\mathrm{E}(P)$ of equivalence relations on P, and $\mathbf{Z}_J$ the set $\mathbf{U}(P)$. If $d' \in \mathbf{Z}_J$ is such that for all $h \geqslant 0$ the quantity $\delta[(Td)(h), r]$ is minimized over $r \in J$ by $r = (Td')(h)$, then $\Delta_1(d, d'')$ is minimized over $d'' \in \mathbf{Z}_J$ by $d'' = d'$; this is a trivial corollary of the expression for Δ_1 as an integral of δ. It might be thought that a stronger result holds, namely that, given d, there exists $d' \in \mathbf{Z}_J$ such that, for all $h \geqslant 0$, the quantity $\delta[(Td)(h), r]$ is minimized over $r \in J$ by $r = (Td')(h)$. This is not so; consider the DC d on six OTU's specified by the following table.

	A	*B*	*C*	*D*	*E*	*F*
A						
B	2					
C	2	2				
D	3	3	1			
E	3	3	3	2		
F	3	3	3	2	2	

The graphs for d at levels 1, 2 are given in Figure 11.1. The relation $(Td)(1)$ is an equivalence relation. The relation $(Td)(2)$ is not. The equivalence relation r such that $\delta[(Td)(2), r]$ is minimal is in this case unique, and is the equivalence relation whose classes are $\{A, B, C\}\{D, E, F\}$ But this relation does not contain $(Td)(1)$, so no Δ_1-best-fitting ultrametric d'

can give the best-fitting equivalence relation at all levels simultaneously in this example. Thus minimization of Δ_1 is not equivalent simply to minimization of δ at all levels, although almost certainly this is a useful first step, and suggests that if absolute minima are wanted Δ_1 may be the easiest Δ to minimize.

Now we give another cautionary example which will serve for both Δ_1 and δ, to show that absolute minimization may well give misleading results. This concerns chains. If $ABCD \ldots I$ is a chain of elements of P, then the

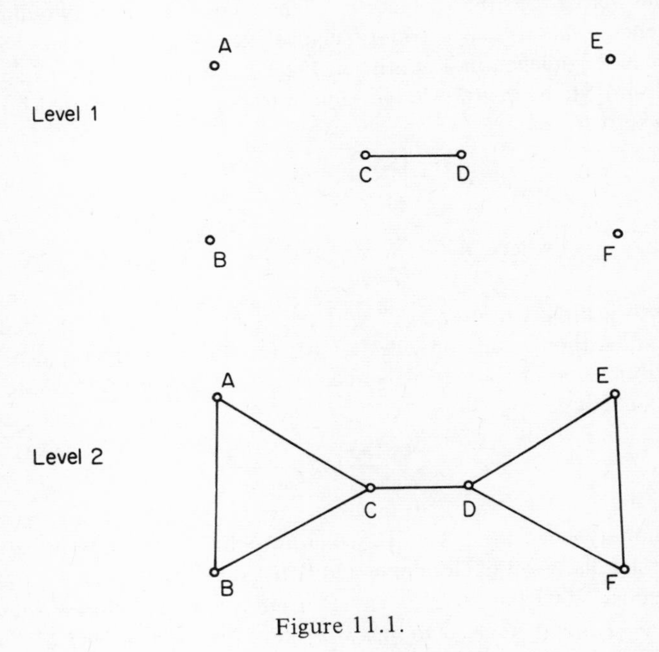

Figure 11.1.

equivalence relation specified by $\{AB\}\{CD\}\{EF\}\{GH\}\{I\}$ is best-fitting. So is $\{A\}\{BC\}\{DE\}\{FG\}\{HI\}$. If now we introduce the extra link AC, the unique best-fitting equivalence relation is specified by $\{ABC\}\{DE\}\{FG\}\{HI\}$. It is undesirable that the decision about which of the closely-linked pairs GH, HI should be regarded as a cluster should be influenced in this way by the link AC by way of considerations of parity of position in the chain. Thus even if we can successfully minimize Δ_1 or δ, the result may be of little value.

Despite the above remarks about the need to treat processes for absolute minimization with caution, there is no reason why we should not use δ to measure goodness-of-fit between relations, in the same sort of way that Δ's and the quantities derived from them are used to compare DC's. We can in particular compare a given relation r_0 with one such as $(Td)(h)$ which we derive ourselves; r_0 might, for example, be the equivalence relation corresponding to some previous author's sectioning of a genus.

The comparison of arbitrary symmetric reflexive relations in terms of their symmetric difference size δ can, of course, be specialized to equivalence relations. The comparison takes as its zero the difference between a relation and itself.

An alternative method of comparing equivalence relations is to compare them in terms of the degree of dependence in the two-dimensional contingency table which a pair of equivalence relations defines. Statistical tests for independence in a contingency table are discussed in Chapter 33 of Kendall and Stuart (1967). Large sample tests are usually based on the fact that the statistic

$$X^2 = \left\{ \sum_{i,j} [n_{ij}^2/n_{i.}\, n_{.j}] - 1 \right\}$$

$$(\text{where } n_{i.} = \sum_j n_{ij} \text{ and } n_{.j} = \sum_i n_{ij})$$

is asymptotically distributed as χ^2 with $(r - 1)(c - 1)$ degrees of freedom in an $r \times c$ table. Although this provides a basis for testing independence, it is not necessarily a good measure of dependence, for which purpose the information-transfer statistic

$$I = \sum_{i,j} \frac{n_{ij}}{n} \log_2 \frac{n\,.\,n_{ij}}{n_{i.}\, n_{.j}}$$

has commonly been used; see Macnaughton-Smith (1965). Some alternative measures are discussed in Goodman and Kruskal (1954).

In practice, we would use χ^2 to *test* whether one partition was dependent on another; I might be used to *measure* their *dependence*; and δ to *measure* their *difference*. The situations in which these measures should be applied in taxonomy are further discussed in Chapter 14.4.

11.5. RANDOM EFFECTS; CLASSIFIABILITY AND STABILITY

In the large majority of cases in which cluster analysis is used there is not only no statistical model involved, but also it is impossible to imagine what such a model might look like. We have already pointed out that this forces us to regard the use of measures of classifiability (such as $\hat{\Delta}$) and measures of stability (such as Δ^* between a d_1 obtained from one set of attributes and a d_2 from another set, or some larger set) as an *ad hoc* procedure.

Nevertheless, there are a few comments which may assist in the investigation of classifiability and stability. In a numerical taxonomic study in which, as we have recommended in Part I, no *a priori* weighting of attributes is carried out, the potential maximum contribution to dissimilarity from each attribute is the same. If we are dealing with the taxonomy of higher organisms it is usually not too difficult to pick out a hundred or more distinct non-constant attributes, and it is an empirical fact that up to this sort of number, dissimilarities go on increasing. There is a supposition, sometimes known as 'Mahalanobis' axiom' (Mahalanobis, Majumdar, and Rao, 1949, p. 253), that under certain circumstances dissimilarity values will actually converge as the number of attributes selected becomes very large, but it is hard to see what such a statement could mean under most practical circumstances, except in the sense that after a time it becomes difficult to think of non-constant attributes which do not effectively duplicate those already selected. What we can investigate to some extent is the degree to which the ratios of dissimilarity values settle down, and the degree to which dissimilarity is variable over different subsets of attributes. The Δ^*'s or π provide appropriate comparison measures, and a suggestion about how they may be used is made in Chapter 14.4.

Sampling errors arising within attributes are not too hard to deal with in a rough-and-ready way. Tables (see Appendix 1) can be computed to provide a guide to the errors expected from estimation of dissimilarity values, and it is thus possible to give some idea of the magnitude of the sampling error when a lot of estimates of this kind are added.

The description of error in the DC may sometimes take the form of a system of confidence intervals for the DC. Thus, rather than just a DC d, we would also have DC's d_L, d_U with $d_L \leqslant d \leqslant d_U$. If D is a subdominant method, then $d_1 \leqslant d_2 \Rightarrow D(d_1) \leqslant D(d_2)$, so $D(d_L) \leqslant D(d) \leqslant D(d_U)$. The nature of cluster methods is such that it is very difficult to give any more precise or powerful kind of error analysis than this.

11.6. VALIDITY OF CLUSTERS

The final topic which we consider in this chapter is the investigation of the validity of individual clusters. There are two ways in which a cluster may be unsatisfactory. It can be only feebly linked internally; and it can be ill-isolated from other OTU's. The obvious example of a feebly-linked cluster is a chain as picked out by the single-link method. The methods B_k provide not merely a technique for analysis into overlapping clusters, but also a simple way of checking the cohesion of single-link clusters–if a cluster picked out by the single-link method $A_S = B_1$ reappears as a cluster picked out by $B_2, B_3, \ldots$ then it is well-linked internally. This method of checking clusters is very easy to carry out, since it involves no extra work over and above that needed anyhow for applying the methods B_k. A very rigorous constraint on a single-link cluster is that it should be a ball-cluster in the sense of Jardine (1969b). A ball-cluster is a subset S of P such that for each $A \in S$ the remaining elements of S are its nearest neighbours. Every ball-cluster is a single-link cluster at some level. The condition that a single-link cluster be a ball-cluster is in part an internal cohesion condition and in part an isolation condition. An even stronger condition is given by van Rijsbergen (1970a). Both conditions are discussed in Chapter 12.4. The isolation of a single-link cluster may be thought of as the 'amount of space around it'; formally, as the difference between the minimum distance between an OTU in the cluster and one outside it, and the level at which the cluster forms. This is simply the difference between the level at which the cluster is added to and the level at which it forms. This criterion of isolation was suggested by Wirth, Estabrook, and Rogers (1966). A generalization of this criterion to deal with clusters in overlapping stratified systems was suggested in Jardine and Sibson (1968a).

If error bounds on the DC are available, then a further check on clusters may be made by investigating whether the clusters for d are also obtained for d_L, d_U, where $d_L \leqslant d \leqslant d_U$; use of this technique can involve considerable extra computation, especially when overlapping cluster methods are used.

CHAPTER 12

Other Approaches to Cluster Analysis

12.1. INTRODUCTION

We have so far focussed attention exclusively on methods which lead from a dissimilarity coefficient to a stratified clustering. Here we shall examine briefly some of the clustering methods described in the literature which have different starting-points and different end-products. The majority of these alternative approaches we show to be unsatisfactory as methods for producing simplified representations of complex data. But it should be emphasized that many of them were proposed as heuristic attempts to solve specific practical problems, and that our criticisms are largely irrelevant to such uses.

The various alternative approaches to cluster analysis can conveniently be grouped into two categories. First, there are methods which go directly from descriptions of the objects to be classified to a clustering without the intermediate calculation of a measure of dissimilarity. Secondly, there are methods which seek to go directly from a dissimilarity measure to an unstratified clustering. Both categories are too heterogeneous to be given generic names.

12.2. CLUSTER ANALYSIS WITHOUT A DISSIMILARITY MEASURE

One special case of this kind of cluster method is well-known. It operates on descriptions of each of the objects by states of quantitative attributes. It is assumed that the data is a mixture of multivariate normal distributions. Clear accounts of methods for sorting out mixtures of multivariate normal distributions are given by Friedman and Rubin (1967) and Sebestyen (1962). Several rapid and sophisticated algorithms are available; see Wolfe (1967, 1968a, 1968b). Many of the earlier methods were limited to small numbers of objects and the case where constant dispersion is assumed; but Wolfe's NORMIX algorithm can be used where dispersion is inconstant for reasonable numbers of parameters and objects. As pointed out by Friedman and Rubin, these algorithms can be described in geometrical terms. The dangers of geometrical and other analogue models in the study of general cluster methods has been mentioned in Chapter 6, but here geometrical models are quite innocuous, because the situation is naturally a geometrical one.

A very large number of *ad hoc* methods have been developed which employ algorithms similar to those used in unscrambling mixtures of multivariate normal distributions, but which are applied to data for which the assumption of multivariate normality is inapplicable. The majority of these methods make use of cohesion measures similar in mathematical form to measures of dispersion, and many of them have no justification beyond that which is obtained by analogy with an intuitively plausible geometrical model. Such methods have been used widely in pattern recognition, and in allocation problems. A clear survey is given by Ball (1966). Where discrete-state attributes are used in such methods it is necessary to code them in quantitative form, for example by disposing them along the real line.

It may be that the methods appropriate to mixtures of multivariate normal distributions can be generalized to deal with less restrictive assumptions about the kind of distribution which is expected, and to a more general class of attributes than those whose states are disposed along the real line. The mathematics described in Chapter 2 suggests an approach to this problem, since the information radius of a set of probability measures can be regarded as a measure of their mutual cohesion.

Another approach to clustering which does not involve intermediate calculation of a dissimilarity measure is that employed in the family of methods generally known as *association analysis.* Methods of association analysis are devised primarily for the classification of sets of individuals each described by a set of binary discrete-state attributes.

Methods of association analysis depend upon measurement of the fit of a partition of a set of individuals in terms of the information loss induced by the partition; see Maccacaro (1958), Orloci (1968), Rescigno and Maccacaro (1961), Wallace and Boulton (1968). The usual approach is to find the bisection of the set which minimizes a measure of induced information loss. Each subset is then bisected by the same criterion, and so on, so that a hierarchic clustering is obtained. Usually a stopping rule is employed to stipulate when further bisection is no longer significant. In so-called *monothetic* association analyses a single binary attribute from the selected set of attributes is used to bisect the set of individuals at each stage. The attribute chosen is that which induces a bisection in such a way that the information loss with respect to the other attributes, or some other related function is minimized. Several authors, including Williams and Lance (1958) and Lance and Williams (1965), have used instead functions based on χ^2. So-called *polythetic* association analyses seek that bisection which minimizes the information loss or some other related function regardless of whether bisection corresponds to the range of the states of any of the selected binary attributes. Such methods have been described by Macnaughton-Smith and coworkers (1964) and Macnaughton-Smith (1965).

Methods of monothetic association analysis have been used in ecology and taxonomy by many authors; see, for example, Williams and Lambert (1959, 1960), Watson, Williams, and Lance (1966). The available methods are unsatisfactory in several respects. First, they are ill-defined. Data can be readily constructed for which no bisection induces a unique minimum information loss. Secondly, partition into *two* subsets at each stage is an arbitrary choice of procedure. Finally, their application in taxonomy is restricted to discrete-state attributes which do not vary within populations. Their use appears to rest upon a confusion between classification and diagnosis. Measures of information loss may be useful in comparing partitions and in evaluating the predictive power of a partition with respect to particular attributes; compare Macnaughton-Smith (1963). Monothetic association analysis produces a hierarchic classification by choosing a diagnostic key based on the available attributes which is in a precise sense optimal. But the production of optimal diagnostic keys is not the primary purpose of classification in ecology or taxonomy.

12.3. SIMPLE CLUSTER ANALYSIS BASED ON A DISSIMILARITY MEASURE

The methods discussed here lead to simple clusterings in which no cluster includes another. Such clusterings are either partitions or coverings in which clusters may overlap. We suggested in the introduction that even when the representation of a dissimilarity coefficient which is required is a simple clustering, it may be preferable to proceed indirectly *via* a stratified clustering. This suggestion is reinforced by the unsatisfactory nature of the available direct methods.

The simplest dissimilarity measure is a nominal measure. Such a measure can be represented as a binary symmetric relation on P (where P is the set of objects), or as a graph whose vertices represent objects. An ordinal dissimilarity measure can be decomposed into an ordinally indexed sequence of nominal dissimilarity measures, one for each ordinal value; a numerical dissimilarity measure can likewise be decomposed into a numerically indexed sequence of nominal dissimilarity measures.

A simple partition method for a nominal dissimilarity measure is to pick out the single-link clusters. This corresponds to picking out the connected components in the related graph. A simple non-partitional covering with arbitrary overlap is obtained by finding the ML-sets. The B_k clusters $(1 < k < p - 1$, where $p = |P|)$ constitute a sequence of simple clusterings intermediate between the clusterings given by the set of all single-link clusters and the clustering given by the set of all ML-sets. Clusterings consisting of progressively more homogeneous clusters are obtained by allowing progressively more overlap. Many other cluster definitions for nominal data could be devised.

Frequently a partition is required in cases where there are either no single-link clusters or single-link clusters would be considered too straggly. Zahn (1964) has suggested that the optimal partition is that which minimizes the cardinal of the symmetric difference between the binary symmetric relation representing the nominal dissimilarity measure and the equivalence relation representing the partition. This minimum is not necessarily achieved uniquely on a single partition. Enumeration of all partitions is obviously unfeasible for even quite small numbers of objects (Rota, 1964), but no algorithm for seeking partitions optimal under this condition is known which avoids the full enumeration. An efficient algorithm for enumeration of partitions is given by Dobeš (1967). Zahn's suggestion is shown in Chapter 11.3 to be capable of leading to counterintuitive results.

A variety of measures of goodness-of-fit of a partition to an ordinal dissimilarity measure have been devised by Lerman (1970) and his colleagues at the University of Paris. Some of these measures are reviewed by Harrison (1968). For example, suppose symmetric functions are defined on pairs of elements of $P \times P$ as follows:

$R[(w, x)(y, z)] = 1$ if w and x occur in the same element of the partition, and y and z occur in different elements; and is otherwise 0.

$S[(w, x)(y, z)] = 1$ if $d(w, x) < d(y, z)$; and is otherwise 0. The goodness-of-fit of a partition to an ordinal dissimilarity measure is then given by:

$$\sum_{q \in P^4} R(r)\, S(q)$$

Several criteria for evaluation of a partition of objects described by a numerical dissimilarity coefficient have been suggested. See, for example, Edwards and Cavalli-Sforza (1965), Lerman (1970). The majority of these methods seek to minimize some function of the root mean square pairwise dissimilarity within elements of the partition and the root mean square pairwise dissimilarity between members of different elements of the partition. They are sometimes called 'variance' methods (Wishart, 1969a).

As with Zahn's partition criterion so with the criteria for ordinal and numerical dissimilarity measures; there is not in general a unique partition on which the measure is optimized, and no algorithm is known which avoids enumeration of all partitions. If a measure of taxonomic distance is used (see Chapter 5.4) the objects can be arranged in a euclidean space. Harding (1967) has shown that in attempting to optimize the 'variance' criteria only partitions which are induced by hyperplanes in this space need be considered. If some algorithm for finding all such partitions could be found, the search for partitions optimal under the criteria might be shortened. Jensen (1969) has shown that it is possible to find partitions optimal under a particular 'variance' criterion by considering the problem as a problem in dynamic programming. In this way a solution is found more rapidly than by enumeration of all partitions. It seems likely that this kind of approach could profitably be used for other criteria of optimality.

Two interesting approaches which yield overlapping clusterings involve searching for individual clusters which have some satisfactory property rather than seeking to optimize some function of the clustering as a whole. Such methods are sometimes called 'clump' methods. One approach due to Needham (1961a) is to seek clusters which satisfy what he called the *GR-clump* definition. Unfortunately no algorithm for finding the set of GR-clumps specified by a dissimilarity coefficient is known. Enumeration of the power set of the set of objects is obviously unfeasible. Another approach suggested by Parker-Rhodes and Jackson (1969), Parker-Rodes and Needham (1960), and Spärck Jones and Jackson (1967) defines a cohesion measure on every subset of the set of objects and seeks subsets which exceed some threshold of the cohesion measure. One of the cohesion measures they suggest for a numerical similarity measure S is

$$\frac{\left\{\sum_{x \in C,\ y \notin C} S(x, y)\right\}^2}{\sum_{x,\ y \notin C} S(x, y) \times \sum_{x,\ y \in C} S(x, y)}.$$

Other suggested measures incorporate weighting factors dependent on the size of the subset C. No algorithm is known which finds all subsets for which the cohesion is greater than some threshold, so that methods which start with an arbitrary partition or simple covering and attempt to improve the cohesion of clusters by re-allocation of objects have to be used.

12.4. ML-SETS AND ASSOCIATED CLUSTERINGS

ML-sets can be defined for nominal, ordinal, or numerical dissimilarity coefficients. The set of all ML-sets defined by a numerical dissimilarity coefficient is a numerically stratified clustering in which arbitrary overlap is permitted. The methods for finding ML-sets (variously called cliques, maximal cliques, and maximal complete subgraphs) proposed by Harary and Ross (1957) and Kuhns (1957) were prohibitively slow for all but the smallest data. Needham (1961a) suggested a more efficient approach to the problem. Moody and Hollis have modified and developed this suggestion and have obtained an algorithm which is outlined in Appendix 5.

If d is a DC, then it is specified by $Td(h)$ for all $h \geqslant 0$, and hence by $ML(Td(h))$ for all $h \geqslant 0$, where, as before, $ML(r)$ is the set of ML-sets for the relation r. The system

$$\bigcup_{h=0}^{\infty} ML(Td(h))$$

of all ML-sets for d will be denoted by $ML(d)$. If with each element of $ML(d)$ we associate its diameter, then it is possible to recover d from this information, so that the clustering $ML(d)$ effects no worthwhile simplification of the data. We can, however, obtain useful information by considering a particular subset of $ML(d)$, the set of ball-clusters for d. Jardine (1969b) has defined a *ball-cluster* to be a set $Q \subseteq P$ satisfying the following condition

For all $A \in Q$ we have

$$\text{L} \qquad \min\{d(A, C) : C \notin Q\} > \max\{d(A, B) : B \in Q\}.$$

It is easy to check that if Q is a ball-cluster, then Q is an ML-set for d at level diam Q, where

$$\operatorname{diam} Q = \max\{d(A, B) : A, B \in Q\}.$$

If we define

$$\operatorname{link}_1 Q = \min\{\max\{d(A, B) : A \in Q\} : B \in Q\}$$

then it is easily seen that if Q is a ball-cluster then it is an ML-set for $A_S(d)$ (that is, a single-link cluster) at level $\text{link}_1\, Q$. Sibson has further shown that if $\text{link}_k\, Q$ is defined by

$$\begin{aligned} \text{link}_k Q &= \text{diam}\ Q \text{ if } |Q| \leqslant k \\ &= \min\{\max\{d(A,B): A \in Q, B \in S\}: S \subseteq Q, |S| = k\} \text{ if } |Q| > k \end{aligned}$$

then if Q is a ball-cluster, it is an ML-set for $B_k(d)$ (a B_k-cluster) at level $\text{link}_k\, Q$. The proof of this result is complicated and will not be given here. It follows from this result that every ball-cluster is a B_k-cluster (at some level depending on k) for all $k \geqslant 1$; as special cases for $k = p - 1$ and $k = 1$ we have the results that a ball-cluster is an ML-set and a single-link cluster. The converse of this result is false.

Thus ball-clusters are closely related to the clusters obtained by the methods B_k, and the process of finding ball-clusters may be regarded as a method for selecting particularly well-isolated and homogeneous clusters from among those picked out by all the methods B_k. The use of ball-clusters in the guessing of phylogenetic branching sequences is discussed in Chapter 14.5.

For some purposes it is desirable to consider a stronger condition than condition L. The following condition has been suggested by van Rijsbergen (1970a).

$$\text{L*} \qquad \min\{d(A,C): A \in Q, C \notin Q\} > \text{diam}\ Q.$$

L* ⇒ L, so every set Q satisfying L* (an L*-cluster) is also a ball-cluster. If Q is an L*-cluster, then it is an ML-set at level diam Q which intersects no other ML-set at this level, and consequently it is a B_k-cluster at level diam Q for all k. The converse of this is trivially true.

Not surprisingly, ball-clusters, and *a fortiori* L*-clusters, are rather infrequent in the kinds of taxonomic data encountered in practice, but when they do occur they are likely to be considered highly significant. Both ball-clusters and L*-clusters can be found by selecting the single-link clusters which satisfy the appropriate conditions. See Appendix 6 and van Rijsbergen (1970b).

12.5. CONCLUSIONS

We conclude that, with one exception, there do not as yet exist, outside the framework we have established in Chapters 6-10, satisfactory methods for the construction of classificatory systems. The exception is the case where the data can be assumed to be a mixture of multivariate normal distributions.

This is less surprising than it appears at first sight. Some of the methods were not originally proposed for this purpose; further, many of the more obviously *ad hoc* methods were proposed for problems where the ideal clustering required is a specification of all ML-sets determined by the data. Until very recently algorithms for finding ML-sets were slow and inefficient. Now that a faster algorithm is available, such methods lose their value for many purposes. It is possible that satisfactory methods will be devised, and the pursuit of them for the applications for which they are needed is far from futile; their proper analysis is likely to be impossible without careful attention to the construction of suitable models.

REFERENCES

Ball, G. H. (1966). Data analysis in the social sciences: what about the details? *Proc. Fall Joint Comput. Conf.*, **27**, 533-559.

Bonner, R. E. (1964). On some clustering techniques. *IBM Jl. Res. Dev.*, **8**, 22-32.

Busemann, H., and P. Kelly (1953). *Projective Geometry and Projective Metrics,* Academic Press, London and New York.

Dobes, I. (1967). Partitioning algorithms. *Inf. Processing Mach.*, **13**, 307-313.

Edwards, A. W. F., and L. L. Cavalli-Sforza (1965). A method for cluster analysis. *Biometrics,* **21**, 362-375.

Estabrook, G. F. (1966). A mathematical model in graph theory for biological classification. *J. theor. Biol.*, **12**, 297-310.

Florek, K., J. Łukaszewicz, J. Perkal, H. Steinhaus, and S. Zubrzycki (1951a). Sur la liaison et la division des points d'un ensemble fini. *Colloquium Math.* **2**, 282-285.

Florek, K., J. Łukaszewicz, J. Perkal, H. Steinhaus, and S. Zubrzycki (1951b). Taksonomia Wrocławska. *Przegl. antrop.*, **17**, 193-207. (In Polish with English summary)

Friedman, H. P., and J. Rubin (1967). On some invariant criteria for grouping data. *J. Am. statist. Ass.*, **62**, 1159-1178.

Goodman, L. A., and W. H. Kruskal (1954). Measures of association for cross-classifications. *J. Am. statist. Ass.*, **49**, 732-764.

Gower, J. C. (1967). A comparison of some methods of cluster analysis. *Biometrics,* **23**, 623-637.

Harary, F., and I. C. Ross (1957). A procedure for clique detection using the group matrix. *Sociometry,* **20**, 205-215.

Harding, E. F. (1967). The number of partitions of a set of N points in K dimensions induced by hyperplanes. *Proc. Edinb. math. Soc.*, **15**, 285-289

Harrison, I. (1968). Cluster analysis. *Metra,* **7**, 513-528.

Hartigan, J. A. (1967). Representation of similarity matrices by trees. *J. Am. statist. Ass.*, **62**, 1140-1158.

Jardine, N. (1969b). Towards a general theory of clustering. *Biometrics,* **25**, 609-610.

Jardine, N. (1970). Algorithms, methods, and models in the simplification of complex data. *Comput. J.*, **13**, 116-117.

Jardine, N., and R. Sibson (1968a). The construction of hierarchic and non-hierarchic classifications. *Comput. J.*, **11**, 117-184.

Jardine, N., and R. Sibson (1968b). A model for taxonomy. *Math. Biosci.*, **2**, 465-482.

Jensen, R. E. (1969). A dynamic programming algorithm for cluster analysis. *J. Ops Res. Soc. Am.*, **7**, 1034-1057.

Johnson, S. C. (1967). Hierarchical clustering schemes. *Psychometrika,* **32**, 241-254.

Kendall, M. G., and A. Stuart (1967). *The Advanced Theory of Statistics, Volume 2,* 2nd ed., Griffin, London.

Kuhns, J. (1957). Work correlation and automatic indexing. *Res. Rep., Ramo Wooldridge Corp.*, Canoga Park, California. Appendix P.

Lance, G. N., and W. T. Williams (1965). Computer programs for monothetic classification (association analysis). *Comput. J.*, **8**, 246-249.

Lance, G. N., and W. T. Williams (1967). A general theory of classificatory sorting strategies. I. Hierarchical systems. *Comput. J.*, **9**, 373-380.

Lerman, I. C. (1970). *Les Bases de la Classification Automatique,* Gauthier-Villars, Paris.

Maccacaro, G. A. (1958). La misura della informazione contenuta nei criteri di classificazione. *Annali Microbiol. Enzimol.*, **8**, 231-239.

Macnaughton-Smith, P. (1963). The classification of individuals by the possession of attributes associated with a criterion. *Biometrics,* **19**, 364-366.

Macnaughton-Smith, P. (1965). *Some Statistical and other Numerical Techniques for Classifying Individuals,* Home Office Res. Unit Rep., Publ. No. 6, H.M.S.O., London.

Macnaughton-Smith, P., W. T. Williams, M. B.Dale, and L. G. Mockett (1964). Dissimilarity analysis: a new technique of hierarchical subdivision. *Nature, Lond.*, **202**, 1034-1035.

McQuitty, L. L. (1957). Elementary linkage analysis for isolating orthogonal and oblique types and typal relevancies. *Educ. Psychol. Measmt.*, **17**, 207-229.

Mahalanobis, P. C., D. N. Majumdar, and C. R. Rao (1949). Anthropometric survey of the United Provinces. *Sankhyā,* **9**, 89-324.

Needham, R. M. (1961a). The application of digital computers to problems of classification and grouping. Ph.D. Thesis, University of Cambridge, England.

Needham, R. M. (1961b). The theory of clumps II. *Cambridge Language Res. Unit. Rep. M.L. 139* (Mimeo)

Ogilvie, J. C. (1968). The distribution of number and size of connected components in random graphs of medium size. *I.F.I.P. Congress, Edinburgh, 1968,* North Holland, Amsterdam. *Volume H,* pp. 89-92.

Orloci, L. (1968). Information analysis in phytosociology: partition, classification and prediction. *J. theor. Biol.*, **20**, 271-284.

Parker-Rhodes, A. J., and D. M. Jackson (1969). Automatic classification in the ecology of the higher fungi. In A. J. Cole (Ed.), *Numerical Taxonomy*. Academic Press, London and New York. pp. 181-215.

Parker-Rhodes, A. J., and R. M. Needham (1960). The theory of clumps. *Cambridge Language Res. Unit, Rep. M.L. 126.* (Mimeo)

Rescigno, A., and G. A. Maccacaro (1961). The information content of biological classifications. In C. Cherry (Ed.), *Information Theory, 4th London Symposium.* Butterworths, London. pp. 437-446.

Rota, G.-C. (1964). The number of partitions. *Am. math. Mon.*, **71**, 498-504.

Sebestyen, G. S. (1962). *Decision-Making Processes in Pattern Recognition,* Macmillan, New York.

Sibson, R. (1970). A model for taxonomy II *Math. Biosci.*, **6**, 405-430.

Sibson, R. (1971). Some observations of a paper by Lance and Williams. *Comput. J.*, **14**, 156-157.

Sneath, P. H. A. (1957). The application of computers to taxonomy. *J. gen. Microbiol.*, **17**, 201-226.

Sneath, P. H. A. (1966a). A comparison of different clustering methods as applied to randomly spaced points. *Classifn Soc. Bull.*, **1**, 2-18.

Sneath, P. H. A. (1969). Evaluation of clustering methods. In A. J. Cole (Ed.), *Numerical Taxonomy*. Academic Press, London and New York. pp. 257-267.

Sokal, R. R., and C. D. Michener (1958). A statistical method for evaluating systematic relationships. *Kans. Univ. Sci. Bull.*, **38**, 1409-1438.

Sokal, R. R., and F. J. Rohlf (1962). The comparison of dendrograms by objective methods. *Taxon,* **11**, 33-40.

Sørensen, T. (1948). A method of establishing groups of equal amplitude in plant sociology based on similarity of species content. *K. danske Vidensk. Selsk. Skr.* (*biol*), **5**, 1-34.

Spärck Jones, K., and D. M. Jackson (1967). Current approaches to classification and clump finding at the Cambridge Language Research Unit. *Comput. J.*. **10**. 29-37.

Van Rijsbergen, C. J. (1970a). A clustering algorithm. *Comput. J.*, **13**, 113-115, (Algorithm 47).

Van Rijsbergen, C. J. (1970b). A fast hierarchic clustering algorithm. *Comput. J.*, **13**, 324-326.

Wallace, C. S., and D. M. Boulton (1968). An information measure for classification. *Comput. J.*, **11**, 185-194.

Watson, L., W. T. Williams, and G. N. Lance (1966). Angiosperm taxonomy: a comparative study of some novel numerical techniques. *J. Linn. Soc.* (*Bot.*), **59**, 491-501.

Williams, W. T., and J. M. Lambert (1959). Multivariate methods in plant ecology. I. Association analysis in plant communities. *J. Ecol.*, **47**, 83-101.

Williams, W. T., and J. M. Lambert (1960). Multivariate methods in plant ecology. II. The use of an electronic digital computer for association analysis. *J. Ecol.*, **48**, 717-729.

Williams, W. T., and G. N. Lance (1958). Automatic subdivision of associated populations. *Nature, Lond.*, **182**, 1755.

Wirth, M., G. F. Estabrook, and D. J. Rogers (1966). A graph theory model for systematic biology, with an example for the Oncidiinae (Orchidaceae). *Syst. Zool.*, **15**, 59-69.

Wishart, D. (1969a). Numerical classification method for deriving natural classes. *Nature, Lond.*, **221**, 97-98.

Wolfe, J. H. (1967). NORMIX: computational methods for estimating the parameters of multivariate normal mixtures of distributions. *Res. Memo SRM 68-2* U.S. Naval Personnel Research Activity, San Diego, California.

Wolfe, J. H. (1968a). NORMIX program documentation. *Res Memo SRM* 69-11, U.S. Naval Personnel Research Activity, San Diego, California.

Wolfe, J. H. (1968b). NORMAP program documentation. *Res. Memo SRM* 69-12, U.S. Naval Personnel Research Activity, San Diego, California.

Zahn, C. T. (1964). Approximating symmetric relations by equivalence relations. *SIAM J. appl. Math.*, **12**, 840-847.

Part III

MATHEMATICAL AND BIOLOGICAL TAXONOMY

'They were obliged to have him with them,' the
Mock Turtle said. 'No wise fish would go anywhere without a porpoise.'

LEWIS CARROLL, *Alice's Adventures in Wonderland*

CHAPTER 13

The Components and Aims of Biological Taxonomy

13.1. THE COMPONENTS OF BIOLOGICAL TAXONOMY

In order to evaluate the rôle in biological taxonomy of the methods developed in Parts I and II we must first examine the kind of classificatory system which biologists use, and the aims and methods of orthodox taxonomy.

The taxonomic or Linnaean hierarchy is an ordinally stratified hierarchic clustering. The disjoint classes at each level are called *taxa*, and the ordinal levels are called *ranks.* The taxa of a given rank constitute a *category*. Thus 'species', 'genus', 'family', 'order', 'class', etc., are the names of categories. Colloquially, taxa belonging to a category are often described by the category name. Thus it is said that *Homo sapiens* is a species, when what is meant strictly is that *Homo sapiens* is a taxon of specific rank, or one belonging to the category 'species'. Where confusion cannot arise we shall employ this lax but convenient usage. Usually it is harmless, but we shall see when we come to discuss the criteria by which taxa of specific rank are delimited that it is not always so. The lowest-rank category which is universally used in the Linnaean hierarchy is *species,* but sometimes *subspecies, variety,* or *form* are used as categories of lower rank. A typical taxonomic hierarchy is shown in Figure 13.1.

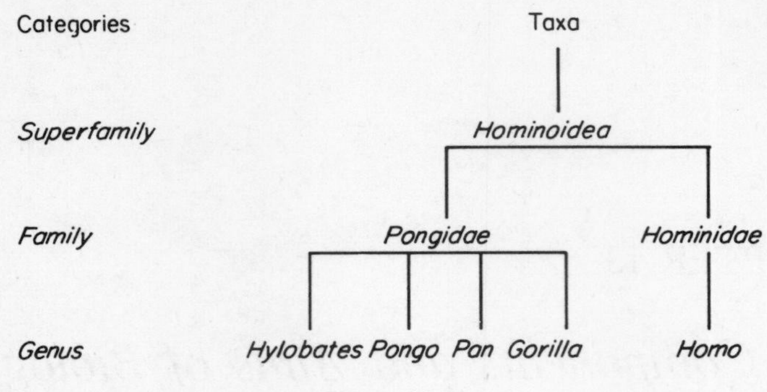

Figure 13.1.

A taxon such as *Hominidae* which includes only a single taxon of the next lower rank is known as a *monotypic* taxon.

Many biologists have supposed that the Linnaean hierarchy raises special logical and philosophical problems; see for example, Buck and Hull (1966), Gregg (1954, 1967). In fact the construction of an adequate set-theoretical model for the taxonomic hierarchy presents no special difficulties. The definition of a taxonomic hierarchy is analogous to the definition of a dendrogram (see Chapter 7.1), but with ordinal rather than numerical indexing of the nested partitions. For details see Jardine, Jardine, and Sibson (1967), and Łuszczewska-Romahnowa (1961). The model given by Łuszczewska-Romahnowa is isomorphic with that of Jardine, Jardine, and Sibson if a definition which precludes taxa monotypic at every rank is dropped; it appears very different for notational reasons. The philosophical problems about criteria for the use of taxon names are no different from the philosophical problems which surround the use of all general terms. The source of much of the confusion about the nature of taxonomic hierarchies is failure to realize that it is possible to give an adequate account of the logical structure of taxonomic hierarchies without also giving an account of the way in which individual organisms are diagnosed into taxa.

Mathematical models for the process of diagnosis have been suggested by Bailey (1965), Ledley and Lusted (1959), and van Woerkom and Brodman (1961). A full account of statistical diagnostic methods used in pattern recognition has been given by Fu (1968). Various non-parametric diagnostic methods were described by Nilsson (1965) and Sebestyen (1962). One of these, the nearest-neighbour assignment rule by which a new object is assigned to the class of its nearest neighbour, is in an obvious way complementary to the single-link clustering method: its properties have been investigated in detail by Cover and Hart (1967) and Peterson

(1970). It is doubtful whether such models are directly applicable as accounts of the way in which the ranges of taxa are extended, because it is doubtful whether the organisms referred to a taxon at a given time can ever be considered as a random sample of its total range. Walters (1961) has shown how the present limits and sizes of flowering-plant genera and families have been largely determined by the historical order in which plants were discovered and described. The construction of efficient diagnostic keys using discrete-state attributes which do not vary within taxa has been treated in detail by Pankhurst (1970), who gives full references to previous work on this subject. The coding of quantitative attributes to facilitate their use in diagnostic keys has been discussed by Jardine and Sibson (1970) The diagnostic criteria for taxa may change as their ranges are extended. This invalidates accounts which assume that taxa can be specified in terms of some fixed set of attribute states or some fixed set of probability measures over attribute states; see Jardine (1969a) for a brief discussion. Clear accounts of the ways in which taxonomic hierarchies are used by biologists are given by Davis and Heywood (1963, Chap. 3) and by Simpson (1961, pp. 18-22).

For the purpose of discussion the practice of biological taxonomy may conveniently, though artificially, be divided into three components, as follows.

(a) The demarcation of taxa of specific rank
(b) The grouping of taxa of specific rank into taxa of supraspecific rank
(c) The grouping of populations within taxa of specific rank into taxa of infraspecific rank

It will be argued that a basis on which taxa of all ranks must be formed is overall resemblance of populations of organisms with respect to many attributes.

The term *population* is used here following Clausen (1951, Chap. 1) and Davis and Heywood (1963, Chap. 11) to denote morphologically homogeneous groups of organisms which are either from a single locality or from a single kind of habitat. These will be taken as the starting point for all classifications. Mayr (1969a) has called such groups *phena,* and Simpson (1961) has called them *demes.* Both terms are unfortunate, having been proposed earlier for quite different specialized uses. The term 'phenon' was used by Camp and Gilly (1943) to describe a particular kind of taxon of specific rank, and by Sokal and Sneath (1963) to describe the clusters in a dendrogram obtained by cluster analysis. The term 'deme' was proposed by Gilmour and Gregor (1939) as a neutral root, which when suitably prefixed describes groups of organisms in the various special purpose genecological and genetic categories. The population concept adopted here is close to Gilmour and Gregor's categories *topodeme* and *ecodeme.* For a fuller discussion of this point see Briggs and Walters (1969, p. 120). This use of the term 'population' might be objected to on the grounds that many biologists use the term only for freely interbreeding groups of organisms. Such groups may be called *Mendelian* populations or *gamodemes.*

The problems raised by the three components of taxonomy are rather different. It will be argued that at the specific and infraspecific level, in addition to certain general problems, there arise many special problems which differ according to the breeding systems and patterns of ecological and geographical variation of the organisms studied. In the delimitation of species, controversy has centred on the issue of whether taxa should be assigned specific rank on grounds of genetic isolation or on grounds of having a certain degree of morphological dissimilarity from other taxa. In supraspecific taxonomy controversy has centred on the issue of whether supraspecific classifications should aim to represent only the relative dissimilarities of taxa of specific rank or whether they should also aim to represent the evolutionary (phylogenetic) relationships between such taxa. In infraspecific taxonomy controversy has centred on the extent to which the Linnaean hierarchy is appropriate for representing patterns of infraspecific variation.

A further controversial issue in classification at the specific and infraspecific levels concerns the relation between the taxa which are recognized and named within the Linnaean hierarchy and the special-purpose or experimental classifications which are based on genetic, cytological, and ecological information about the populations. The reader is referred to Davis and Heywood (1963, Chap. 13) for a detailed and balanced discussion of this issue.

13.2. THE DEMARCATION OF TAXA OF SPECIFIC RANK

Biologists often maintain that taxa of specific rank are in some sense more real or more basic than taxa of other ranks. There is some common-sense justification for this view. The most homogeneous kinds of plant and animal which the layman can discriminate usually correspond, more or less, to taxa of specific rank, generic rank, or some intermediate rank. Thus the common primrose in Britain corresponds to a single taxon of specific rank (*Primula vulgaris*), and the British elms are precisely the British members of the taxon *Ulmus* of generic rank. The discrimination of kinds by laymen is rarely consistent, being much more acute when applied to organisms which are in some way useful or unusual; see Berlin, Breedlove, and Raven (1966), Briggs and Walters (1969, p. 14). But this is not a peculiarity of laymen.

Taxonomists also tend to be more acute when dealing with the useful or the unusual. There is no such regular correspondence between the families and orders of flowering plants recognized by taxonomists and any groups or kinds which would be recognized by a laymen. There is a simple explanation for the fact that the diagnostic criteria for taxa of specific rank are often both more numerous and more clear-cut than are the diagnostic criteria for taxa of other ranks, namely that many taxa of specific rank consist of populations which are genetically isolated from populations of all other taxa of specific rank, but which may exchange genes with each other. Infraspecific taxa commonly consist of populations which are not genetically isolated from populations of other infraspecific taxa within the same species. Supraspecific taxa commonly include populations which are genetically isolated from each other.

Many taxonomists have suggested that barriers to potential gene-flow should be used to define species; see, for example, Dobzhansky (1941, Chap. 11), Mayr (1969b). These so-called *biological* species definitions are frequently contrasted favourably or unfavourably with so-called *morphological* species definitions. It seems that the various species definitions cover two distinct issues which may have become confused by the ambiguity of the term 'species' when it is used both to refer to individual taxa of specific rank and to denote the set of all taxa of specific rank. One issue concerns the way in which individual taxa of specific rank should be delimited. The other issue arises only if it is supposed that all taxa are clusters based on the relative dissimilarities of populations with respect to many attributes. It may then be asked what kind of criteria should be used to decide which of these clusters should be accorded specific rank. The biological species definition, if seriously applied to the first issue, would demand that minimal sets of genetically isolated populations be recognized as taxa of specific rank regardless of any other attributes of the organisms concerned. Fortunately, no-one has ever tried to apply this definition. It would have the absurd consequence that in groups of asexual organisms each individual would be assigned to a different taxon of specific rank. In sexually reproducing organisms morphologically homogeneous assemblages of populations would often have to be divided into several taxa of specific rank, despite the fact that, as first shown by Müntzing (1938), partial or complete intersterility of populations within what is usually regarded as a good species may be controlled by simple genetic effects. Even those taxonomists who accept the necessity for 'sibling' species are reluctant to allocate populations to distinct taxa of specific rank when there is no significant morphological divergence between populations.

The dispute appears more reasonable if the biological species definition is applied to the second issue, that of deciding which clusters of populations should be accorded specific rank. By the morphological criterion the extent of the morphological isolation of a cluster is used to decide whether it merits specific rank; by the biological criterion clusters whose component populations are genetically isolated from populations of other clusters are given specific rank regardless of the magnitude of their relative morphological isolation. The criteria based on genetic isolation or barriers to potential gene-flow are by no means clear-cut. Gene-flow may be prevented or restricted not only by intrinsic genetically determined mechanisms, but also by geographical, ecological and seasonal barriers, and in higher plants by the discrimination of pollinating agents; see Clausen (1951, Chap. 8), Stebbins (1950, Chap. 6). Intrinsic genetically determined barriers between populations may vary from complete and permanent intersterility to marginally reduced viability or fertility of hybrid offspring; see Clausen (1951, Chap. 8), Davis and Heywood (1963, Chap. 13) for detailed discussions. Populations which cannot exchange genes directly may do so indirectly *via* other populations. Genetic barriers may be asymmetrical when the first generation progeny can backcross with one parent only. And finally, the potential gene-flow criterion is inapplicable to asexually reproducing organisms.

All the above considerations militate against the strong version of the biological species definition, that barriers to gene-flow should be used as the sole criterion for delimitation of taxa of specific rank; and also against the weaker version, that barriers to gene-flow should be used to decide which clusters of populations, grouped according to other criteria, merit specific rank. One justification which is often given for attaching great weight to barriers to potential gene-flow is that it is minimal genetically isolated groups of populations which are the basic evolutionary units: at a given time they are the cross-sections of the branches of an evolutionary tree. But this is a spurious justification, since it is the actual rather than the potential barriers to gene-flow which determine what at a given time are the evolutionary units, the *gene-pools.* This point was argued in more detail by Ehrlich and Raven (1969), who cited evidence for actual amounts of gene-flow between isolated but interfertile populations. Further, it is impossible to determine whether or not populations at present separated only by extrinsic genetic barriers will at some future date exchange genes. The fact that a consistent biological species definition would have to be based on all barriers to gene-flow, rather than on intrinsic barriers alone, has been emphasized by Davis and Heywood (1963, Chap. 13). Unfortunately, evidence for intrinsic barriers is much more easily obtained than are estimates of the naturally occurring amounts of gene-flow.

In practice, as pointed out by Clausen (1951) and Clausen, Keck, and Hiesey (1941), there are not just two but at least four distinct criteria which have been used in deciding which groups of populations should be accorded specific rank: (a) dissimilarity in morphological and other attributes; (b) difference in ecological and/or geographical range, (c) degree of interfertility revealed by ease of hybridization, and fertility and vigour of second generation offspring, (d) cytological difference as indicated by chromosome number and degree of pairing between chromosomes at meiosis in hybrids. To this list it is nowadays necessary to add (e) serological differences (see, for example, Hawkes and Tucker, 1968); and (f) extent of DNA hybridization (see, for example, Kohne, 1968). Use of the criteria (c)-(f) has often been justified on the grounds that they provide a better measure of evolutionary relationship than does morphological divergence. This argument is of dubious validity. Thus Snyder (1951, cited in Briggs and Walters, 1969) gives evidence that poor chromosome pairing between populations of *Elymus glaucus* (a North American grass) may be caused by small structural differences in chromosomes or by specific genes. To these technical criteria for specific rank should be added the criterion which in practice has been used to delimit the majority of species: namely, a good species is what a good taxonomist believes to be a good species. Priority in proposing this criterion is uncertain.

Decisions about which groups of populations should be accorded specific rank are straightforward when these criteria are concordant. When they differ some kind of compromise has to be adopted. Clusters of populations which are separated by only a small morphological dissimilarity may be recognized as distinct taxa of specific rank if there is an intrinsic genetic barrier or a definite difference in ecological-geographical range. Conversely, clusters of populations separated by a larger morphological dissimilarity may be assigned to different taxa of subspecific rank only if there is no intrinsic genetic barrier and little or no difference in ecological-geographical range. But how the two criteria are balanced in the compromise seems to be largely a matter of personal taste. The tendency amongst zoologists to state that they give great weight to genetic criteria may arise in part from the fact that morphological discontinuities and genetic barriers between animal populations are more usually concordant than are morphological discontinuities and genetic barriers between plant populations. Acquiescence in one of the many biological species definitions may therefore have relatively little effect on zoological taxonomic practice. Botanists are faced with a situation where concordance is relatively rare; acquiescence in a biological species definition would render taxonomy at the specific and infraspecific level quite intractable. They are therefore forced to be more critical of genetic criteria, see Walters (1962). Botanists who hold that the biological species definition is correct usually apply this belief in practice by giving great weight to chromosome number in the delimitation of species. The reasons for this are obscure: there is abundant evidence that differences in chromosome number frequently fail to coincide with intrinsic barriers to gene-flow.

If, as is supposed here, taxa of specific rank must at least satisfy the condition of being discrete clusters of populations based on the distribution of many attributes, then partitional methods of clustering will be useful in delimiting taxa of specific rank just as they are useful in delimiting taxa of higher rank. The methods described in Chapter 11 for calculating the distortion imposed by a hierarchic clustering may sometimes provide a guide in deciding which taxa should be assigned specific rank. Where a hierarchic clustering of a set of populations imposes very high distortion this may indicate that their differentiation is best considered to be infraspecific. This kind of use of partitional clustering in biological taxonomy is discussed in detail in the next section. If potential or actual barriers to gene-flow are to be used together with morphological criteria to decide which clusters of populations are to be given specific rank, a further use for partitional methods becomes apparent. They can be used to investigate the extent to which groups of populations isolated by actual or potential genetic barriers coincide with clusters of populations obtained by automatic classification on the basis of morphological criteria.

13.3. SUPRASPECIFIC CLASSIFICATION

The opinions of taxonomists on the purposes of supraspecific classification show a range of variation which is suprisingly wide given the remarkable extent to which there is a consensus of opinion about the adequacy of many of the accepted classifications.

There are two main aims of supraspecific classification which appear to be acceptable to almost all taxonomists.

(a) To group taxa of specific rank into taxa of higher categories in ways which convey as accurately as possible information about their component populations.

(b) To provide a consistent system of names for groups of organisms so that information about the groups can be communicated.

The conventional system of Latin nomenclature for taxa, which is outlined in the most recent versions of the *International Code of Botanical Nomenclature* and the *International Code of Zoological Nomenclature* is still generally accepted. Suggestions for substituting a numerical indexing system for the conventional Latin nomenclature have been made by Sokal and Sneath (1963, Chap. 10). Uninomial nomenclature for taxa of specific rank has been discussed by Cain (1959b) and Michener (1964). Such systems will doubtless come to be used increasingly for mechanized retrieval of information about organisms. But for purposes of general communication they are unlikely ever to gain acceptance, since an enormous initial loss of facility in the communication of information would ensue.

The serious area of dispute concerns the way in which aim (a) is to be fulfilled. What kind of information should supraspecific classifications seek to represent? There are three well-marked extreme views—the *conservative* view, the *cladistic* view, and the *phenetic* view—and a number of intermediate views.

Conservatism

On the conservative view classifications should be based as far as conscience allows on previous classifications. New taxa should be fitted into existing classifications so as to disrupt them as little as possible. A justification for this view is that frequent alterations of classifications in attempts to make them satisfy aim (a) more adequately result in many nomenclatural changes so that their performance of aim (b) is seriously impaired. The deleterious effects of alterations on the communication of information are particularly severe when the name of a taxon of supraspecific rank is retained, but its range is repeatedly changed. A good example of this is given by the fern family Polypodiaceae whose range has been revised at least a dozen times in the last thirty years. A further justification for the conservative view can be given by those who believe that the pursuit of a classification which is optimal in the way it satisfies aim (a) is the pursuit of a myth. Moderate exponents of this view have included Gilmour and Walters (1964) and Walters (1965). The majority of professional taxonomists probably have secret conservative sympathies.

Cladism

On this view the sole aim of supraspecific classification is to express as accurately as possible the evolutionary relations of taxa of specific rank. This is achieved by ensuring as far as possible:

(a) That every taxon be *strictly monophyletic.* A taxon is said to be strictly monophyletic if its component populations are all and only the descendants of a single ancestral population.

(b) That the ranks of taxa be decided by the relative times since their component populations diverged from a common ancestral population. The cladistic criterion of classification is contrasted with two weaker criteria of consistency with phylogeny in Figures 13.2, 13.3, and 13.4.

This is the view of classification which has been advocated by Hennig (1965, 1967) and Stensiö (1963). Application of the cladistic criterion in classifying groups which include both recent and fossil organisms results in classifications which may be completely discordant with the relative dissimilarities of populations of organisms. Thus the populations 1 and 2 shown in Fig. 13.2 are assigned to different taxa whereas 2 and 3 are assigned to a single taxon of the same rank. If evolutionary rates had been reasonably constant 1 and 2 would be much less dissimilar than are 2 and 3.

The application of the strict monophyly criterion to both living and fossil populations appears to be an unsatisfactory method of classification, because it results in the allocation of fossil populations to taxa without regard to their relative dissimilarities to other living and fossil populations. When the position in the evolutionary branching sequence of fossil populations is not known they must, *faute de mieux*, be classified on grounds of relative dissimilarity. The cladistic view requires that if their evolutionary position is subsequently determined drastic reclassification may be necessary. An extreme example of this is demonstrated in Stensiö (1963). The order Porolepiformes of primitive Devonian fishes includes two groups generally recognized as superfamilies, the Osteolepoideae and the Holoptychoideae. The morphological dissimilarity between the two groups is roughly comparable to that which is found between other pairs of superfamilies within the same order of fishes. Stensiö believed that one group, the Holoptychoideae, is ancestral to the anuran amphibians (frogs and toads), and that the other is ancestral to the remaining living amphibians as well as all other tetrapods. He therefore assigned the two groups to different classes, despite the fact that their difference is vastly smaller than that which separates other taxa which are conventionally recognized as classes by zoologists. Pichi-Sermolli (1959) has shown how cladism has led to chaos in the classification of ferns.

Perhaps the most obvious objection to the cladistic theory is that representation of evolutionary branching sequences is most adequately performed by evolutionary trees: to use classifications for this purpose seems otiose. Recent cladists do not commit the error of identifying evolutionary trees with classifications. But the cladistic approach appears to have evolved from this naive fallacy.

The application of the strict monophyly criterion to living populations only does not necessarily produce such absurd classifications. It is discussed more fully below.

Pheneticism

On this view the sole aim of taxonomy is to produce classifications which reflect as accurately as possible the relative similarities or dissimilarities of populations without regard to their evolutionary relationships. The relative similarities or dissimilarities of organisms are measured or assessed intuitively on the basis of many attributes without *a priori* differential weighting of attributes or attribute states. This is the view of classification frequently said to have been pioneered by Adanson (1763)*.

Clear statements of the phenetic view of classification have been given by Bather (1927), Bigelow (1956), Bremekamp (1939), Gilmour (1937, 1940), Gilmour and Turrill (1941), and Thompson (1952). The phenetic view of taxonomy has been presented as the basis for the use of numerical methods in taxonomy by Sneath (1962) and Sokal and Sneath (1963).

Unfortunately the term 'phenetic' has acquired several additional connotations since its introduction in the above sense by Cain and Harrison (1960). Some authors have failed to realize that renunciation of *a priori* weighting of attributes is not the same thing as renunciation of all weighting of attributes. Various grounds for precise numerical *a posteriori* weighting have been discussed in Chapter 4. Even where the assessment of relative dissimilarities is intuitive, weighting according to the relative variabilities within populations is probably standard practice amongst taxonomists. It is at least part of what is meant when taxonomists write of 'good' and 'bad' attributes. Some authors, in rejecting the myth of an ideal classification based on phylogeny, have reinstated the myth of an ideal phenetic classification which is ever more nearly approximated as more and more attributes are used as data for automatic classification. The basis for this belief is the hypothesis stated by Rao (1948), and, in a rather different form, by Sokal and Sneath (1963, Chap. 5), that similarities and dissimilarities between populations of organisms can be considered as parameters which are estimated progressively more accurately as the number of attributes considered is increased. Sokal and Sneath base this hypothesis on the nexus hypothesis. By the nexus hypothesis each attribute is supposed to be partially determined by many genes; and conversely each gene is supposed partially to determine many attributes. Hence as progressively more attributes are considered, progressively more genes are sampled in a random manner, providing that attributes are selected without bias. The notions of sampling, randomness, and bias involved here are highly metaphorical. We suggest that whether or not different attributes, or attributes selected according to different rules, yield similar relative dissimilarities is entirely a matter for empirical investigation.

*It has been stated repeatedly that Adanson was the founder of phenetic classification. However, Guédès (1967) has shown conclusively that whilst Adanson did advocate the use of many attributes in classification rather than just those thought to be 'essential' or of functional importance, he did not explicitly formulate a doctrine of *a priori* equal weighting. Guédès shows that the misunderstanding of Adanson originated with the adverse criticisms of his method by Cuvier (1807) and De Candolle (1813).

Likewise, whether or not as more attributes are considered the relative magnitudes of dissimilarities between populations undergo drastic alterations is a matter for empirical investigation. The importance of such investigations and some ways in which they may be carried out are discussed in Chapter 14.3.

It is perhaps the prevalence of these fallacious versions of the phenetic view of classification amongst certain advocates of the use of numerical methods in taxonomy which has led to such extreme reactions against numerical taxonomy as are expressed in Kiriakoff (1962) and Mayr (1965).

The following two views of the aims of supraspecific classification are intermediate between cladism and pheneticism.

Phylogenetic weighting

Several authors (Cain, 1959a; 1962; Cain and Harrison, 1960; Mayr, 1965, 1968) have suggested that classifications should be based on the phylogenetic affinity or genetic affinity of populations. Such affinities are estimated on the basis of the distributions of attribute states amongst populations, but they differ from phenetic similarities because attribute states are differentially weighted according to their supposed relative importances as phylogenetic indicators.

This kind of weighting is quite distinct from *a posteriori* weighting to deal with conditional definition and relative variability, which we have shown to be consistent with a purely phenetic approach. Mayr (1969a, p. 218) defines this kind of weighting as 'a method for determining the phyletic information content of a character'. However, the majority of the criteria which he suggests for such weighting are in fact the same as those which arise naturally in the calculation of phenetic dissimilarity, although the purpose is quite different. Cain (1959a) shows that this approach was clearly formulated by Darwin (1859).

This approach to classification, rather than the cladistic approach, is what the majority of taxonomists seem to have had in mind when stating that classification should be based on phylogenetic principles. We shall therefore discuss it in some detail. There are three main difficulties. First, the meanings of phylogenetic and genetic affinity, and of phylogenetic importance, require clarification. Secondly, whatever definition of phylogenetic importance is adopted, there appear to be considerable difficulties in estimating it for attribute states when fossil evidence for the evolutionary history of a group of organisms is lacking. And thirdly, even if objective methods of phylogenetic weighting can be found, the usefulness of the resultant classifications is questionable.

If a biologist wishes to understand the evolution of a group of organisms there are two distinct features of phylogeny which he will need to determine. First, the branching sequence of the evolutionary tree must be established. Secondly, the amount of divergence which has occurred since each branching must be determined. Cladistic classification aims solely at the representation of the former aspect of phylogeny. Phenetic classification aims solely at the representation of the second aspect of phylogeny: phenetic classification does, therefore, have an evolutionary basis because the observed dissimilarities of populations of organisms have supposedly been established by evolutionary divergence of populations. Similarities or dissimilarities calculated with phylogenetic weighting will be a function both of the relative times of branching in evolution and of the subsequent amounts of phenetic divergence of populations; *cf.* Mayr (1969a, p. 200). But such a compromise is a compromise only in the derogatory sense of the term. Similarities or dissimilarities calculated in this way will be unknown and variable functions of phenetic similarity or dissimilarity and time since branching in evolution. The ambiguous nature of the notion of phylogenetic similarity or affinity was emphasized by Gilmour (1940). A classification based upon phylogenetic affinity will, in attempting to convey both kinds of information, convey no information.

Even if the calculation of phylogenetically weighted similarities or dissimilarities were worthwhile, the estimation of phylogenetic weights for attributes would present considerable difficulties. Two distinct, but frequently confused, basic criteria of phylogenetic importance have been suggested–*primitiveness* and *conservatism.* An attribute state or range of states is said to be primitive in a group of organisms if it was present in the common ancestor of the group. The degree of primitiveness will depend upon the time since divergence (evolutionary age) of the group of organisms. An attribute state or range of states is conservative if it occurs in all descendants of the population in which it arose. Many authors have confused conservatism with primitiveness; it is clear, however, that a conservative attribute state may be of recent or ancient origin. A further confusion is involved in the frequent assumption that a given attribute state can unambiguously be said to be conservative or have a particular degree of primitiveness. In fact an attribute state which is conservative or primitive in one group of organisms may not be so in other groups. Davis and Heywood (1963, p. 189) mention an excellent example of this. It is generally assumed that in the flowering plants orthotropous ovules are primitive and campylotropous ovules advanced (see Bocquet and Bersier, 1960). Corner (1962) has presented convincing evidence that in the figs (Moraceae) the reverse is true. The abundant evidence for the parallel evolutionary development of closely similar structures should be adequate warning against the construction of evolutionary trees for morphological, cytological, and molecular structures, as if they had a life of their own apart from the organisms in which they occur. But the practice is still widespread. Clear examples of evidence for parallel evolution of similar or identical morphological features are given in Inger (1967), Michener (1949), and Throckmorton (1965).

The usual grounds for determining primitiveness and conservatism depend either upon prior classification or upon the interpretation of a fossil record; the latter may itself depend heavily upon taxonomic judgements. Attribute states are likely to be primitive if they are widely distributed amongst otherwise dissimilar present-day populations. Conservative attribute states are likely to be those which are diagnostic for taxa thought to be strictly or approximately monophyletic.

If either primitiveness or conservatism is to be used in constructing classifications, precise methods must be found for deciding which attribute states are primitive or conservative, prior to the construction of supraspecific classifications; otherwise, as Sokal and Sneath (1963, Chap. 2) have pointed out, circular inference will be involved. Recently a variety of such methods have been suggested, both qualitative (Inger, 1967; Maslin, 1952; Stebbins, 1967) and quantitative (Cain and Harrison, 1960; Farris, 1966; Kluge and Farris, 1969; Sporne, 1948, 1969; Wilson, 1965). If such methods can be justified they will be of obvious value in attempts to reconstruct evolutionary branching sequences for groups of organisms which lack a fossil record. They may also be useful in the approach to supraspecific classification outlined in the next section, whereby phenetic classifications are modified *post facto* to improve their consistency with phylogeny. For the reasons given above it is doubtful if the use of such methods as a basis for exact phylogenetic weighting of attribute states is worthwhile.

Phylogenetic modification

Phylogenetic weighting of attributes does not seem to be a viable method of ensuring that supraspecific classifications reflect the phylogeny of the organisms classified. An alternative approach, which has been discussed by Mackerras (1964) and Simpson (1961), is to modify the results obtained by phenetic methods so as to ensure consistency with what can be inferred about the evolutionary branching sequence of the populations studied.

The relation between phenetic classification and the evolutionary branching sequences of populations is determined by evolutionary rates of divergence, provided that the rate of divergence is estimated using the same dissimilarity measure and the same attributes as are used in constructing the phenetic classification. Since adequate methods for the measurement of evolutionary rates have not been developed, very little is known about them. Some of the ways in which they might be investigated are discussed in Chapter 14.5. Since such investigations have not been carried out, the following discussion is of necessity hypothetical.

The strongest assumption about evolutionary rates which can be made is that they are constant. On this assumption the dissimilarities between present-day populations would be monotone with the times since their divergence. They would therefore be ultrametric, since the times of divergence of populations in an evolutionary tree form an ultrametric. The fact that the dissimilarities between present-day populations are rarely ultrametric refutes the hypothesis of constancy of evolutionary rates in terms of known measures of dissimilarity.

As pointed out earlier, the cladistic requirement that all taxa in a hierarchy which covers both recent and fossil organisms be strictly monophyletic would not hold for a phenetic classification even if the evolutionary rates had been constant. However, if only present-day populations were classified, any reasonable phenetic method would produce strictly monophyletic supraspecific taxa if evolutionary rates had been constant. Provided that evolutionary rates have not been wildly inconstant it is probable that highly isolated clusters found by a phenetic method will be strictly monophyletic. Unfortunately, at least in the groups of organisms which have been studied by numerical phenetic methods, it appears that an adequate supraspecific classification would rarely be possible if only highly isolated clusters were recognized as taxa.

A criterion more moderate than the cladistic criterion of consistency between hierarchic classifications and phylogeny is that the taxa of every supraspecific rank should partition present-day organisms into strictly monophyletic groups. It follows from the considerations above that phenetic classifications may have to be modified if this criterion is to be fulfilled. It is probable that such modifications will be less drastic the more closely the dissimilarities between the present-day populations approximate to ultrametric structures (see Chap. 14.5). It is interesting to note that the classification of present-day mammals which was generally adopted when little was known of their fossil record differs relatively little from, for example, Simpson's (1945) classification. Examination of Simpson's classification shows that it largely obeys the strong consistency criterion given above.

A much weaker and more plausible assumption about evolutionary rates is the hypothesis that in evolution populations never show overall convergence. More precisely, if two populations are dissimilar by an amount h at a time t, measured with respect to a given large set of attributes, their descendants will not be dissimilar by an amount less than h measured with respect to the same set of attributes at any time t' later than t. It was shown in Jardine (1971) that if the non-convergence hypothesis is correct, then phenetic classifications obtained by the single-link method of cluster analysis will always satisfy the criterion of consistency with phylogeny proposed by Beckner (1959) and Simpson (1961). The Beckner-Simpson criterion of consistency is as follows.

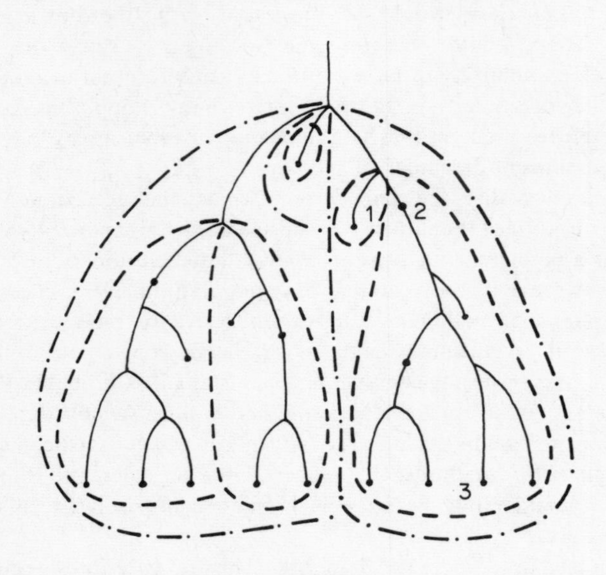

Figure 13.2. A purely cladistic classification. Bold points indicate populations of specific rank; – – – delimits taxa of a given supraspecific rank; – . – . – delimits taxa of a higher rank. Note that populations 2 and 3 belong to the same taxon of the lower supraspecific rank, whereas population 1 belongs to a different taxon of the same rank. Phenetic classifications will rarely satisfy this criterion.

A taxon of a given rank is said to be *minimally monophyletic* if its component populations are directly descended from populations referred to not more than one taxon of the same rank. A taxonomic hierarchy is consistent with phylogeny if all its taxa of supraspecific rank are minimally monophyletic. The Beckner-Simpson criterion is applicable only to classifications which cover both fossil and present-day populations.

The operation of the two criteria of consistency is illustrated and contrasted with the cladistic criterion for classification in Figures 13.2, 13.3, 13.4, and 13.5.

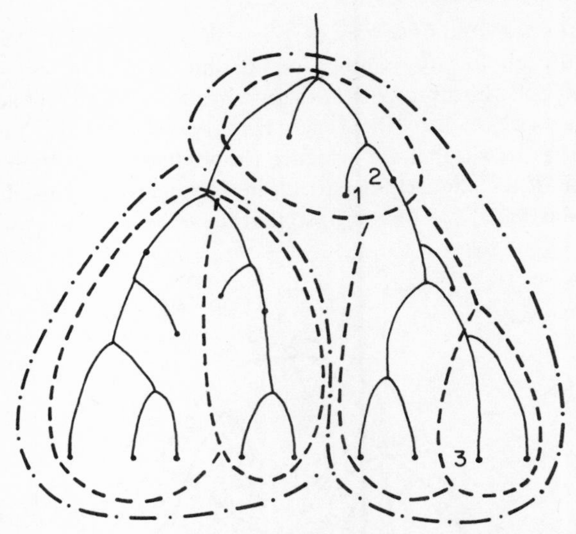

Figure 13.3. A classification which satisfies the requirement that taxa of each supraspecific rank should partition present-day populations into strictly monophyletic groups. Phenetic classifications may or may not satisfy this criterion. For explanation of symbols see Figure 13.2

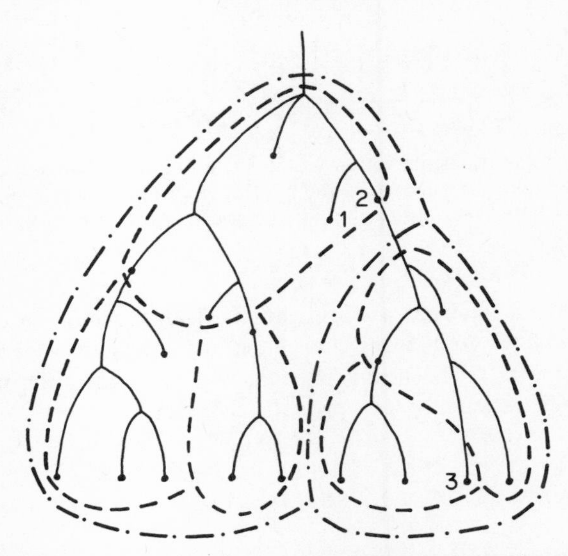

Figure 13.4. A classification which satisfies the Beckner-Simpson criterion of consistency between classification and phylogeny. Phenetic classifications will usually satisfy this criterion. For explanation of symbols see Figure 13.2

The *post facto* modification of phenetic classifications to ensure consistency with phylogeny is fully compatible with the use of numerical phenetic methods in supraspecific taxonomy. It has the great advantage over the phylogenetic weighting approach that it leaves open to the taxonomist the decision as to whether the evidence for phylogeny is sufficient to warrant modification of the phenetic classification. Further, it is applicable only when there is a substantial fossil record, so that it can be applied only on sufficient

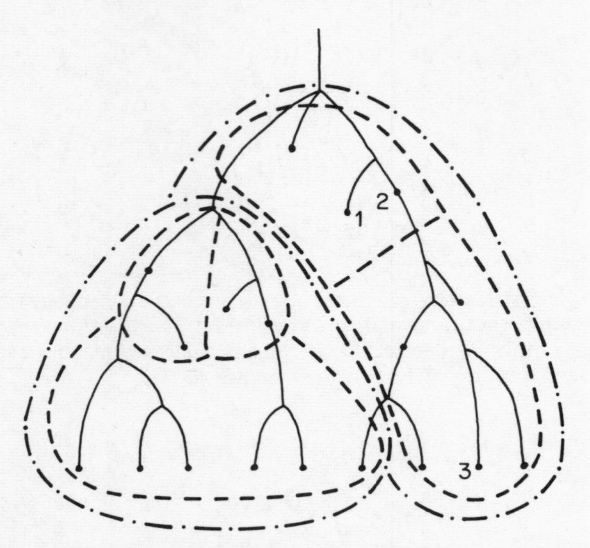

Figure 13.5. A classification which violates the Beckner-Simpson criterion of consistency between classification and phylogeny. This kind of classification will usually violate also any phenetic classification. For explanation of symbols see Figure 13.2

evidence. It is arguable that a classification which violates the criterion is seriously misleading. And finally, since on the available evidence overall convergence appears to be a rare phenomenon, its application may rarely involve any modification of purely phenetic classifications.

13.4. INFRASPECIFIC CLASSIFICATION

At the end of the nineteenth century many taxonomists, especially botanists, used regularly five or six infraspecific categories. Thus in Rouy and Foucauld's *Flore de France* (1893-1913), subspecies, proles, varieties, subvarieties, forms, and subforms were recognized. Today many botanists consider that the only generally useful infraspecific category is the subspecies (see Valentine and Löve, 1958). The only infraspecific category allowed by the *International Rules of Zoological Nomenclature* is the subspecies.

The kinds of groups of populations which are usually accorded subspecific rank by botanists are: (a) groups which are not isolated by intrinsic genetic barriers, but which show well-marked morphological differences correlated with well-marked differences in ecological or geographical range; (b) groups which are wholly or partially isolated from other groups by intrinsic genetic barriers but which are not definitely discriminable by any single gross morphological attribute. Zoologists tend by and large to recognize as subspecies groups which are less well-differentiated both morphologically and geographically. By the well known '75% rule' many zoologists recognize as distinct subspecies ecological or geographical races which are sufficiently well-differentiated to allow correct identification of at least 75% of specimens. As many recent authors (see, for example, Wilson and Brown, 1953) have pointed out, the 75% rule is absurd. For example, if linear combinations of attributes obtained by discriminant function analysis are used as diagnostic criteria, it would allow the allocation of the inhabitants of the more remote and inbred East Anglian villages to distinct subspecies. Hedberg (1958) has similarly attempted to provide rules stipulating minimum amounts of dissimilarity to justify the recognition of distinct taxa of subspecific rank for geographically isolated groups of plants.

Groups of type (b) are usually recognized by zoologists as sibling species. Groups of taxa which would be considered as subspecies of a single species by botanists are commonly recognized as species lying in a single superspecies or species group by zoologists. The species group or aggregate is less commonly used by botanists.

The basic difficulty in infraspecific classification is that the patterns of ecological and geographical variation which arise within taxa of specific rank are frequently poorly represented by any kind of hierarchic classification. This is revealed by the fact that dissimilarity coefficients between populations frequently show very poor hierarchic structure. A few of the patterns of variation which arise are described below. Many of these patterns cannot adequately be expressed by allocation of populations to subspecies and varieties. The list could be greatly extended, and it is emphasized that several of these kinds of variation may coexist within the same species.

(a) *Geographical differentiation.* There may be concordant differentiation in several attributes of populations from different parts of the geographical range of a species. Often intermediate populations may occur at the junctions of the parts of the range. Whether or not geographical subspecies are recognized depends upon the extent to which intergradation occurs and on the extent of the differentiation involved.

(b) *Ecotypic differentiation.* There may be concordant differentiation in several attributes of populations from different kinds of habitat within the range of a species. This is generally called *ecotypic* variation. For a concise discussion see Briggs and Walters (1969, Chap. 10). Often ecotypic variation is accompanied by differences in the geographical range of the ecotypes, as in *Potentilla glandulosa* which was investigated by Clausen, Keck, and Hiesey (1940). The study of the genetic bases for ecotypic differentiation is often called genecology, and is reviewed by Heslop-Harrison (1964). Ecotypes may show various degrees of intergradation. Where there is substantial intergradation the pattern of variation may be better regarded as ecoclinal.

(c) *Clinal variation.* Here there is a gradient in some attribute or group of attributes which varies more or less continuously with geographical location (a *topocline*), or with an environmental gradient (an *ecocline*), or with both. It is probable that many taxa of varietal and subspecific rank in fact represent segments of clines. Likewise many of the groups recognized as ecotypes by botanists may result from uneven sampling of populations which do in fact form an ecocline. See Briggs and Walters (1969, pp. 165-170) for a discussion of this point. Clinal variation may cut across well-marked geographical discontinuities in other attributes. Furthermore, clinal variation in different attributes may be discordant; that is, different attributes or groups of attributes may vary in different geographical directions or along different ecological gradients.

(d) *Discordant variation.* This kind of variation involves geographically or ecologically correlated discontinuities which do not coincide for different attributes or groups of attributes. This situation appears to be very common indeed. Good zoological examples are given by Gillham (1956), Highton (1962), and Sokal and Rinkel (1963). Highton showed that geographically disjoint populations of the salamander *Plethodon jordani* show clear-cut differences in skin colouration which many zoologists would use to create subspecific taxa, but that there is discordance between the various attributes. Gillham (1956) described the confusion which has arisen in the taxonomy of butterflies from attempts to represent discordant patterns of infraspecific variation by allocation to geographical subspecies.

Two special cases create particular difficulties for taxonomists. Sometimes several morphological characters show concordant geographically or ecologically correlated discontinuities over part of the range, and discordant variation in other parts of the range of a species. North European populations of *Sagina apetala* (Caryophyllaceae) can be referred to two population groups which show many slight morphological differences, well-marked ecological preferences, and slight differences in geographical range. In southern Europe there are well-marked discontinuities in morphology which are correlated with geographical location, but these are discordant with respect to the attributes which differentiate the northern populations. It is a moot point whether the northern population groups should be given taxonomic recognition.

Another awkward special case occurs when concordant and geographically correlated discontinuities in many attributes occur, so that well-marked population groups can be recognized, but topoclinal or ecoclinal variation in other attributes cuts across the population groups. This is the situation with topoclinal variation in *Pinus nigra* (Davis and Heywood, 1963, p. 380) and with ecoclinal variation in the Caucasian populations of the genus *Onobrychis* (Papilionaceae); see Sinskaja (1960).

(e) *Reticulate variation.* Here local populations show well-marked differences which are uncorrelated with ecological and geographical factors and which are so distributed that they cannot be combined to form morphologically homogeneous groups. In plants this pattern of infraspecific variation arises frequently when apomixis or autogamy occurs; see Briggs and Walters (1969, Chap. 8). In this situation some authors have thought it correct to give taxonomic recognition to every local population. Thus Almquist (1921) recognized dozens of microspecies within *Capsella bursa-pastoris* (Shepherd's Purse), and Jordan (1873) recognized numerous microspecies within the *Erophila verna* aggregate (Whitlow Grass). In both species plants are usually self-pollinated, but populations with new combinations of character states may arise because of occasional cross-pollination. The *Erophila verna* aggregate illustrates the complex situation, quite frequent in flowering plants, where chromosomal races showing different geographical ranges are superimposed on a local pattern of reticulate variation.

Reticulate variation should be carefully distinguished from the slight apparent differentiation of small local populations which arises because of sampling errors even in species which are not internally differentiated. This phenomenon emphasizes the danger of basing infraspecific taxa on small samples.

(f) *Genotypic mimicry of phenotypic modification.* Frequently in plant populations an awkward situation occurs. There is a well-marked morphological discontinuity which separates a particular ecotype from other populations. But the ecotype turns out to be inhomogeneous. In some populations the differences are solely the result of phenotypic modification whereas in others the modifications are genetically determined. A good example is the group of prostrate maritime forms of *Sarothamnus scoparius* (broom) in Britain, which are sometimes recognized as subspecies *maritimus.*

(g) *Variation in variability*. Populations from different parts of the range of a species often show differences in variability with respect to particular attributes or groups of attributes. Sometimes this is due to the availability of a wider range of suitable environments in some parts of the range than in others, so that correspondingly different amounts of phenotypic modification occur. Sometimes, however, the extent to which attributes are environmentally modifiable is itself under genetic control. This raises the question of whether phenotypic modifiability itself should be considered as an attribute which should be used in infraspecific taxonomy.

With all these kinds of infraspecific variations the difficulties of hierarchic classification are clearly revealed if attempts are made to use methods of automatic classification. The dissimilarity coefficients between populations based on many attributes usually show imperfect hierarchic structure, so that the hierarchic dendrogram obtained by the single-link method represents them with considerable distortion (see Chap. 7). Variation of kind (a), where intergradation between otherwise well-marked geographical clusters of populations occurs, can sometimes be elucidated by using non-hierarchic cluster methods. Variation of kind (d) can be elucidated by comparison of the hierarchic and non-hierarchic clusterings obtained from dissimilarity coefficients based on different sets of attributes. Clinal variation may, under certain circumstances, be revealed by chains of overlapping clusters obtained by non-hierarchic cluster methods. But cluster analysis is not a natural method for the investigation of clinal variation, although the measurement of dissimilarity between populations with respect to selected attributes and groups of attributes is a prerequisite for its study. The methods appropriate for seeking clinal variation given a dissimilarity coefficient on a set of populations are scaling methods. The combination of hierarchic and non-hierarchic cluster methods with nonmetric scaling methods provides a powerful and flexible tool for describing even the most complex patterns of variation.

The rôle of cluster analysis and scaling methods in the investigation of infraspecific variation is discussed in more detail in Chapter 14.6. The potential usefulness of such methods lies in the fact that the accurate description of patterns of infraspecific variation is a prerequisite for their explanation in terms of geographical and ecological isolation, genetic differentiation, and hybridization. For discussions of the factors which produce the various patterns of infraspecific variation the reader is referred to books by Briggs and Walters (1969), Dobzhansky (1951), and Stebbins (1950).

CHAPTER 14

Automatic Classification as a Research Tool

14.1. INTRODUCTION

The discussion in the last chapter leads us to suggest some of the kinds of taxonomic problem which may most profitably be investigated by numerical methods. In other words, we shall be concerned with the first stage in any application of numerical taxonomy, the selection of a profitable field of enquiry and of methods appropriate to the chosen problem. Discussion of the details of procedure once a problem and appropriate methods have been selected will be outlined in the final chapter and the appendices. The methods of automatic classification which we have developed are designed to describe as accurately as possible the patterns of differentiation of groups of organisms. Delimitation and naming of taxa is only one of the many taxonomic activities for which such methods may be useful. Our view of the aims of taxonomy is in many ways conservative. The use of methods of automatic classification to construct new taxonomic hierarchies is, in our view, justifiable only under certain circumstances, and under other circumstances may be futile or positively deleterious in its effects. We suggest that applications in taxonomy which are not aimed directly at the creation of new taxa probably represent the most valuable applications of methods of automatic classification. Amongst the uses which we shall discuss in subsequent sections are: the investigation of the congruence between morphological divergence and genetic isolation of populations; the investigation of the concordance of classifications based upon different selections of attributes, the internal stability of classifications, and their predictive utility; the investigation of evolutionary rates and of the extent to which information about present-day populations can be used to infer phylogenetic branching sequences, the description of patterns of infraspecific variation which are more complex than can be represented within the Linnaean hierarchy, and the investigation of the relations between such complex patterns and geographical and ecological factors. The use of methods of automatic classification to tackle these problems depends upon the methods for evaluating the results of automatic classification described in Chapter 11.

14.2. AUTOMATIC CLASSIFICATION AND THE CREATION OF TAXA

Prima facie the case for using numerical methods of automatic classification to construct taxonomic hierarchies for all kinds of organisms can be made to appear quite strong. The case rests upon the following assumptions. First, the purely phenetic approach to taxonomy must be accepted. Secondly, it must be assumed that automatic classification based upon an adequate selection of populations and attributes can be guaranteed to produce optimal phenetic classifications.

These two assumptions have been advocated by Sokal and Sneath (1963). They argue that the phenetic approach is superior to all phylogenetic approaches on the grounds that for the vast majority of organisms all kinds of phylogenetic classification rest upon speculation and circular inference. The conservative view of taxonomy they ignore. Their main positive reason for advocating phenetic classifications is that 'The ideal taxonomy is that in which the taxa have the greatest content of information and which is based on as many characters as possible' (1963, p. 50). The secondary assumption that numerical phenetic methods can approach such ideal classifications is defended on the hypothesis that as successively more attributes are used so the classifications obtained tend towards a limiting ideal classification. In our view the question of the stability of classifications as the number of attributes used is increased, or as the kinds of attribute used are varied, is entirely a matter for empirical investigation. An amusing discussion of the role in taxonomy of the myth of an ideal classification was given by Johnson (1968).

It is, however, our contention that given a purely phenetic aim for taxonomy (or the approach described in Chapter 13.3 in which phenetic classifications are modified to ensure consistency with phylogeny), the range of appropriate methods of classification is much more limited than has hitherto been realized. In Parts I and II it has been shown that simple and intuitively plausible constraints determine measures of D-dissimilarity as pairwise measures of taxonomic dissimilarity between populations, and determine the single-link method as the only appropriate hierarchic cluster method. The range of appropriate non-hierarchic cluster methods has likewise been shown to be limited. In other words, whilst pursuit of optimal phenetic classifications may be pursuit of a chimaera, pursuit of optimal phenetic methods of classification is reasonable.

Some of the circumstances under which we consider it to be justifiable to use methods of automatic classification to create new taxonomic hierarchies are as follows.

(a) When no previous supraspecific classification has been established; or when the existing classifications are known to be based on only a small fraction of the existing populations from a group of organisms. This is the situation which holds for many groups of micro-organisms, and for the majority of groups of organisms which are poorly represented in the northern temperate region and other regions where collection of specimens is a convenient and relatively comfortable occupation.

(b) When existing classifications are the subject of dispute between different workers, or are generally acknowledged to be provisional and incomplete.

(c) When a minor modification or modifications to an existing classification leads to a marked improvement in the accuracy with which it represents the dissimilarities between populations.

Finally, we suggest that numerical phenetic methods should only be used to construct new classifications on the basis of the most exhaustive possible descriptions of a wide range of populations in the group of organisms studied*. The use of numerical phenetic methods in taxonomy at the specific and infraspecific levels should be combined with experimental study of the genetics, cytology, and environmental modifiability of populations.

When the existing classifications command a substantial measure of agreement, regardless of whether they are thought to be phenetic, artificial, or phylogenetic classifications, drastic revision by numerical phenetic methods is ill-advised. Likewise, revision on the basis of incomplete data is ill-advised. Incomplete data is perhaps even more damaging to the application of numerical phenetic methods than it is to intuitive methods. In the intuitive construction of classificatory systems taxonomists often make use of forms

* For example, Burtt, Hedge, and Stevens (1970) argued that unsatisfactory groupings were obtained by El Gazzar, Watson, Williams, and Lance (1968) in their numerical taxonomic study of *Salvia* because the selection both of populations and of attributes was inadequate. They emphasized also the importance of checking the concordance of classifications based on varied selections of attributes.

of inference which have no obvious parallel in automatic methods. Thus the properties of specimens which are incomplete may be guessed by analogical inferences: inferences of the form 'the specimen has attribute states *a, b, c,* etc., so probably it has also attribute state *d* since in related specimens attribute states *a, b,* and *c* usually concur with attribute state *d*'. Similarly, where specimens are available from only part of the geographical range of a group of organisms it is often possible in intuitive taxonomy to make useful guesses about the likely effects of discovery of further specimens. For example, one might be able to guess on biogeographical or climatic grounds that some at least of the subgroups found in the available material were fairly completely represented by the available specimens. When methods of automatic classification are used the taxonomist is strictly limited to available data. The fitting-together conditions (see Chap. 9) guarantee that for certain kinds of cluster methods inclusion of further OTU's will, under certain circumstances, leave parts of the existing classification intact, but no such guarantee of stability can be given when further attributes are considered.

In the construction of supraspecific classificatory systems the representation of the pattern of variation in a group of organisms which is most useful is the hierarchic dendrogram. In infraspecific classification non-hierarchic systems and various kinds of scaling may also be useful.

Given a hierarchic dendrogram, how should clusters be selected as taxa? An obvious requirement is that the ranks of taxa in the derived taxonomic hierarchy should be monotone with the levels of the corresponding clusters in the dendrogram. Subject to this general constraint a variety of methods are available:

(a) Fixed levels may be chosen in the dendrogram, and the clusters at consecutively higher levels then recognized as taxa of consecutively higher ranks. Sokal and Sneath (1963) suggested this method and called the selected levels *phenon* levels.

(b) Fixed levels may be chosen in such ways as to maximize the average isolation of clusters, or so as to ensure that no cluster at a level has less than some minimum allowed amount of isolation (cf. Holloway and Jardine, 1968).

(c) The range of cluster levels may be dissected into a number of intervals, and the most isolated clusters in each interval may be recognized as taxa of a given rank.

(d) Cluster selection may be carried out as in (a)-(c), but using either measures of the homogeneity of clusters, or composite measures of goodness of a cluster based upon both isolation and homogeneity. See Chapter 11.

There is a strong case for considering only clusters which have more than some minimum amount of isolation as candidates for recognition as taxa. If a cluster in a single-link dendrogram has isolation I then the cluster can be destroyed by alteration of a single pairwise dissimilarity by an amount I. The minimum value for I will be determined in each case by the confidence limits on the original dissimilarities. But this kind of error, arising from sampling of specimens from OTU's, is not the only kind which we have to guard against in choosing clusters if we are to produce stable classifications. A very much more difficult problem is posed by the need to ensure stability of a classification with respect to the attributes chosen. Methods for investigating this kind of stability are discussed in Sections 4 and 5. These considerations will under most circumstances lead to a choice of minimum permissible isolation of clusters which are to be recognized as taxa very much larger than the minimum imposed by the need to guard against errors arising from sampling of OTU's.

Quite apart from these internal constraints there are a variety of external constraints which may guide the way in which methods of automatic classification are used in the creation or revision of taxonomic hierarchies. One such constraint has already been mentioned in Chapter 13.3. It may be thought desirable to select for recognition as taxa those clusters which are thought to be monophyletic groups of populations. Other things being equal, if a choice between clusters which are polyphyletic and clusters which are monophyletic is available the choice of the latter is indicated. For reasons outlined in Section 5 it is reasonable to suppose that the more isolated and homogeneous is a cluster of present-day populations the more likely is it to be monophyletic.

Further external constraints are of a more directly practical nature. If clusters are found which correspond to taxa recognized in previous classifications this constitutes strong grounds for retaining the taxa. If one choice of clusters for taxonomic recognition would lead to drastic revision of existing classifications whereas another choice would require only minor modifications, the latter choice is indicated. Two constrasting situations which require particular care are: (a) when the clustering obtained suggests that an existing taxon should be broken up into several taxa of the same rank; (b) when the clustering obtained suggests that several existing taxa should be amalgamated into a single taxon of the same rank. In both situations there may be grounds for retaining the existing classification if it partitions the taxa involved into groups of roughly equal sizes. For purposes of diagnosis and communication of information about organisms very large taxa and very small taxa are both inconvenient. The International Codes of Zoological and Botanical Nomenclature recognize as principal categories species, genera, families, etc., and as subordinate categories subfamilies, tribes, subgenera, sections, subsections, etc. For purposes of diagnosis and communication of information about organisms it is the principal ranks which are mainly used. We suggest, therefore, that whenever possible the results of automatic phenetic classification should be expressed by revision or creation of taxa of subordinate ranks rather than revision of taxa of the principal ranks.

A final word of caution is needed about the use of automatic phenetic methods to create new taxa or taxonomic hierarchies. There are no *a priori* grounds on which the stability of a classification can be guaranteed under extension of the domain of objects, or under change or increase in the number of attributes used. The investigation of stability is obviously a prerequisite for the valid application of automatic phenetic classification to the creation of new taxa. Hitherto no analytic techniques for the investigation of stability have been available. It is therefore concluded that the majority of the taxonomic revisions which have been carried out in the light of the results of automatic classification must be regarded as premature, at least where the revision has been claimed as an improvement.

14.3. THE STABILITY OF CLASSIFICATIONS

It is not possible to undertake meaningful investigations of the stability of the results produced by a classificatory method unless the cluster method employed has certain properties. The investigation of stability under alteration of the attributes used can be undertaken for stratified cluster methods only if the methods induce continuous transformations of the data. If cluster methods which induce discontinuous transformations of the data are used (and this includes all methods which produce simple clusterings), then it is very difficult to determine whether a difference between results based on different attributes represents a genuine difference in the structure of the data, or is a byproduct of a discontinuity in the transformation induced by the method. The investigation of stability under extension of the range of objects classified is likewise very difficult if the cluster method used does not satisfy the fitting-together condition (see Chapter 9.4) or some similar condition. We should expect a classification to be altered if objects intermediate in their attributes between the objects in two well-marked clusters are introduced, but the use of methods which allow that introduction of objects which clearly fall within one cluster may disrupt other clusters would pose great difficulties.

Both in investigation of stability of classifications under change in selection of attributes, and under extension of range, we need to be able to compare clusterings on the same set of objects. The measures of discordance between clusterings described in Chapter 11 are suitable for this purpose.

Taxonomists, whether numerical or orthodox, usually start by selecting the attributes which are good discriminators and which are easily observed. There may be a tendency to select first those attributes which discriminate well some of the groups of OTU's which are expected to occur in the resultant classification. Whilst this is to be deplored it may be very hard to avoid. Subsequently, attributes which are less easily observed, or which are poorer discriminators, will be selected. Thus there is no sense in which taxonomists' selections of attributes can be considered as random selections from some predetermined set of attributes. It follows that the view that dissimilarity is necessarily a parameter which is estimated with increasing accuracy as the number of attributes used is increased is ill-founded, as is the corollary that as the number of attributes is increased so the resultant classification must tend to some limiting 'ideal' classification. This illusion may be fostered by the fact that as the number of attributes already considered is increased so the maximum proportional effect on a dissimilarity coefficient arising from addition of a further given number of attributes will decrease. It is, nevertheless, possible to devise *ad hoc* ways of investigating whether or not a given selection of attributes is adequate for classificatory purposes. Suppose that we have a dissimilarity coefficient and resultant classification based on a given set of attributes. We may then look at the distribution of values of the discordances between the dissimilarity coefficients obtained from pairs of subsets of the set of attributes selected at random. The mean and variance of values of discordances between pairs of DC's based on random bisections of a set of attributes provide measures of the *internal stability* of the DC based on the whole set of attributes. There can be no guarantee that the internal stability will improve as the total number of attributes is increased. However, low internal stability would be a clear indicator that more attributes should be considered.

The internal stability of a classification is related to two features which biologists have generally felt to characterize 'natural' classifications: 'information content', and 'predictive power'. A classification with high internal stability is one in which different selections of reasonable size from the set of attributes initially selected are expected to give rise to fairly similar classifications. The clusters of OTU's have high information content in the sense that it will be possible to make many true generalizations about them. There can be no guarantee that a classification with high internal stability will serve as a basis for successful predictions about the way in which further attributes will discriminate the OTU's. However, if the further attributes are considered to be in some biological sense of the same kind as those already selected, it is reasonable to base predictions on the classification. A classification with low internal stability will in no case serve as a basis for reasonable predictions. In general we should expect that infraspecific classifications will be more likely to have low internal stability even for large selections of attributes than will supraspecific classifications, but much more extensive investigations will be needed before any firm generalization can be made.

It may be felt that the above remarks are disappointingly vague. The vagueness arises from the fact that it does not in general make sense to regard the attributes selected by a taxonomist as a random sample from some larger set of attributes. Only in certain special cases, for example if the attributes were based on randomly selected homologous segments of DNA base-pair sequences, would it be reasonable to regard the internal stability of a classification as indicating something about the probable behaviour of the classification when further attributes are considered.

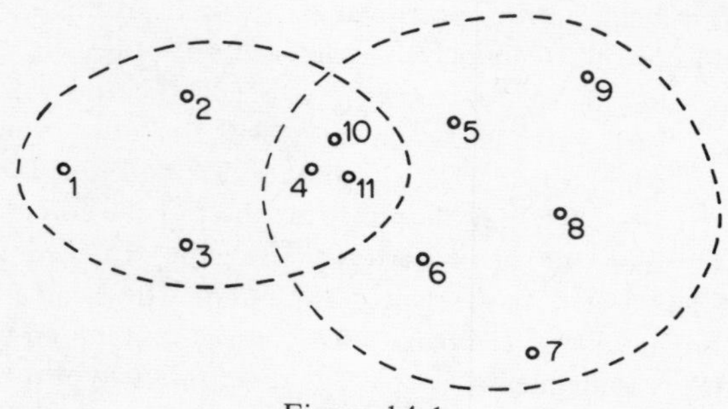

Figure 14.1.

The investigation of the stability of a clustering as the range of objects is extended poses very different problems. When the number of attributes is increased the general rule about induced change is that it can strike anywhere. When the range is extended it is possible to say something about the ways in which particular kinds of extension of the range will affect the clustering.

(a) *Single-link clustering.* The conditions under which new clusters will be created within existing clusters can easily be worked out, a formal version would be tedious. Existing clusters will be amalgamated whenever objects intermediate between one or more of their members are introduced. The fitting-together condition guarantees that neither the creation of new clusters amongst the OTU's added, nor the amalgamation of clusters by intermediate OTU's, will affect other clusters.

(b) *Non-hierarchic stratified clustering.* Two sequences of non-hierarchic stratified cluster methods B_k and C_u were shown to be satisfactory within the axiomatic framework of Chapter 9. B_k operates by restriction on the size of the permitted overlap between clusters. C_u operates by a restriction on the diameter of the permitted overlap between clusters which is proportional to the level of the cluster. B_k is likely under certain circumstances to be more unstable under extension of range than is C_u (see Chapter 8.7). Since the main application of non-hierarchic methods of cluster analysis is in the study of patterns of infraspecific variation, an example from this field is chosen to illustrate the difference. Suppose we have two well-marked geographical races which differ in many morphological characters, but which intergrade in a small area where their ranges overlap. To facilitate the discussion let us further make the assumption that the dissimilarity between local populations within each race, and in the intermediate zone, is roughly related to their geographical separation, so that the distribution of points representing local populations can be taken both as a map and as a rough indication of dissimilarity between populations within each race and within the intermediate zone. See Figure 14.1. Suppose that populations 1-9 are selected. Single-link gives no satisfactory representation of the pattern of variation. Both B_2 and $C_{0.3}$ (say) find {1, 2, 3, 4} and {4, 5, 6, 7, 8, 9} as clusters at some level. Suppose, however, that the range is extended by selecting also populations 10 and 11. B_2 will no longer find distinct clusters including {1, 2, 3} and {5, 6, 7, 8, 9} respectively. $C_{0.3}$ will however find clusters {1, 2, 3, 4, 10, 11} and {4, 10, 11, 5, 6, 7, 8, 9} at some level.

In other words, whenever there are well-marked groups with intermediates, C_u is likely to produce clusterings which are more stable as the range is extended, because it is less vulnerable to alteration in the number of OTU's intermediate between clusters. As has been pointed out, C_u pays for this greater stability by requiring stronger assumptions about the significance of the underlying dissimilarity coefficient than does B_k.

The investigation of the stability of the various acceptable cluster methods when applied to various kinds of taxonomic data is a major outstanding field for practical investigation. Until such investigations have been carried out the intepretation of all the published results of applications of cluster analysis in taxonomy must be considered provisional. Crovello (1968, 1969), Rohlf (1963), and Sokal and Michener (1967) investigated the concordance of clusterings under change in selection of attributes and OTU's. The interpretation of these results is difficult, because some of the cluster methods used are discontinuous.

14.4. COMPARISON OF CLASSIFICATIONS BASED ON DIFFERENT CRITERIA

There are two main kinds of comparison between classifications which are of interest. First, we may wish to compare the clusterings obtained on the same set of OTU's for different selections of attributes. Secondly, we may wish to compare the clustering obtained by an automatic phenetic method with a partition of the set of OTU's induced by some such external criterion as geographical distribution or genetic isolation.

The extent to which clusterings of a set of OTU's based on different sets of attributes are concordant is intimately related to the stability of clusterings. High discordance between different sets of attributes will lead to internal instability. Often taxonomists are interested in discordance for other reasons. The extent of discordance between phenetic classifications based upon, for example, morphological and biochemical attributes is in itself of interest.

Also of interest is the extent of the discordance between classifications based on sets of attributes selected from different functional complexes. A functional complex can loosely be regarded as a set of attributes so related that it is reasonable to suppose that selection pressures which lead to change in one of the attributes are likely to lead to compensating changes in the others. They have been investigated by Olson and Miller (1958), who presented detailed and convincing evidence that functional complexes of quantitative attributes are often indicated by high correlations within populations. Olson (1964) and several other authors have suggested that evolutionary changes in different functional complexes may be discordant, a pattern of evolutionary change which is sometimes called *mosaic evolution*; see Le Gros Clark (1950) and Mayr (1964). The difficulty with this view is that the notion of a functional complex is rather imprecise. Different functional complexes cannot be assumed to be sets of attributes which induce discordant classifications, or form distinct systems of attributes mutually correlated within populations, but there are great difficulties in finding any independent definition of community or difference in function.

If a sufficiently large set of attributes is available it is a simple matter to test whether particular subsets of the set of attributes, either of different kinds, or from different functional complexes, show significantly higher discordances than would be expected from randomly selected subsets of the total set of attributes. See Appendix 7 for an example of such a test.

Sometimes the situation arises where we wish to compare a phenetic clustering based on many attributes with a clustering based upon some criterion which gives rise directly to a dissimilarity coefficient on the set of populations. Criteria which give rise directly to a dissimilarity coefficient on a set of populations include measures of extent of DNA hybridization, measures of serological affinity, and measures of degree of chromosome pairing in hybrid offspring. It is important to note that some of these kinds of dissimilarity coefficient may, despite their numerical form, have at best ordinal significance, so that such clustering methods as C_u may be inappropriate, and the use of numerical discordance measures to compare the resultant clusterings may be ill-advised.

The case where a phenetic clustering must be compared with the partition of the set of populations by some other criterion arises frequently. At least three distinct situations can arise.

(a) *Testing independence.* We may wish to know whether a phenetic grouping of populations is independent of its grouping by some other criterion or set of criteria. This is the question involved when we ask, for example, whether the morphological differentiation of a set of population is affected by the kind of habitat. The appropriate methods are tests of independence for contingency tables (χ^2 for example) which fall outside the scope of this book.

(b) *Measurement of dependence.* We may wish to find out the extent to which the partitioning of a set of populations derived from a phenetic classification is dependent upon some criterion, or we may wish to compare the extents of dependence of a partition derived from a phenetic classification on various criteria. For example, we might wish to know to what extent a partition on morphological grounds is dependent on kind of habitat; or whether the partition is affected more by soil type than by geographical distribution. For this purpose a measure of departure from independence in a contingency table is appropriate. Such measures include measures of information gain and the various predictive functions discussed by Goodman and Kruskal (1954). They pointed out that the widespread use of functions of χ^2 as measures of dependence rests on a confusion between a test of independence and a measure of departure from independence.

(c) *Measurement of congruence.* A rather different situation arises when we wish to measure the extent to which a partition of a set of populations departs from perfect fit with the partition induced by some other criterion. For example, we may wish to know how well a partition derived from a phenetic classification fits the taxa of a given rank recognized by some previous taxonomist on intuitive grounds. Or we may wish to know how well a partition derived from a phenetic classification fits a partition based on a chromosome number or on intrinsic genetic barriers. Here what is required is some measure of congruence such as the symmetric difference between the equivalence relations which correspond to the partitions. See Chapter 11.3.

It is difficult to give clear-cut guidance about the situations in which measures of dependence (b) and congruence (c) are appropriate. Roughly speaking, measures of dependence are appropriate when the question at issue is 'To what extent is the phenetic grouping *influenced* or *affected by* factors x, y, etc.?' Measures of congruence are appropriate when the question at issue is 'Does the phenetic grouping *correspond* to the grouping given by criterion *x?*', or 'Is the phenetic grouping *accounted for, determined by,* or *caused by* factor *x?*' Often when taxonomists ask 'Is a classification by one set of criteria correlated with a classification by some other criterion?', it is difficult to decide what kind of question is at issue.

In interpreting the results of comparisons it is important to be aware that in certain cases, notably when conditional definition of attributes arises or when an overall size factor is involved, the way in which attributes are selected to convey a given body of observations on a given set of OTU's may affect the resultant classification. See Chapter 4.3.

14.5. CLASSIFICATION AND PHYLOGENY

The relationship between phenetic classifications of populations of present-day organisms and their phylogeny cannot be investigated directly when there is no fossil record. There are, however, various indirect ways in which the relationship can be investigated.

If evolutionary rates with respect to a measure of divergence based on a given set of attributes had been constant the dissimilarities between present-day populations, based on the same set of attributes, would be ultrametric. The converse is, of course, untrue. The fact that a dissimilarity coefficient based on a particular attribute or set of attributes is ultrametric does not necessarily imply that evolutionary divergence with respect to that set of attributes has proceeded at a constant rate. To take a trivial counterexample, a dissimilarity coefficient based on a single binary attribute which does not vary within OTU's will always have perfect ultrametric structure. If, however, it were found that present-day dissimilarities with respect to some set of attributes were very close to ultrametric structure for a wide variety of populations, it would be hard to think of any alternative plausible explanation.

The extent to which certain kinds of attribute, and certain methods of selecting attributes, lead to dissimilarity coefficients which are relatively close to ultrametric structure could conclusively refute or provide partial support for the frequent claims that some kinds of attribute are better phylogenetic indicators than are others. The closeness of a dissimilarity coefficient to ultrametric structure can be measured by the distortion induced on it by the subdominant ultrametric (see Chapter 11). There has been a strong tendency in the past for each new kind of taxonomic information made available by some technical advance to be hailed as an important indicator of phylogenetic affinity. This has happened in turn with serological data, first proposed as a phylogenetic indicator by Reichert and Brown (1909); with protein amino-acid sequence data (Buettner-Janusch and Hill, 1965); with DNA hybridization data (Schildkraut and coworkers, 1961); and most recently with DNA nucleotide-pair frequency data (Bellett, 1967). Several authors have suggested that eventual knowledge of the whole nucleotide sequence for the DNA of various organisms will provide the basis for the ultimate taxonomy. Jardine has found (unpublished work) that both amino-acid sequence data and nucleotide-pair frequency data sometimes yield dissimilarity coefficients very much closer to ultrametric structure than is usual for other kinds of data. But very much more extensive investigation is needed before assertions about the relative reliability of different kinds of information as phylogenetic indicators can be made with any confidence. It was argued in Chapter 13.3 that even if it can be shown that some kinds of attribute are better phylogenetic indicators than others, this is no justification for differential weighting of attributes in the measurement of dissimilarity between populations.

A large number of other ways of inferring the phylogenetic branching sequence from information about present-day organisms have been proposed, often under the title 'cladistic analysis'. The majority of these methods depend upon the minimum evolution hypothesis, whereby it is supposed that the observed differences have been established by the minimum number of mutations or unit morphological or biochemical changes. Such methods as that of Camin and Sokal (1965) depend upon the further assumption that it is possible to arrange the states of each discrete-state attribute in order of primitiveness. Both assumptions appear to be untenable, at least in these simple versions, since both are refuted if extensive parallel evolution has occurred. There is substantial evidence that this is so: a few of the many studies which suggest this are cited. Throckmorton (1965) has given convincing evidence for the parallel evolution of identical structures in the genitalia of *Drosophila*, and Inger (1967) has presented convincing evidence for parallel evolution in the phylogeny of frogs. Osborn (1893, 1902) gave extensive evidence for the parallel evolution of homologous cusps and folds in mammalian teeth.

Several authors have suggested the use of cluster analysis to suggest which phylogenetic branching sequences are plausible; see Blackith and Blackith (1968), Edwards and Cavalli-Sforza (1964, 1965), Fitch and Margoliash (1967), Taylor and Campbell (1969). Edwards and Cavalli-Sforza's method was mentioned in Chapter 12.2; the other authors cited have used average-linkage cluster analysis. Whilst it is obvious that, provided that evolutionary rates have not been wildly inconstant, highly homogeneous and isolated clusters of present-day populations found by any method are likely to be monophyletic, the assumption that hierarchic clusterings are intimately related to phylogenetic branching sequences is quite unjustified. The assumption rests on a naïve confusion between geometrical diagrams showing the inclusion relation of clusters in hierarchic stratified clusterings and geometrical representations of evolutionary trees.

Jardine, van Rijsbergen, and Jardine (1969) showed that the assumption that highly homogeneous clusters found by the ball-cluster method (described in Chapter 12.4) are monophyletic is consistent with the assumption that evolutionary rates are often locally constant. That is

$$d(A, B) < d(B, C) \Leftrightarrow t(A, B) < t(B, C)$$

where t is the time since divergence*. Dissimilarity coefficients on present-day populations often yield very few ball-clusters, but in the cases investigated the monophyly of the clusters found is consistent with what can be inferred about phylogeny on independent grounds. The hypothesis that rates have often been locally constant is suggested by the observation that in groups of organisms with a substantial fossil record, horses for example, particular groups of related lineages often show rates of divergence characteristically different from those shown by other groups of related lineages. If rates of divergence are partly under direct genetic control, or if internal selective factors predominate in determining rates of divergence, it would be expected that rates would be more similar the more similar the genomes of the populations considered.

The fact that inferences from dissimilarities between present-day populations to phylogenetic branching sequences depend upon the assumptions which are made about evolutionary rates has been emphasized also by Kirsch (1969). He suggested a stochastic model for serological divergence in phylogeny.

One obvious check on the validity of inferences about phylogenetic branching sequences from the results of cluster methods is provided by the fact that evolutionary relationships of adult and juvenile specimens of the same population must be identical. If a method of cluster analysis yields clusters which are incongruent when attributes of juvenile and adult forms are used, as happened in Rohlf's (1963) comparison of adult and larval mosquitoes, inferences about phylogenetic branching sequences are clearly invalidated.

We conclude that use of dissimilarities between present-day populations as a basis for inference of phylogenies will remain premature until more is known about rates of evolutionary divergence with respect to the various

* It may be more reasonable to consider evolutionary rates in terms of number of generations elapsed rather than time elapsed. It is, however, difficult to find a reasonable way of estimating past replication rates.

kinds of attributes. Study of dissimilarities in groups of organisms for which the branching sequence and the approximate times of branching are established by independent fossil evidence may yield such information, compare Sarich and Wilson (1967).

14.6. PATTERNS OF INFRASPECIFIC VARIATION

Some of the many patterns of infraspecific variation which arise were discussed in Chapter 13.4. It seems to be usual that the dissimilarities between infraspecific populations based on morphological attributes show relatively poor hierarchic structure. The extent to which dissimilarities depart from hierarchic structure may conveniently be measured by the distortion imposed by the subdominant ultrametric (see Chapter 11.2). Poor hierarchic structure may be the result of one of the many patterns of variation which cannot adequately be represented by a hierarchic classification, or it may indicate that there is no significant differentiation between populations.

Both the choice of OTU's and the subsequent strategy of analysis depend crucially upon the pattern of infraspecific variability which is expected. If the K-dissimilarity measure is used an obvious first step in analysis of a dissimilarity coefficient is to check that some at least of the pairwise dissimilarities are statistically significant.

The various techniques which may be used if the dissimilarity coefficient turns out to be significant will be discussed separately for each of the patterns of infraspecific variation (a)-(e) outlined in Chapter 13.4.

(a) and (b) *Geographical and ecotypic differentiation.* The methods of overlapping stratified clustering B_k and C_u are appropriate for this kind of investigation. C_u requires strict numerical significance for the dissimilarity measure, whereas B_k is equivariant under order-isomorphisms of the data, and is hence applicable to ordinal dissimilarity measures. However, as pointed out earlier, C_u is in a precise sense more stable under varied selection of OTU's and for this reason its use in combination with the strictly numerical K-dissimilarity measure can be of value.

In using non-hierarchic methods it is important to find a suitable compromise between the increased accuracy of representation and the increased complexity of the resultant clustering as the overlap criterion is relaxed. The normalized distortion measures described in Chapter 11 can be used to find out at what stage relaxation of the overlap criterion ceases to produce a useful increase in accuracy of representation. In evaluating the significance of particular clusters the measures of isolation and homogeneity of clusters described in Chapter 11 and in Jardine and Sibson (1968a) are useful.

The application of B_k to the study of geographical differentiation in the flowering plant species *Sagina apetala* (Caryophyllaceae) was described in Jardine and Sibson (1968a). The application of B_k to a study of the differentiation of local human populations is described in Appendix 7.

(c) *Clinal variation.* When the kind of clinal variation involved can be guessed in advance so that the OTU's can be disposed along the relevant climatic, altitudinal, edaphic, or geographical gradients, it is simple to test whether the data shows the existence of a cline. It is necessary to show only that the dissimilarities of other OTU's from each OTU are monotone, or approximately monotone, with their distances from it along the gradient.

Often we may wish to find out first whether there are apparent trends in the data, and then seek an ecological or geographical gradient which fits these trends. Clustering methods are not strictly appropriate for such investigations, although where non-hierarchic methods produce chains of overlapping clusters, clinal variation might be suspected. The various techniques of factor analysis are of limited applicability here, because, whilst the mean values of the character states of populations are expected to be more or less monotone with position on a geographical or ecological gradient, there are no grounds for supposing that they will in general vary in a linear fashion. The extensive data collected by Gregor (1938, 1939) suggests that clinal variation is generally non-linear.

The methods of non-metric multidimensional scaling developed by Kruskal (1964a, 1964b), Shepard (1962), and Shepard and Carroll (1966) have properties which suggest that they may be more appropriate for the study of clinal variation. Given as data a dissimilarity coefficient on a set of objects, the non-metric multidimensional scaling algorithm devised by Kruskal (1964b) seeks that disposition of points representing the objects in euclidean space of a given dimension, for which a monotone regression measure between the interpoint distances and the dissimilarity measure is maximized. This method has been applied to the dating of human artefacts (Doran and Hodson, 1966), and to the exploration of the relations between the distribution of animal species on islands and the geographical position of the islands (Holloway and Jardine, 1968). Its use for the investigation of geographical trends of variation at the infraspecific level has been suggested by Sneath (1966b).

In the study of clinal variation scaling to one dimension is needed. For technical reasons the Kruskal algorithm cannot find the best-fitting one-dimensional disposition, because in this case the optimization procedure becomes trapped in local minima. The one-dimensional disposition has to be recovered by inspection of the disposition in the best-fitting two-dimensional representation. Related algorithms which have been developed by Guttman and Lingoes under the title 'Smallest Space Analysis' use a different monotone regression measure and a different optimization procedure. They appear to avoid the technical difficulties mentioned above, see Guttman (1968) for a full discussion and references.

(d) *Discordant variation.* Discordant variation at the infraspecific level may be investigated using the methods for comparison of clusterings described previously. It is important to check that the dissimilarity coefficient for each of the attributes or sets of attributes concerned is statistically significant.

(e) *Reticulate variation.* This pattern of variation is indicated when the dissimilarity coefficients between local populations are statistically significant, but no significant structure is revealed either by non-hierarchic clustering or by the scaling methods appropriate for the investigation of clinal variation which are described above. It should be noted that discordant clustering or discordant clinal variation with respect to different sets of attributes may produce an overall effect of reticulate variation. We conjecture that many of the cases of reticulate variation described in the literature are in fact of this kind.

14.7. DATA SIMPLIFICATION AND HYPOTHESIS TESTING

The applications of automatic classification and scaling which we have discussed aim to describe patterns of differentiation of organisms prior to their explanation in terms of evolutionary processes and ecological, geographical, and genetic factors. They are best regarded in this context as techniques for simplifying information in ways which may suggest explanatory hypotheses and facilitate their testing. When specific hypotheses about patterns of differentiation are available quite different methods may be appropriate. For example, if infraspecific populations separated by a particular ecological or geographical barrier are thought to be differentiated, the various techniques or discriminant analysis may be used to test the hypothesis. Rao (1952, Chap. 8) should be consulted for mathematical details, and Ashton, Healy, and Lipton (1957) and Pritchard (1960) for examples of the application of discriminant functions to the study of infraspecific variation. The information radius K can be used as the basis for a discriminant test.

Many of the statistical tests which taxonomists currently use to test hypotheses about the differentiation of populations may be criticized in detail on the grounds that they involve assumptions about normality and constant dispersion of distributions which are unwarranted by the data. But they in no way conflict with the methods offered here. They are complementary to methods of automatic classification and scaling, since they are methods for checking hypotheses against data, whereas methods of automatic classification and scaling aim to simplify data in ways which will suggest hypotheses.

CHAPTER 15

The Application of Methods of Automatic Classification

15.1. INTRODUCTION

The first stage in any application of automatic classification in biological taxonomy is the selection of OTU's and attributes. Previous discussion has already indicated that the choice of strategy in the representation of data is constrained both by certain general requirements about the properties of methods of data representation and by the particular kind of taxonomic problem investigated. The situation which arises in the selection of data is more fluid. Both the purpose of the investigation and the strategy of analysis to be used impose certain constraints, but these are, at least at the present stage of development of the subject, very imprecise. There are few rules; the most that we can do is to offer some guidelines. It appears that fairly clear guides to the selection of OTU's can be given, but that selection of relevant attributes is highly problematical.

15.2. SELECTION OF OTU'S

It has been suggested that the basic units of classification are topodemes or ecodemes: that is, morphologically homogeneous populations, each drawn from either a single locality, or a single kind of habitat. This obviously does not imply that topodemes or ecodemes should always be selected as OTU's in every application of automatic classification.

In any profitable application a statement of the purpose of the investigation will provide some external constraints on the selection of OTU's. Two simple examples illustrate this. If the aim of the investigation is to study the extent of concordance between chromosome number and morphological differentiation within a taxon, the OTU's selected should all lie within the taxon, and the individuals within each OTU should have the same chromosome number. Further, the OTU's should each be morphologically as homogeneous as possible, and the OTU's with a given chromosome number should be selected from as wide a range of locations and kinds of habitat as possible, since these factors also may be correlated with morphological differentiation. If the aim of the investigation is to study the differentiation of taxa of specific rank within a taxon of generic rank, the OTU's should all lie within the genus, and in no case should a single OTU include specimens which would be referred to more than one taxon of specific rank by even the most ardent 'splitter'. If the taxa of specific rank are believed to be geographically or ecogeographically differentiated, several OTU's selected from as wide as possible a range of locations and habitats should be used to represent each taxon of specific rank.

In general, therefore, the selection of OTU's is always conditioned by previous partitions of the domain of individuals by external criteria. These external criteria are of two kinds. First, there are criteria which delimit the range of individual organisms from which OTU's may be selected. A particular study may investigate, for example, the genus *Sagina*, or the genus *Sagina* in Europe, or the genus *Sagina* at high altitudes. Secondly, there are criteria which partition the individuals within the range delimited by criteria of the first kind. A particular study may investigate taxa of specific rank within the genus *Sagina*, or ecological and geographical differentiation within the species *Sagina apetala.* The general rule is that OTU's must be homogeneous with respect to all relevant external criteria. This rule is not, alas, usually applicable in this simple form, because certain relevant external criteria may be impossible to apply with precision. For example, as pointed out in Chapter 13.2, the criterion of genetic isolation is often relevant, but is difficult to apply in a precise and non-arbitrary way. Similarly, if populations from a particular taxon are to be investigated, difficulties may arise when there is no general agreement about the range of the taxon. Further difficulties may arise in deciding which external criteria are relevant. Suppose it has been decided that OTU's should be sampled from as wide a range of habitats as possible. The secondary question arises, as to which are the relevant climatic, edaphic, and other variables which differentiate habitats. The only general rule which can be offered to deal with these difficulties is caution. If it is doubtful whether a particular external constraint is relevant, it should be assumed to be relevant. The subsequent analysis of data may reveal that a constraint assumed relevant was irrelevant, it cannot reveal that a constraint not assumed relevant was relevant. If the range of a taxon is doubtful it is best to interpret it in a wide sense. Subsequent analysis may reveal that the OTU's definitely included in the taxon form a cluster which excludes those dubiously referred to the taxon. If it is doubtful whether two populations form a single morphologically homogeneous population, it is generally preferable to treat them as distinct OTU's rather than lump them together as a single OTU.

15.3. SELECTION OF ATTRIBUTES

Restriction of attributes to those which show a particular kind of distribution within OTU's is undesirable. The widespread use of association measures, and of methods of association analysis, has resulted in the use in biological taxonomy of classification strategies based only on discrete-state attributes which are invariant within OTU's, or worse, to the sampling of OTU's by single representative specimens on the hopeful assumption that the attribute states recorded are constant within OTU's. The use of normalized taxonomic distance measures is strictly valid only if all the attributes selected are normally distributed with constant variance in all OTU's. Use of measures of D-dissimilarity, of which K-dissimilarity is an example, removes the need for selection of attributes which show particular families of distributions within OTU's and emphasizes the need for adequate sampling.

There remain a number of criteria for the selection of attributes, not all of which turn out to provide clear-cut guidance in practice.

Relevance

Attributes should be relevant in the sense that their corresponding character states should discriminate some of the OTU's selected. Measures of D-dissimilarity are unchanged by irrelevant attributes. The collection of information and subsequent processing for irrelevant attributes is a waste of effort, but in large studies it may be impossible to check the relevance of all selected attributes in advance.

Homology

If attributes of parts of organisms are selected it is necessary that the parts themselves should be homologous in all the organisms studied. Thus if petal colour is one of the selected attributes, the parts called petals must be homologous in every plant considered. The notion of biological homology has recently been the subject of considerable dispute. The usual textbook definition states that parts of organisms are homologous if they are descended from a single part in a common ancestor. This definition is logically circular; it also fails to provide any practical criterion for the determination of homologies. It has been criticized on these grounds by Boyden (1947) and Zangerl (1948). Woodger (1937, p. 137) expressed this point with great clarity:

> Some authors have written as though homologies of parts could not be determined until the phylogenetic relations of their owners have been determined. But the determination of such phylogenetic relations presupposes the establishment of at least some homologies quite independently of all such considerations. In other words, we must possess some *criteria* of homology which the earlier morphologists also possessed before phylogenetic questions were considered at all. There is a primary sense of 'homology' which we all use intuitively and upon which all the more sophisticated senses of the word depend. Just because it is used intuitively we do not stop to analyse it, and are scarcely aware that we are using it.

Still less satisfactory is the definition of homology as 'essential' similarity which is sometimes offered, usually by biologists working with organisms which have a poor fossil record.

In higher animals which have a relatively stable adult phase the criterion of homology which is in practice used by comparative anatomists is correspondence in relative position. A computer program described in Jardine and Jardine (1967) has been used successfully to study the homologies of fish skulls which are too complex for correspondence in relative position to be matched successfully by eye; see Jardine (1969c). Additional criteria in determining homologies are similarity in composition of parts and similarity in developmental origin. Where these criteria conflict with correspondence in relative position it appears that the latter criterion must be considered basic. With plants, where there is no stable adult phase, it is very much more difficult to find workable criteria of homology.

Sattler (1966, 1967) has emphasized the inadequacy of the criteria used in the comparative morphology of higher plants. Here homologies of parts are allegedly determined by deciding to which of a number of fundamental categories such as 'caulome' and 'phyllome' they belong. There is no general agreement about these categories, rival categorizations include the classical caulome-phyllome theory (see Eames, 1961; Esau, 1965), the telome theory (see Zimmermann, 1965), and the gonophyll theory (see Melville, 1962, 1963) Further, the categories seem to have been carried over from pre-Darwinian Platonist morphology, and few objective criteria for assignment of parts to categories are available. The interpretation of these fundamental categories as parts of an ancestral flowering plant is, in the absence of a fossil record, dubious in the extreme, and does nothing to avoid the practical difficulties in deciding to which category a particular organ should be referred. It does, however, seem possible that embryological investigation may render such decisions less arbitrary (see Wardlaw, 1965). Both Sattler (1966), and Sokal and Sneath (1963), have proposed to avoid these difficulties by using a quantitative concept of homology by which parts may vary in degree of homology according to their similarity in terms of shared attributes. But this runs counter to the habitual use of the term homology by biologists for an all-or-none relation rather than for a relation of degree. It appears absurd to say that the 'flower' of *Potamogeton* is partially homologous with the flower, and partially homologous with the infloresence, of other flowering plants. It would be better to admit that its homology cannot be reliably established. The only way in which these difficulties can be avoided is to recognize that it may be impossible to establish homologies between widely dissimilar plants. This places a corresponding restriction on the scope of methods of automatic classification, for it implies that morphological comparisons between, for example, different orders of flowering plants may be strictly meaningless. An example of a numerical taxonomic study whose results are rendered suspect by use of attributes of doubtfully homologous parts is Hamann's (1961) study of the Farinosae.

Redundancy

The investigation of redundancy amongst attributes has been discussed in Chapter 4.3. It will suffice here to reiterate some of the main conclusions. It was pointed out that there has been substantial confusion between redundancy and at least three senses in which biologists have used the term 'correlation'. Correlation between attributes may mean statistical correlation within populations; it may mean concordance between the ways in which attributes discriminate populations; and it may mean functional correlation. Both statistical correlation within populations and concordance may provide clues to the functional correlations of attributes. It is obvious that an attribute can be considered redundant only if it is *both* statistically correlated within populations with another attribute or set of attributes *and* concordant with the same attribute or set of attributes, in the populations studied. The converse does not hold: it is quite possible that attributes which are statistically correlated and concordant within a given set of populations may cease to be so when further populations are considered. In this case any assumption of redundancy would have been premature. Unfortunately many writers have assumed that statistical dependence within populations indicates that attributes are redundant for taxonomic purposes. They have, in effect, confused redundancy in describing a given OTU, with redundancy relevant to the classification of a set of OTU's.

It is possible to give only some tentative general guides to ways in which study of the statistical correlations and concordances of attributes may be used to improve on initial selection of attributes.

(a) If attributes are both statistically correlated within OTU's and concordant, this may indicate redundancy. In general, attributes should only be eliminated from the list if a sound biological explanation for the statistical correlation and concordance can be found. For example, if two leaf shape indices were selected and found to be both statistically correlated and concordant, studies of leaf growth might indicate that one of the indices was truly redundant. The relations between growth rates and statistical correlations between sizes of adult parts are discussed in detail by Cock (1966), who emphasizes that whilst mutually dependent growth rates may explain statistical correlations between sizes of adult parts they cannot be used to infer them.

(b) If attributes are statistically correlated within OTU's, but are not concordant, redundancy is not indicated. If the attributes do not individually give good discrimination it may indicate that they should be compounded to form a single attribute. Suppose that leaf shape is a good discriminator between the selected OTU's: that is, leaf shape is fairly constant within OTU's, and differs widely between OTU's. If leaf length and leaf breadth were selected as attributes they would turn out to be statistically correlated within OTU's, and would give poor and discordant discrimination. Replacement of the two attributes by their ratio would yield an attribute which gave good discrimination.

Much further research on the statistical correlations and concordances of attributes is needed before it will be possible to give any definite rules about their use in the selection of attributes for taxonomic purposes.

15.4. ESTIMATION OF CHARACTER STATES

Two main questions have to be settled in order to obtain adequate estimates of the character states of each OTU for each attribute. First, a reasonable sample of specimens from each OTU must be selected. Secondly, an appropriate family of distributions to describe the states of each character must be chosen.

The choice of samples for OTU's will often be severely restricted by the available material. Ideally, specimens should be randomly selected. If the

OTU is itself thought to be heterogeneous according to geographical location or habitat the sample should include specimens from the whole geographical or ecological range of the OTU. However, if the differentiation within an OTU is well-marked it is better, as mentioned previously, to partition the group into several OTU's. It is obvious, however, that any selection of OTU's must inevitably tolerate some inhomogeneity of OTU's. For example, in a study of differentiation of genera within a family it would be reasonable to select as OTU's taxa of specific rank, but impossibly arduous to insist on use of several ecodemes and topodemes from each species as OTU's. The special problems which arise in obtaining representative samples of extinct populations from the available fossil assemblages are discussed in Johnson (1960), Miller and Kahn (1962, Appendix B), and Olson and Miller (1955).

A further serious problem concerns the value in taxonomy of growth or rearing of samples of OTU's under standard conditions. The difficulty is that conditions which are optimal for some of the populations studied may be lethal or sublethal for others. Should phenotypic variability, and range of conditions under which organisms are viable, themselves be considered as attributes relevant to classification in taxonomy? Ideally one might suppose that samples from every OTU should be grown or reared under a whole range of standard conditions. This would be prohibitively laborious in practice. Even if it were feasible it is hard to see how such observations could be used to eliminate the phenotypic component in morphological and other variations, for phenotypic variability is itself partially under genetic control. Indeed, the widespread notion that morphological variation can be partitioned into genotypic and phenotypic components is seriously misleading.

The selection of an appropriate family of distributions presents no problems for discrete-state attributes. For quantitative attributes the situation is more awkward. In biological taxonomy, when the OTU's are ecodemes or topodemes, it is sometimes a good approximation to assume that an attribute varies within OTU's according to a normal distribution; or, more commonly, that its logarithm does. When one or more of the OTU's are inhomogeneous or when the distribution follows no obvious pattern it is necessary to dissect its range and to record the relative frequencies of values in each interval of the range for each OTU. The estimation of information radius for pairs of character states is discussed in Appendix 1. It is shown that information radius for discrete-state attributes can reasonably be estimated by first estimating probabilities by relative frequency, and then calculating the information radius between these estimated values. Values of information radius for numerically scaled attributes can be estimated by dissection of the range.

15.5. STRATEGIES FOR THE ANALYSIS OF TAXONOMIC DATA

Once OTU's and attributes have been selected, an appropriate method of analysis for the data must be chosen. The selection of an appropriate method depends crucially on the kind of taxonomic problem under investigation. The methods which are appropriate for several kinds of problem have been discussed in Chapter 14, and the inadvisability of blind use of automatic methods prior to a clear formulation of a problem has been stressed. Our aim in this final section is to show how the various analytic methods which we have described fit together, so that once the investigator has decided what he wants to do he can select an efficient strategy for doing it.

In Tables 15.1 and 15.2 we outline a general strategy for analysis of taxonomic data. Table 15.1 covers the calculation of dissimilarity coefficients and their evaluation. Table 15.2 covers cluster analysis and evaluation of results.

The major computational steps indicated in the table are discussed either in previous chapters or in appendices. But a few details of the procedure outlined require comment.

(a) *File inversion.* It is necessary in calculating *D*-dissimilarities for each attribute to have the character states of every OTU listed separately for each attribute. Unfortunately, except in very small-scale studies when the taxonomist can collate the entire material in one place, it is much more convenient for the taxonomist to record his data separately for each OTU. The botanist, for example, may have to travel between herbaria or botanic gardens to collect information about the relevant specimens and thus be forced to record his data in this way. Since the estimation of character states may involve a number of decisions by the taxonomist, it is best first to estimate character states for each OTU separately, and then to invert the file before computing *D*-dissimilarities.

(b) *Conditional definition.* The method described in Chapter 3.3, by which weights are introduced to deal with conditional definition, is strictly correct. However, some taxonomists may prefer to use a less tedious but somewhat *ad hoc* method. One such method involves adding a dummy state 'not comparable' to the list of states of any attribute which is conditionally defined. This state is recorded whenever the specimen fails to take whatever attribute states the attribute in question is conditional upon. The attribute is then handled in the same way as any other discrete-state attribute. This approach is obviously inapplicable to quantitative attributes.

(c) *Stability of classifications.* We emphasize again that the value of a numerical taxonomic study is increased if the options for investigation of

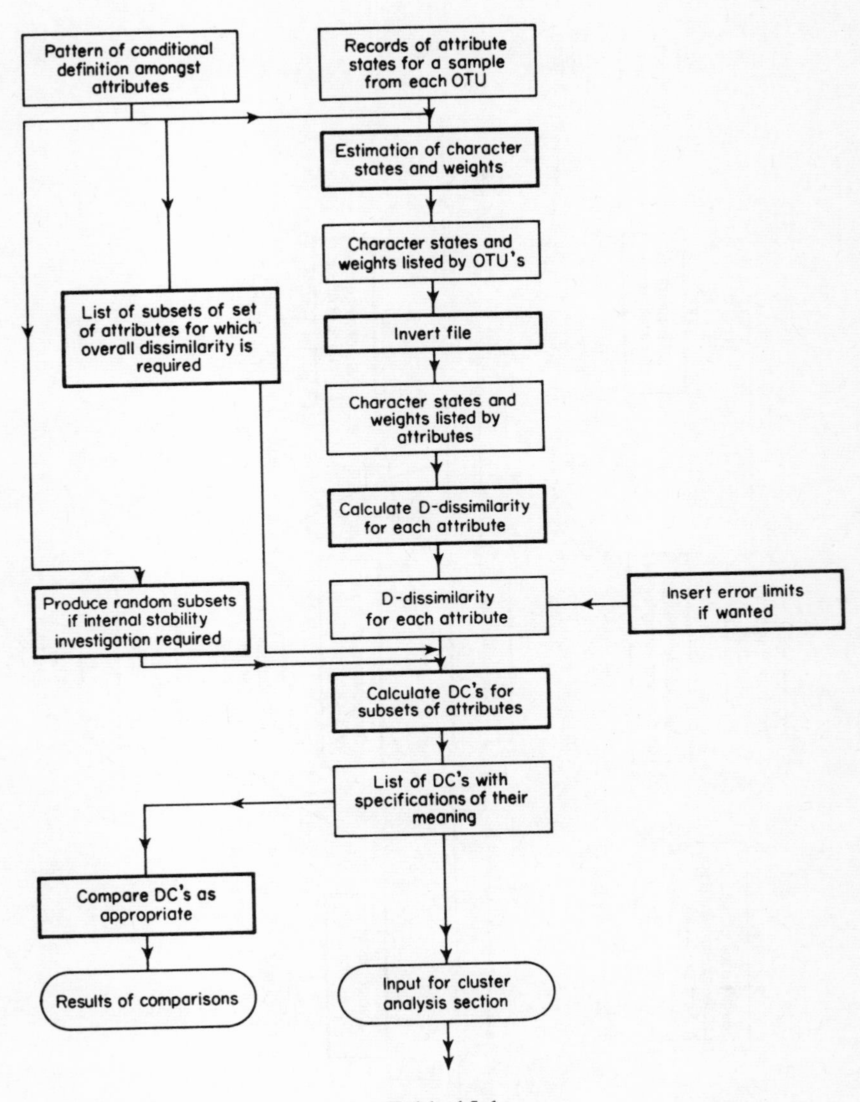
Pattern of conditional definition amongst attributes
Records of attribute states for a sample from each OTU
Estimation of character states and weights
Character states and weights listed by OTU's
List of subsets of set of attributes for which overall dissimilarity is required
Invert file
Character states and weights listed by attributes
Calculate D-dissimilarity for each attribute
Produce random subsets if internal stability investigation required
D-dissimilarity for each attribute
Insert error limits if wanted
Calculate DC's for subsets of attributes
List of DC's with specifications of their meaning
Compare DC's as appropriate
Results of comparisons
Input for cluster analysis section

Table 15.1.

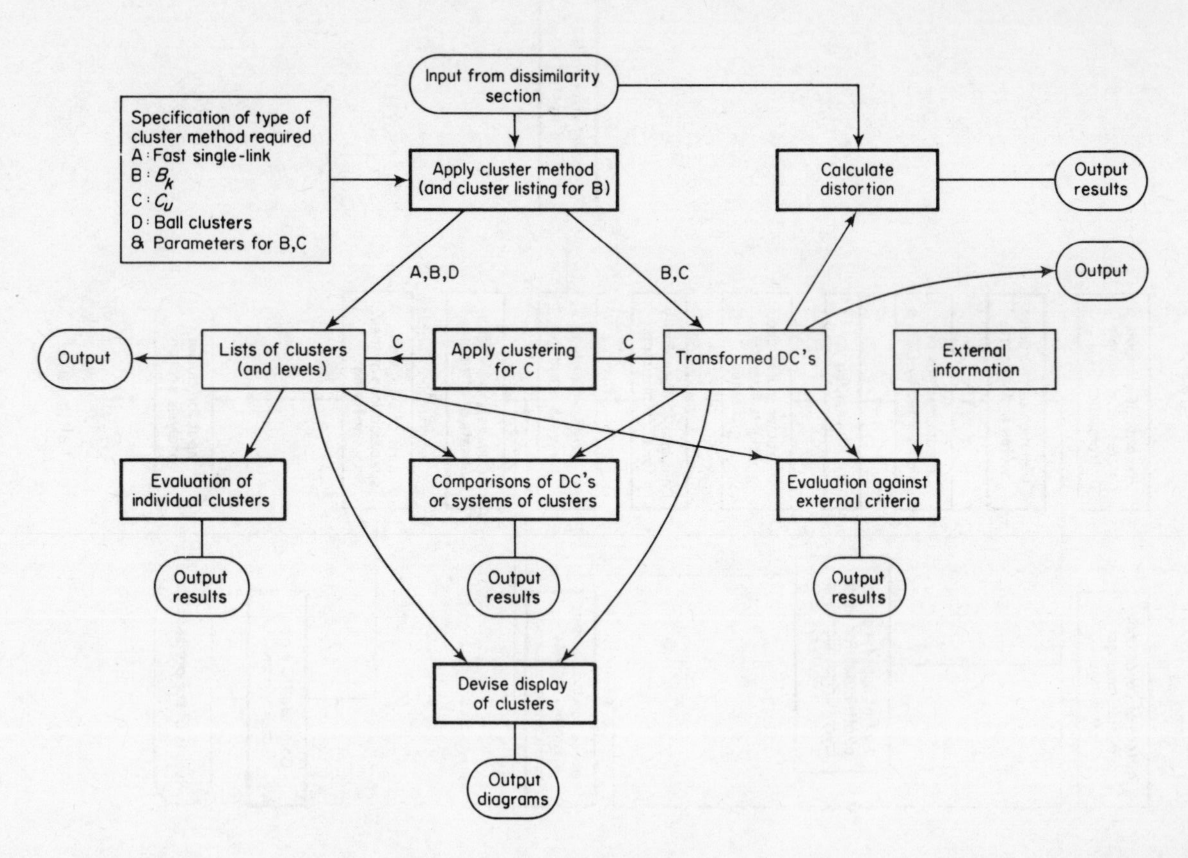
Specification of type of cluster method required
A : Fast single-link
B : B_k
C : C_u
D : Ball clusters
& Parameters for B,C
Input from dissimilarity section
Apply cluster method (and cluster listing for B)
Calculate distortion
Output results
A,B,D
B,C
Output
Output
Lists of clusters (and levels)
C
Apply clustering for C
C
Transformed DC's
External information
Evaluation of individual clusters
Comparisons of DC's or systems of clusters
Evaluation against external criteria
Output results
Output results
Output results
Devise display of clusters
Output diagrams

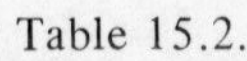
Table 15.2.

internal stability are used. It is inadvisable ever to assume internal stability, however large the initial set of attributes.

(d) *Fast hierarchic clustering.* For large problems computer times for the full procedure may be long. It may often be advisable to use the fast single-link method to get an initial idea of the structure of the data.

(e) *Display of results.* Dendrograms in tree-form are the most convenient representation when only hierarchic clusterings are involved. Users should beware of the psychological effects of different dispositions of OTU's along the base of a dendrogram (see Figures 7.1, 7.2, 7.3). A dendrogram can often be made to look as if it supports an existing classification by grouping together at the base of the dendrogram OTU's belonging to the same classes in the existing classification, even when the fit is in fact very poor. This is amply illustrated in the literature. Tree-diagrams for non-hierarchic clusterings are hopelessly complicated, but lists of clusters at each level are not readily grasped. A good method of display is to represent the clusters at selected levels in graphical form (see Figure 7.4 and Appendix 7). In order to do this it is necessary to obtain a disposition of points representing OTU's in two dimensions which minimizes the tangling between clusters. This is conveniently done by applying non-metric multidimensional scaling to the output dissimilarity coefficient. If the dissimilarities less than or equal to the current level are to be represented by edges of the graph the disposition should be modified to eliminate collinearity between points. Again, users should beware of the psychological effects. It is often hard to remember that the disposition of points representing OTU's is only a convenient format; use of multidimensional scaling for this purpose is purely presentational. Only when clinal variation is being investigated can the results be interpreted meaningfully, and then the method is applied to the input dissimilarity coefficient rather than the dissimilarity coefficient output by a cluster method.

Methods of automatic classification which yield a simple partition or covering, or which operate directly on data specifying the attribute states of individual organisms, fall outside the scope of the treatment offered here. It has been shown in Chapter 12 that the majority of such methods are unsatisfactory for the analysis of taxonomic data. An exception is the methods for sorting mixtures of multivariate normal distributions which may be useful in the study of infraspecific variation, using quantitative attributes which can be assumed to be normally distributed in the groups sought. Wolfe (1968b) described the application of one such method to the analysis of a mixed population of *Iris setosa, I. versicolor*, and *I. virginica.*

REFERENCES

Adanson, M. (1763). *Familles des Plantes,* Vincent, Paris.

Almquist, E. (1921). Studien über *Capsella,* II. *Acta Horti Bergen,* **7**, 41-93.

Ashton, E. H., M. J. R. Healy, and S. Lipton (1957). The descriptive use of discriminant functions in physical anthropology. *Proc. R. Soc. (Ser. B),* **146**, 552-572.

Bailey, N. T. J. (1965). Probability methods of diagnosis based on small samples. In *Mathematics and Computer Science in Biology and Medicine,* M.R.C., H.M.S.O., London. pp. 103-110.

Bather, F. A. (1927). Biological classification: past and future. *Q. Jl geol. Soc. Lond.,* **83**, Proc. lxii-civ.

Beckner, M. (1959). *The Biological Way of Thought,* Columbia University Press, New York.

Bellett, A. J. D. (1967). Numerical classification of some viruses, bacteria and animals according to nearest-neighbour base sequence frequency. *J. molec. Biol.,* **27**, 107-112.

Berlin, B., D. E. Breedlove, and P. H. Raven (1966). Folk taxonomies and biological classification. *Science, N.Y.,* **54**, 273-277.

Bigelow, R. S. (1956). Monophyletic classification and evolution. *Syst. Zool.,* **5**, 145-146.

Blackith, R. E., and R. M. Blackith (1968). A numerical taxonomy of orthopteroid insects. *Aust. J. Zool.,* **16**, 111-131.

Bocquet, G., and J. D. Bersier (1960). La valeur systématique de l'ovule: developpements teratologiques. *Archs. Sci. (Genève)* **13**, 475-496.

Boyden, A. (1947). Homology and analogy. A critical review of the meanings and implications of these concepts in biology. *Am. Midl. Nat.,* **37**, 648-669.

Bremekamp, C. E. B. (1939). Phylogenetic interpretations and genetic concepts in taxonomy. *Chronica Bot.,* **5**, 398-403.

Briggs, D., and S. M. Walters (1969). *Plant Variation and Evolution,* Weidenfeld and Nicolson, London.

Buck, R. C., and D. L. Hull (1966). The logical structure of the Linnean hierarchy. *Syst. Zool.,* **15**, 97-111.

Buettner-Janusch, J., and R. L. Hill (1965). Molecules and monkeys. *Science, N.Y.,* **147**, 836-842.

Burtt, B. L., I. C. Hedge, and P. H. Stevens (1970). A taxonomic critique of recent numerical studies in Ericales and Salvia. *Notes R. bot. Gdn Edinb.,* **30**, 141-158.

Cain, A. J. (1959a). Deductive and inductive processes in post-Linnaean taxonomy. *Proc. Linn. Soc. Lond.,* **170**, 185-217.

Cain, A. J. (1959b). Function and taxonomic importance. In A. J. Cain (Ed.), *Function and Taxonomic Importance,* Systematics Association, Publ. No. 3, London. pp. 5-19.

Cain, A. J. (1962). The evolution of taxonomic principles. In G. C. Ainsworth and P. H. A. Sneath (Eds.), *Microbiol Classification,* 12th Symposium of the Society for General Microbiology, Cambridge University Press. pp. 1-13.

Cain, A. J., and G. A. Harrison (1958). An analysis of the taxonomist's judgement of affinity. *Proc. zool. Soc. Lond.*, **131**, 184-187.

Cain, A. J., and G. A. Harrison (1960). Phyletic weighting. *Proc. zool. Soc. Lond.*, **135**, 1-31.

Camin, J. H., and R. R. Sokal (1965). A method for deducing branching sequences in phylogeny. *Evolution, Lancaster, Pa.*, **19**, 311-326.

Camp, W. H., and C. L. Gilly (1943). The structure and origin of species. *Brittonia*, **4**, 323-385.

Candolle, A. P. de (1813). *Théorie Elémentaire de la Botanique*, Deterville, Paris.

Clausen, J. (1951). *Stages in the Evolution of Plant Species*, Cornell University Press, Ithaca, New York.

Clausen, J., D. D. Keck, and W. M. Hiesey (1940). *Experimental Studies on the Nature of Species. I. Effects of Varied Environments on Western North American Plants*, Carnegie Inst. Washington, Publ. No. 520.

Clausen, J., D. D. Keck, and W. M. Hiesey (1941). Experimental taxonomy. *Carnegie Inst. Washington Year Book*, **40**, 160-170.

Cock, A. G. (1966). Genetical aspects of metrical growth and form in animals. *Q. Rev. Biol.*, **41**, 131-190.

Corner, E. J. H. (1962). The classification of Moraceae. *Gdns' Bull. Singapore*, **19**, 187-252.

Cover, T. M., and P. E. Hart (1967). Nearest-neighbor pattern classification. *IEEE Trans. Inf. Theory*, **13**, 21-27.

Crovello, T. J. (1968). The effects of changes of OTU's in a numerical taxonomic study. *Brittonia*, **20**, 346-367.

Crovello, T. J. (1969). Effects of changes of characters and number of characters in numerical taxonomy. *Am. Midl. Nat.*, **81**, 68-86.

Cuvier, G. (1807). Eloge historique de Michel Adanson. *Mém. Inst. natn. Fr., Cl. Sci. math. phys.*, **7**, 159-188.

Darwin, C. (1859). *On the Origin of Species by Means of Natural Selection*, 1st ed. Murray, London.

Davis, P. H., and V. H. Heywood (1963). *Principles of Angiosperm Taxonomy*, Oliver and Boyd, Edinburgh and London.

Dobzhansky, T. (1941). *Genetics and the Origin of Species*, revised ed. Columbia University Press, New York.

Dobzhansky, T. (1951). *Genetics and the Origin of Species*, 3rd ed. Columbia University Press, New York.

Doran, J. E., and F. R. Hodson (1966). A digital computer analysis of palaeolithic flint assemblages. *Nature, Lond.*, **210**, 688-689.

Eames, A. J. (1961). *Morphology of the Angiosperms*, McGraw-Hill, New York.

Edwards, A. W. F., and L. L. Cavalli-Sforza (1964). Reconstruction of phylogenetic trees. In V. H. Heywood and J. McNeill (Eds.), *Phenetic and Phylogenetic Classification*, Systematics Association, Publ. No. 6, London. pp. 67-76.

Edwards, A. W. F., and L. L. Cavalli-Sforza (1965). A method for cluster analysis. *Biometrics*, **21**, 362-375.

Ehrlich, P. R., and P. H. Raven (1969). Differentiation of populations. *Science, N.Y.*, **165**, 1228-1232.

El Gazzar, A., L. Watson, W. T. Williams, and G. N. Lance (1968). The taxonomy of *Salvia*: a test of two radically different numerical methods. *J. Linn. Soc.* (*Bot.*), **60**, 237-250.

Esau, K. S. (1965). *Plant Anatomy,* 2nd ed. Wiley, New York.

Farris, J. S. (1966). Estimation of conservatism of characters by consistency within biological populations. *Evolution, Lancaster, Pa.,* **20**, 587-591.

Fitch, W. M., and E. Margoliash (1967). Construction of phylogenetic trees. *Science, N.Y.,* **155**, 279-284.

Fu, K. S. (1968). *Sequential Methods in Pattern Recognition and Machine Learning,* Academic Press, London and New York.

Gillham, N. W. (1956). Geographic variation and the subspecies concept in butterflies. *Syst. Zool.,* **5**, 110-120.

Gilmour, J. S. L. (1937). A taxonomic problem. *Nature, Lond.,* **139**, 1040-1042.

Gilmour, J. S. L. (1940). Taxonomy and philosophy. In J. S. Huxley (Ed.), *The New Systematics,* Oxford University Press, Oxford. pp. 461-474.

Gilmour, J. S. L., and J. W. Gregor (1939). Demes: a suggested new terminology. *Nature, Lond.,* **144**, 333-334.

Gilmour, J. S. L., and W. B. Turrill (1941). The aim and scope of taxonomy. *Chronica Bot.,* **6**, 217-219.

Gilmour, J. S. L., and S. M. Walters (1964). Philosophy and classification. In W. B. Turrill (Ed.), *Vistas in Botany, IV,* Pergamon Press, Oxford. pp. 1-22.

Goodman, L. A., and W. H. Kruskal (1954). Measures of association for cross-classifications. *J. Am. statist. Ass.,* **49**, 732-764.

Gregg, J. R. (1954). *The Language of Taxonomy,* Columbia University Press, New York.

Gregg, J. R. (1967). Finite Linnaean structures. *Bull. math. Biophys.,* **29**, 191-206.

Gregor, J. W. (1938). Experimental taxonomy. II. Initial population differentiation in *Plantago maritima* L. of Britain. *New Phytol.,* **37**, 15-49.

Gregor, J. W. (1939). Experimental taxonomy. IV. Population differentiation in North American and European sea plantains allied to *Plantago maritima* L. *New Phytol.,* **38**, 293-322.

Guédès, M. (1967). La méthode taxonomique d'Adanson. *Revue Hist. Sci.,* **20**, 361-386.

Guttman, L. (1968). A general nonmetric technique for finding the smallest coordinate space for a configuration of points. *Psychometrika,* **33**, 469-506.

Hamann, U. (1961). Merkmalbestand und Verwandtschaftsbeziehungen der Farinosae. Ein Beitrag zum System der Monokotyledonen. *Willdenowia,* 2, 639-768.

Hawkes, J. G., and W. G. Tucker (1968). Serological assessment of relationships in a flowering plant family (Solanaceae). In J. G. Hawkes (Ed.), *Chemotaxonomy and Serotaxonomy,* Systematics Association, Special Volume No. 2. Academic Press, London and New York. pp. 77-88.

Hedberg, O. (1958). The taxonomic treatment of vicarious taxa. *Upps. Univ. Årsskr.*, **6**, 186-195.

Hennig, W. (1965). Phylogenetic systematics. *A. Rev. Ent.*, **10**, 97-116.

Hennig, W. (1967). *Phylogenetic Systematics.* Illinois University Press, Urbana. Translated by D. D. Davis and R. R. Zangerl.

Heslop-Harrison, J. (1964). Forty years of genecology. *Adv. ecol. Res.*, **2**, 159-247.

Highton, R. (1962). Revision of North American salamanders of the genus *Plethodon. Bull. Fla St. Mus. biol. sci.*, **61**, 235-367.

Holloway, J. D., and N. Jardine (1968). Two approaches to zoogeography: a study based on the distributions of butterflies, birds and bats in the Indo-Australian area. *Proc. Linn. Soc. Lond.*, **179**, 153-188.

Inger, R. F. (1967). The development of a phylogeny of frogs. *Evolution, Lancaster, Pa.*, **21**, 369-384.

Jardine, C. J., N. Jardine, and R. Sibson (1967). The structure and construction of taxonomic hierarchies. *Math. Biosci.*, **1**, 173-179.

Jardine, N. (1969a). A logical basis for biological classification. *Syst. Zool.*, **18**, 37-52.

Jardine, N. (1969c). The observational and theoretical components of homology: a study based on the morphology of the dermal skull roofs of rhipidistian fishes. *Biol. J. Linn. Soc.*, **1**, 327-361.

Jardine, N. (1971). The application of Simpson's criterion of consistency to phenetic classifications. *Syst. Zool.*, **20**, 70-73.

Jardine, N., and C. J. Jardine (1967). Numerical homology. *Nature, Lond.*, **216**, 301-302.

Jardine, N., C. J. van Rijsbergen, and C. J. Jardine (1969). Evolutionary rates and the inference of evolutionary tree forms. *Nature, Lond.*, **224**, 185.

Jardine, N., and R. Sibson (1968a). The construction of hierarchic and non-hierarchic classifications. *Comput. J.*, **11**, 177-184.

Jardine, N., and R. Sibson (1970). Quantitative attributes in taxonomic descriptions. *Taxon* **19**, 862-870.

Johnson, L. A. S. (1968). Rainbow's end: the quest for an optimal taxonomy. *Proc. Linn. Soc. N.S.W.*, **93**, 8-45.

Johnson, R. G. (1960). Models and methods for analysis of the mode of formation of fossil assemblages. *Bull. Geol. Soc. Am.*, **71**, 1075-1086.

Jordan, A. (1873). Remarques sur le fruit de l'existence en société à l'état sauvage des espèces végétales affines. *Bull. Ass. fr. Avanc. Sci.*, 2.

Kiriakoff, S. G. (1962). On the neo-Adansonian school. *Syst. Zool.*, **11**, 180-185.

Kirsch, J. A. W. (1969). Serological data and phylogenetic inference: the problem of rates of change. *Syst. Zool.*, **18**, 296-311.

Kluge, A. J., and J. S. Farris (1969). Quantitative phyletics and the evolution of anurans. *Syst. Zool.*, **18**, 1-32.

Kohne, D. E. (1968). Taxonomic applications of DNA hybridisation techniques. In J. G. Hawkes (Ed.), *Chemotaxonomy and Serotaxonomy.* Systematics Association Special Volume No. 2. Academic Press, London and New York. pp. 117-130.

Kruskal, J. B. (1964a). Multidimensional scaling by optimising goodness-of-fit to a nonmetric hypothesis. *Psychometrika,* **129**, 1-27.

Kruskal, J. B. (1964b). Nonmetric multidimensional scaling: a numerical method. *Psychometrika,* **129**, 115-129.

Ledley, R. S., and L. B. Lusted (1959). Reasoning foundations of medical diagnosis. *Science, N.Y.,* **130**, 9-21.

Le Gros Clark, W. E. (1950). New palaeontological evidence bearing on the evolution of the Hominoidea. *Q. Jl geol. Soc. Lond.*, **105**, 225-264.

Łuszczewska-Romahnowa, S. (1961). Classification as a kind of distance function. Natural classifications. *Studia Logica,* **12**, 41-66.

Mackerras, I. M. (1964). The classification of animals. *Proc. Linn. Soc. N.S.W.,* **88**, 324-335.

Maslin, T. P. (1952). Morphological criteria of phylogenetic relationships. *Syst. Zool.,* **1**, 49-70.

Mayr, E. (1964). The taxonomic evaluation of fossil hominoids. In S. L. Washburn (Ed.), *Classification and Human Evolution,* Methuen, London. pp. 332-345.

Mayr, E. (1965). Numerical phenetics and taxonomic theory. *Syst. Zool.,* **14**, 73-97.

Mayr, E. (1968). Theory of biological classification. *Nature, Lond.,* **220**, 545-548.

Mayr, E. (1969a). *Principles of Systematic Zoology,* McGraw-Hill, New York.

Mayr, E. (1969b). The biological meaning of species. *Biol. J. Linn. Soc.,* **1**, 311-320.

Melville, R. (1962). A new theory of the angiosperm flower. I. The gynoecium. *Kew Bull.,* **16**, 1-50.

Melville, R. (1963). A new theory of the angiosperm flower. II. The androecium. *Kew Bull.,* **17**, 1-66.

Michener, C. D. (1949). Parallelism in the evolution of saturnid moths. *Evolution, Lancaster, Pa.,* **3**, 129-141.

Michener, C. D. (1964). The possible use of uninomial nomenclature to increase the stability of names in biology. *Syst. Zool.,* **13**, 182-190.

Miller, R. L., and J. S. Kahn (1962). *Statistical Analysis in the Geological Sciences,* Wiley, New York.

Müntzing, A. (1938). Sterility and chromosome pairing in infraspecific *Galeopsis* hybrids. *Hereditas,* **24**, 117-188.

Nilsson, N. J. (1965). *Learning Machines–Foundations of Trainable Pattern-Classifying Systems,* McGraw-Hill, New York.

Olson, E. C. (1964). Morphological integration and meaning of characters. In V. H. Heywood, and J. McNeill (Eds.), *Phenetic and Phylogenetic Classification,* Systematics Association, Publ. No. 6, London. pp. 115-121.

Olson, E. C., and R. L. Miller (1955). The statistical stability of quantitative properties as a fundamental criterion for the study of environments. *J. Geol.,* **63**, 376-387.

Olson, E. C., and R. L. Miller (1958). *Morphological Integration,* Chicago University Press.

Osborn, H. F. (1893). The rise of the mammalia in North America. *Proc. Am. Ass. Advmt. Sci.,* **42**, 188-227.

Osborn, H. F. (1902). Homoplasy as a law of latent or potential homology. *Am. Nat.* **36**, 259-271.

Pankhurst, R. J. (1970). A computer program for generating diagnostic keys. *Comput. J.*, **12**, 145-151.

Peterson, D. W. (1970). Some convergence properties of a nearest-neighbor decision rule. *IEEE Trans. Inf. Theory,* **16**, 26-31.

Pichi-Sermolli, R. E. G. (1959). Pteridophyta. In W. B. Turrill (Ed.), *Vistas in Botany. A Volume in Honour of the Bicentenary of the Royal Botanic Gardens, Kew,* Pergamon Press, London. pp. 421-493.

Pritchard, N. M. (1960). *Gentianella* in Britain. 2. *Gentianella septendrionalis* (Druce) E. F. Warburg. *Watsonia,* **4**, 218-237.

Rao, C. R. (1948). The utilization of multiple measurements in problems of biological classification. *Jl R. statist. Soc.* (*Ser. B*), **10**, 159-193.

Rao, C. R. (1952). *Advanced Statistical Methods in Biometric Research,* Wiley, New York.

Reichert, E. T., and A. P. Brown (1909). *The Differentiation and Specification of Corresponding Protein and other Vital Substances in Relation to Biological Classification and Organic Evolution,* Carnegie Inst. Washington, Publ. No. 116.

Rohlf, F. J. (1963). Congruence of larval and adult classifications in *Aedes* (Diptera: Culicidae). *Syst. Zool.*, **12**, 97-117.

Sarich, V. M., and A. C. Wilson (1967). Rates of albumen evolution in primates. *Proc. nat. Acad. Sci. U.S.A.*, **58**, 142-148.

Sattler, R. (1966). Towards a more adequate approach to comparative morphology. *Phytomorphology,* **16**, 417-429.

Sattler, R. (1967). Petal inception and the problem of pattern recognition. *J. theor. Biol.*, **17**, 3-39.

Schildkraut, C. L., J. Marmur, and P. Doty (1961). The formation of hybrid DNA molecules and their use in studies of DNA homologies. *J. molec. Biol.*, **3**, 595-617.

Sebestyen, G. (1962). *Decision-Making Processes in Pattern Recognition,* Macmillan, New York.

Shepard, R. N. (1962). The analysis of proximities: multidimensional scaling with an unknown distance function. I. *Psychometrika,* **27**, 125-140; II. 219-246.

Shepard, R. N., and J. D. Carroll (1966). Parametric representation of nonlinear data structures. In P. R. Krishnaiah (Ed.), *Multivariate Analysis.* Academic Press, London and New York. pp. 561-592.

Simpson, G. G. (1945). The principles of classification and a classification of mammals. *Bull. Am. Mus. nat. Hist.*, **85**, 1-350.

Simpson, G. G. (1961). *Principles of Animal Taxonomy,* Columbia University Press, New York.

Sinskaja, E. N. (1960). Investigations in the composition of ecotypical and varietal populations. A brief survey of some of our works published in Russian. *Rep. Scott. Pl. Breed. Stn.*, **1960**.

Sneath, P. H. A. (1962). The construction of taxonomic groups. In G. C. Ainsworth and P. H. A. Sneath (Eds.), *Microbial Classification.* Cambridge University Press. pp. 289-332.

Sneath, P. H. A. (1966b). Estimating concordance between geographical trends. *Syst. Zool.,* **15**, 250-252.

Snyder, L. A. (1951). Cytology of inter-strain hybrids and the probable origin of variability in *Elymus glaucus. Am. J. Bot.,* **38**, 195-202.

Sokal, R. R., and C. D. Michener (1967). The effects of different numerical techniques on the phenetic classification of bees of the complex (Megachilidae). *Proc. Linn. Soc. Lond.,* **178**, 59-74.

Sokal, R. R. and R. C. Rinkel (1963). Geographical variation of alate *Pemphigus populi-transversus* in eastern North America. *Kans. Univ. Sci. Bull.,* **44**, 467-507.

Sokal, R. R., and P. H. A. Sneath (1963). *Principles of Numerical Taxonomy,* Freeman, San Francisco and London.

Sporne, K. R. (1948). Correlation and classification in dicotyledons. *Proc. Linn. Soc. Lond.,* **160**, 40-47.

Sporne, K. R. (1969). The ovule as an indicator of evolutionary status in angiosperms. *New Phytol.,* **68**, 555-566.

Stebbins, G. L. (1950). *Variation and Evolution in Plants,* Columbia University Press, New York.

Stebbins, G. L. (1967). Adaptive radiation and trends of evolution in higher plants. In T. Dobzhansky, M. K. Hecht, and W. C. Steere (Eds.), *Evolutionary Biology,* Vol. 1. North Holland, Amsterdam. pp. 101-142.

Stensiö, E. A. (1963). The brain and cranial nerves in fossil lower craniate vertebrates. *Skr. norske Vidensk-Akad.,* N.S. **13**, 5-120.

Taylor, R. J., and D. Campbell (1969). Biochemical systematics and phylogenetic interpretations in the genus *Aquilegia. Evolution, Lancaster, Pa.,* **23**, 153-162.

Thompson, W. R. (1952). The philosophical foundations of systematics. *Can. Ent.,* **84**, 1-16.

Throckmorton, L. H. (1965). Similarity *versus* relationship in *Drosophila. Syst. Zool.,* **14**, 221-236.

Valentine, D. H., and A. Löve (1958). Taxonomic and biosystematic categories. *Brittonia,* **10**, 153-179.

Van Woerkom, A. J., and K. Brodman (1961). Statistics for a diagnostic model. *Biometrics,* **17**, 299-318.

Walters, S. M. (1961). The shaping of angiosperm taxonomy. *New Phytol.,* **60**, 74-84.

Walters, S. M. (1962). Generic and specific concepts in the European flora. *Preslia,* **34**, 207-226.

Walters, S. M. (1965) 'Improvement' versus stability in botanical classification. *Taxon,* **14**, 6-12.

Wardlaw, C. W. (1965). *Organization and Evolution in Plants,* Longmans, London.

Wilson, E. O. (1965). A consistency test for phylogenies based on contemporary species. *Syst. Zool.,* **14**, 214-220.

Wilson, E. O., and W. L. Brown (1953). The subspecies concept and its taxonomic implications. *Syst. Zool.,* **2**, 97-111.

Wolfe, J. H. (1968b). NORMAP program documentation. *Res. Memo SRM* 69-12, U.S. Naval Personnel Research Activity, San Diego, California.

Woodger, J. H. (1937). *The Axiomatic Method in Biology,* Cambridge University Press.

Zangerl, R. R. (1948). The methods of comparative anatomy and its contribution to the study of evolution. *Evolution, Lancaster, Pa.,* 2, 351-374.

Zimmermann, W. (1965). *Die Telomtheorie,* Fortschritte der Evolutionsforschung, Band I. Gustav Fischer, Stuttgart.

APPENDICES

APPENDIX 1 Estimation of Information Radius

The information radius

$$K\begin{bmatrix} \mu_1 & \mu_2 \\ w_1 & w_2 \end{bmatrix}$$

is, for fixed w_1, w_2, a function of the pair of probability measures μ_1, μ_2. In practice it is determined by estimation, and the usual statistical problem of investigating the behaviour of the estimation process arises. The presence of the log function in the explicit expression for K makes analytical treatment of methods of estimating it very difficult to carry out, although A. Gordon has been able to obtain some results in this direction. In order to obtain some idea of the accuracy of estimation processes for K, Gordon and Sibson have carried out some computer-based studies in a simple case, namely with $w_1 = w_2$, and μ_1, μ_2 probability measures on a two-point space $X = \{x_1, x_2\}$. If μ_1, μ_2 are specified by $\{p_1(x_1), p_1(x_2)\}, \{p_2(x_1), p_2(x_2)\}$, then

$$K = \tfrac{1}{2}\left\{p_1(x_1)\log_2 \frac{2p_1(x_1)}{p_1(x_1)+p_2(x_1)} + p_2(x_1)\log_2 \frac{2p_2(x_1)}{p_1(x_1)+p_2(x_1)} + p_1(x_2)\log_2 \frac{2p_1(x_2)}{p_1(x_2)+p_2(x_2)} + p_2(x_2)\log_2 \frac{2p_2(x_2)}{p_1(x_2)+p_2(x_2)}\right\}$$

Since $p_1(x_1) + p_1(x_2) = p_2(x_1) + p_2(x_2) = 1$, this may be regarded as a function $k[p_1(x_1), p_2(x_1)]$ where

$$k(x, y) = \tfrac{1}{2}\left\{x\log_2 \frac{2x}{x+y} + y\log_2 \frac{2y}{x+y} + (1-x)\log_2 \frac{2(1-x)}{2-x-y} + (1-y)\log_2 \frac{2(1-y)}{2-x-y}\right\}$$

Take a sample of size n_1 for the distribution μ_1, and one of size n_2 for μ_2, and suppose that the samples are divided between x_1 and x_2 as $n_1 = r_{11} + r_{12}$, $n_2 = r_{21} + r_{22}$. The estimation process considered was to estimate $k[p_1(x_1), p_2(x_1)]$ by $k(r_{11}/n_1, r_{21}/n_2)$, and tables showing the bias (difference between the sample mean and the true value) and the expected error (square root of the expected squared deviation of the sample value from the true value) were prepared for $n_1 = n_2 = n$ for values 5 (5) 30 (10) 100 and for $p_1(x_1)$, $p_2(x_1)$ for values 0.0 (0.1) 1.0. Tables for $n = 15, 25, 50, 100$ are given here. In this particular case the figures suggest that sample size should not be allowed to fall below 15, and should if possible be at least 25. The gains in taking larger samples probably do not justify the extra work involved. Some modified estimation processes which were investigated showed no effective reduction in the expected error. It is interesting to note that the bias is always positive, and that it makes only a fairly small contribution to the expected error.

The first eight tables are in four sets of two, the sets being for $n = 15, 25, 50, 100$, and the tables within each set being for bias and expected (root mean square) error. Since each table is symmetrical about the principal diagonal, only the lower half is given; symmetry also occurs about the other diagonal. The probability values are given to the left and below for each table. The final table gives true values for $k[p_1(x_1), p_2(x_1)]$. Computation was done at the Cambridge University Mathematical Laboratory on the Titan computer.

In dealing with multistate attributes when it is not convenient to use a computer, and for certain statistical tests, tables of $-x\log_2 x$ for $x \in [0, 1]$ are useful. The following table computed by C. J. van Rijsbergen gives values of $-x\log_2 x$ for $x = 0.000\ (0.001)\ 0.999$ in the customary style for four-figure tables.

In Appendix 2 tables of information radius for normal distributions are given, and can be used for estimation by using samples to estimate the parameters of the normal distributions involved and then looking up the information radius between normal distributions with these parameters. If the distributions over a numerical attribute are not of known form, the only reasonable way of estimating information radius is by partitioning the range and treating the resultant grouped values in the same way as a multistate attribute. This technique is also of value in many cases even when the form of the distributions is known. If μ_1, μ_2 are probability measures on the line, and μ_1^G, μ_2^G are the multistate probability measures obtained as a result of grouping, then $K(\mu_1, \mu_2) \geqslant K(\mu_1^G, \mu_2^G)$, and $K(\mu_1^G, \mu_2^G) \to K(\mu_1, \mu_2)$ as the fineness of the grouping is increased. Since the estimation process for $K(\mu_1^G, \mu^G)$ has small positive bias, this will tend to correct the reduction in K produced by grouping. Little is yet known in detail about the effects of grouping on the estimation of K, but probably reasonably large samples are needed to obtain good results from this technique.

2-STATE BIAS FOR SAMPLE SIZE 15

0.	0.										
0.1	0.0013	0.0299									
0.2	0.0028	0.0295	0.0276								
0.3	0.0044	0.0301	0.0272	0.0261							
0.4	0.0063	0.0314	0.0278	0.0261	0.0255						
0.5	0.0084	0.0331	0.0291	0.0268	0.0257	0.0254					
0.6	0.0110	0.0353	0.0309	0.0281	0.0266	0.0257	0.0255				
0.7	0.0140	0.0382	0.0334	0.0302	0.0281	0.0268	0.0261	0.0261			
0.8	0.0181	0.0420	0.0369	0.0334	0.0309	0.0291	0.0278	0.0272	0.0276		
0.9	0.0237	0.0474	0.0420	0.0382	0.0353	0.0331	0.0314	0.0301	0.0295	0.0299	
1.0	0.0000	0.0237	0.0181	0.0140	0.0110	0.0084	0.0063	0.0044	0.0028	0.0013	0.0000
	0.	0.1	0.2	0.3	0.4	0.5	0.6	0.7	0.8	0.9	1.0

2-STATE DEVIATIONS ABOUT TRUE VALUE FOR SAMPLE SIZE 15

0.	0.										
0.1	0.0429	0.0470									
0.2	0.0622	0.0604	0.0478								
0.3	0.0782	0.0817	0.0568	0.0455							
0.4	0.0927	0.1030	0.0760	0.0536	0.0444						
0.5	0.1064	0.1233	0.0975	0.0719	0.0522	0.0441					
0.6	0.1194	0.1426	0.1192	0.0937	0.0706	0.0522	0.0444				
0.7	0.1317	0.1612	0.1410	0.1170	0.0937	0.0719	0.0536	0.0455			
0.8	0.1423	0.1787	0.1623	0.1410	0.1192	0.0975	0.0760	0.0568	0.0478		
0.9	0.1430	0.1900	0.1787	0.1612	0.1426	0.1233	0.1030	0.0817	0.0604	0.0470	
1.0	0.0000	0.1430	0.1423	0.1317	0.1194	0.1064	0.0927	0.0782	0.0622	0.0429	0.0000
	0.	0.1	0.2	0.3	0.4	0.5	0.6	0.7	0.8	0.9	1.0

2-STATE BIAS FOR SAMPLE SIZE 25

0.	0.										
0.1	0.0008	0.0171									
0.2	0.0016	0.0166	0.0154								
0.3	0.0026	0.0171	0.0155	0.0150							
0.4	0.0037	0.0179	0.0159	0.0152	0.0149						
0.5	0.0050	0.0190	0.0167	0.0156	0.0151	0.0149					
0.6	0.0064	0.0203	0.0177	0.0163	0.0155	0.0151	0.0149				
0.7	0.0081	0.0218	0.0190	0.0174	0.0163	0.0156	0.0152	0.0150			
0.8	0.0102	0.0238	0.0209	0.0190	0.0177	0.0167	0.0159	0.0155	0.0154		
0.9	0.0134	0.0269	0.0238	0.0218	0.0203	0.0190	0.0179	0.0171	0.0166	0.0171	
1.0	0.0000	0.0134	0.0102	0.0081	0.0064	0.0050	0.0037	0.0026	0.0016	0.0008	0.0000
	0.	0.1	0.2	0.3	0.4	0.5	0 6	0.7	0.8	0.9	1.0

2-STATE DEVIATIONS ABOUT TRUE VALUE FOR SAMPLE SIZE 25

0.	0.										
0.1	0.0329	0.0289									
0.2	0.0476	0.0408	0.0270								
0.3	0.0598	0.0589	0.0360	0.0261							
0.4	0.0708	0.0759	0.0526	0.0340	0.0259						
0.5	0.0811	0.0917	0.0700	0.0498	0.0333	0.0258					
0.6	0.0908	0.1066	0.0871	0.0674	0.0490	0.0333	0.0259				
0.7	0.0997	0.1206	0.1038	0.0856	0.0674	0.0498	0.0340	0.0261			
0.8	0.1069	0.1335	0.1198	0.1038	0.0871	0.0700	0.0526	0.0360	0.0270		
0.9	0.1088	0.1432	0.1335	0.1206	0.1066	0.0917	0.0759	0.0589	0.0408	0.0289	
1.0	0.0000	0.1088	0.1069	0.0997	0.0908	0.0811	0.0708	0.0598	0.0476	0.0329	0.0000
	0.	0.1	0.2	0.3	0.4	0.5	0.6	0.7	0.8	0.9	1.0

2-STATE BIAS FOR SAMPLE SIZE 50

	0.	0.1	0.2	0.3	0.4	0.5	0.6	0.7	0.8	0.9	1.0
0.	0.										
0.1	0.0004	0.0077									
0.2	0.0008	0.0077	0.0074								
0.3	0.0013	0.0080	0.0075	0.0074							
0.4	0.0018	0.0084	0.0077	0.0074	0.0073						
0.5	0.0024	0.0090	0.0081	0.0076	0.0074	0.0073					
0.6	0.0031	0.0096	0.0086	0.0080	0.0076	0.0074	0.0073				
0.7	0.0040	0.0103	0.0092	0.0085	0.0080	0.0076	0.0074	0.0074			
0.8	0.0049	0.0112	0.0101	0.0092	0.0086	0.0081	0.0077	0.0075	0.0074		
0.9	0.0062	0.0125	0.0112	0.0103	0.0096	0.0090	0.0084	0.0080	0.0077	0.0077	
1.0	0.0000	0.0062	0.0049	0.0040	0.0031	0.0024	0.0018	0.0013	0.0008	0.0004	0.0000

2-STATE DEVIATIONS ABOUT TRUE VALUE FOR SAMPLE SIZE 50

	0.	0.1	0.2	0.3	0.4	0.5	0.6	0.7	0.8	0.9	1.0
0.	0.										
0.1	0.0230	0.0135									
0.2	0.0334	0.0243	0.0129								
0.3	0.0419	0.0383	0.0212	0.0127							
0.4	0.0495	0.0508	0.0341	0.0199	0.0127						
0.5	0.0567	0.0621	0.0468	0.0322	0.0194	0.0127					
0.6	0.0634	0.0726	0.0591	0.0452	0.0317	0.0194	0.0127				
0.7	0.0695	0.0825	0.0710	0.0582	0.0452	0.0322	0.0199	0.0127			
0.8	0.0742	0.0914	0.0821	0.0710	0.0591	0.0468	0.0341	0.0212	0.0129		
0.9	0.0749	0.0978	0.0914	0.0825	0.0726	0.0621	0.0508	0.0383	0.0243	0.0135	
1.0	0.0000	0.0749	0.0742	0.0695	0.0634	0.0567	0.0495	0.0419	0.0334	0.0230	0.0000

2-STATE BIAS FOR SAMPLE SIZE 100

0.	0.										
0.1	0.0002	0.0037									
0.2	0.0004	0.0038	0.0037								
0.3	0.0006	0.0039	0.0037	0.0036							
0.4	0.0009	0.0041	0.0038	0.0037	0.0036						
0.5	0.0012	0.0044	0.0040	0.0038	0.0037	0.0036					
0.6	0.0016	0.0047	0.0043	0.0040	0.0038	0.0037	0.0036				
0.7	0.0020	0.0050	0.0046	0.0042	0.0040	0.0038	0.0037	0.0036			
0.8	0.0024	0.0055	0.0050	0.0046	0.0043	0.0040	0.0038	0.0037	0.0037		
0.9	0.0030	0.0060	0.0055	0.0050	0.0047	0.0044	0.0041	0.0039	0.0038	0.0037	
1.0	0.0000	0.0030	0.0024	0.0020	0.0016	0.0012	0.0009	0.0006	0.0004	0.0002	0.0000
	0.	0.1	0.2	0.3	0.4	0.5	0.6	0.7	0.8	0.9	1.0

2-STATE DEVIATIONS ABOUT TRUE VALUE FOR SAMPLE SIZE 100

0.	0.										
0.1	0.0162	0.0064									
0.2	0.0235	0.0157	0.0063								
0.3	0.0295	0.0260	0.0134	0.0063							
0.4	0.0348	0.0349	0.0231	0.0125	0.0063						
0.5	0.0398	0.0430	0.0323	0.0217	0.0121	0.0063					
0.6	0.0445	0.0505	0.0410	0.0311	0.0213	0.0121	0.0063				
0.7	0.0488	0.0575	0.0494	0.0404	0.0311	0.0217	0.0125	0.0063			
0.8	0.0521	0.0637	0.0573	0.0494	0.0410	0.0323	0.0231	0.0134	0.0063		
0.9	0.0524	0.0681	0.0637	0.0575	0.0505	0.0430	0.0349	0.0260	0.0157	0.0064	
1.0	0.0000	0.0524	0.0521	0.0488	0.0445	0.0398	0.0348	0.0295	0.0235	0.0162	0.0000
	0.	0.1	0.2	0.3	0.4	0.5	0.6	0.7	0.8	0.9	1.0

2-STATE TRUE VALUES OF K

	0.	0.1	0.2	0.3	0.4	0.5	0.6	0.7	0.8	0.9	1.0
0.	0.										
0.1	0.0519	-0.0000									
0.2	0.1080	0.0144	0.0000								
0.3	0.1692	0.0468	0.0097	-0.0000							
0.4	0.2365	0.0913	0.0349	0.0079	0.0000						
0.5	0.3113	0.1468	0.0731	0.0303	0.0073	-0.0000					
0.6	0.3958	0.2141	0.1245	0.0667	0.0290	0.0073	-0.0000				
0.7	0.4934	0.2958	0.1912	0.1187	0.0667	0.0303	0.0079	0.0000			
0.8	0.6100	0.3973	0.2781	0.1912	0.1245	0.0731	0.0349	0.0097	-0.0000		
0.9	0.7583	0.5310	0.3973	0.2958	0.2141	0.1468	0.0913	0.0468	0.0144	0.0000	
1.0	1.0000	0.7583	0.6100	0.4934	0.3958	0.3113	0.2365	0.1692	0.1080	0.0519	-0.0000

		0	1	2	3	4	5	6	7	8	9
000		0000	0100	0179	0251	0319	0382	0443	0501	0557	0612
010		0664	0716	0766	0814	0862	0909	0955	0999	1043	1086
020		1129	1170	1211	1252	1291	1330	1369	1407	1444	1481
030		1518	1554	1589	1624	1659	1693	1727	1760	1793	1825
040		1858	1889	1921	1952	1983	2013	2043	2073	2103	2132
050		2161	2190	2218	2246	2274	2301	2329	2356	2383	2409
060		2435	2461	2487	2513	2538	2563	2588	2613	2637	2662
070		2686	2709	2733	2756	2780	2803	2826	2848	2871	2893
080		2915	2937	2959	2980	3002	3023	3044	3065	3086	3106
090		3127	3147	3167	3187	3207	3226	3246	3265	3284	3303
100		3322	3341	3359	3378	3396	3414	3432	3450	3468	3485
110		3503	3520	3537	3555	3571	3588	3605	3622	3638	3654
120		3671	3687	3703	3719	3734	3750	3766	3781	3796	3811
130		3826	3841	3856	3871	3886	3900	3915	3929	3943	3957
140		3971	3985	3999	4012	4026	4040	4053	4066	4079	4092
150		4105	4118	4131	4144	4156	4169	4181	4194	4206	4218
160		4230	4242	4254	4266	4278	4289	4301	4312	4323	4335
170		4346	4357	4368	4379	4390	4401	4411	4422	4432	4443
180		4453	4463	4474	4484	4494	4504	4514	4523	4533	4543
190		4552	4562	4571	4581	4590	4599	4608	4617	4626	4635
200		4644	4653	4661	4670	4678	4687	4695	4704	4712	4720
210		4728	4736	4744	4752	4760	4768	4776	4783	4791	4798
220		4806	4813	4820	4828	4835	4842	4849	4856	4863	4870
230		4877	4883	4890	4897	4903	4910	4916	4923	4929	4935
240		4941	4947	4954	4960	4966	4971	4977	4983	4989	4994
250		5000	5006	5011	5016	5022	5027	5032	5038	5043	5048
260		5053	5058	5063	5068	5072	5077	5082	5087	5091	5096
270		5100	5105	5109	5113	5118	5122	5126	5130	5134	5138
280		5142	5146	5150	5154	5158	5161	5165	5169	5172	5176
290		5179	5182	5186	5189	5192	5196	5199	5202	5205	5208
300		5211	5214	5217	5220	5222	5225	5228	5230	5233	5235
310		5238	5240	5243	5245	5247	5250	5252	5254	5256	5258
320		5260	5262	5264	5266	5268	5270	5272	5273	5275	5277
330		5278	5280	5281	5283	5284	5286	5287	5288	5289	5291
340		5292	5293	5294	5295	5296	5297	5298	5299	5299	5300
350		5301	5302	5302	5303	5304	5304	5305	5305	5305	5306
360		5306	5306	5307	5307	5307	5307	5307	5307	5307	5307
370		5307	5307	5307	5307	5307	5306	5306	5306	5305	5305
380		5305	5304	5304	5303	5302	5302	5301	5300	5300	5299
390		5298	5297	5296	5295	5294	5293	5292	5291	5290	5289
400		5288	5286	5285	5284	5283	5281	5280	5278	5277	5275
410		5274	5272	5271	5269	5267	5266	5264	5262	5260	5258
420		5256	5255	5253	5251	5249	5246	5244	5242	5240	5238
430		5236	5233	5231	5229	5226	5224	5222	5219	5217	5214
440		5211	5209	5206	5204	5201	5198	5195	5193	5190	5187
450		5184	5181	5178	5175	5172	5169	5166	5163	5160	5157
460		5153	5150	5147	5144	5140	5137	5133	5130	5127	5123
470		5120	5116	5112	5109	5105	5102	5098	5094	5090	5087
480		5083	5079	5075	5071	5067	5063	5059	5055	5051	5047
490		5043	5039	5034	5030	5026	5022	5017	5013	5009	5004

		0	1	2	3	4	5	6	7	8	9
500		5000	4996	4991	4987	4982	4978	4973	4968	4964	4959
510		4954	4950	4945	4940	4935	4930	4926	4921	4916	4911
520		4906	4901	4896	4891	4886	4880	4875	4870	4865	4860
530		4854	4849	4844	4839	4833	4828	4822	4817	4811	4806
540		4800	4795	4789	4784	4778	4772	4767	4761	4755	4750
550		4744	4738	4732	4726	4720	4714	4708	4702	4696	4690
560		4684	4678	4672	4666	4660	4654	4648	4641	4635	4629
570		4623	4616	4610	4603	4597	4591	4584	4578	4571	4565
580		4558	4551	4545	4538	4532	4525	4518	4511	4505	4498
590		4491	4484	4477	4471	4464	4457	4450	4443	4436	4429
600		4422	4415	4408	4401	4393	4386	4379	4372	4365	4357
610		4350	4343	4335	4328	4321	4313	4306	4298	4291	4283
620		4276	4268	4261	4253	4246	4238	4230	4223	4215	4207
630		4199	4192	4184	4176	4168	4160	4152	4145	4137	4129
640		4121	4113	4105	4097	4089	4080	4072	4064	4056	4048
650		4040	4031	4023	4015	4007	3998	3990	3982	3973	3965
660		3956	3948	3940	3931	3923	3914	3905	3897	3888	3880
670		3871	3862	3854	3845	3836	3828	3819	3810	3801	3792
680		3783	3775	3766	3757	3748	3739	3730	3721	3712	3703
690		3694	3685	3676	3666	3657	3648	3639	3630	3621	3611
700		3602	3593	3583	3574	3565	3555	3546	3537	3527	3518
710		3508	3499	3489	3480	3470	3460	3451	3441	3432	3422
720		3412	3403	3393	3383	3373	3364	3354	3344	3334	3324
730		3314	3305	3295	3285	3275	3265	3255	3245	3235	3225
740		3215	3204	3194	3184	3174	3164	3154	3144	3133	3123
750		3113	3102	3092	3082	3072	3061	3051	3040	3030	3020
760		3009	2999	2988	2978	2967	2956	2946	2935	2925	2914
770		2903	2893	2882	2871	2861	2850	2839	2828	2818	2807
780		2796	2785	2774	2763	2752	2741	2731	2720	2709	2698
790		2687	2676	2665	2653	2642	2631	2620	2609	2598	2587
800		2575	2564	2553	2542	2530	2519	2508	2497	2485	2474
810		2462	2451	2440	2428	2417	2405	2394	2382	2371	2359
820		2348	2336	2325	2313	2301	2290	2278	2266	2255	2243
830		2231	2219	2208	2196	2184	2172	2160	2149	2137	2125
840		2113	2101	2089	2077	2065	2053	2041	2029	2017	2005
850		1993	1981	1969	1957	1944	1932	1920	1908	1896	1884
860		1871	1859	1847	1834	1822	1810	1797	1785	1773	1760
870		1748	1736	1723	1711	1698	1686	1673	1661	1648	1636
880		1623	1610	1598	1585	1572	1560	1547	1534	1522	1509
890		1496	1484	1471	1458	1445	1432	1420	1407	1394	1381
900		1368	1355	1342	1329	1316	1303	1290	1277	1264	1251
910		1238	1225	1212	1199	1186	1173	1159	1146	1133	1120
920		1107	1093	1080	1067	1054	1040	1027	1014	1000	0987
930		0974	0960	0947	0933	0920	0907	0893	0880	0866	0853
940		0839	0826	0812	0798	0785	0771	0758	0744	0730	0717
950		0703	0689	0676	0662	0648	0634	0621	0607	0593	0579
960		0565	0552	0538	0524	0510	0496	0482	0468	0454	0440
970		0426	0412	0398	0384	0370	0356	0342	0328	0314	0300
980		0286	0271	0257	0243	0229	0215	0201	0186	0172	0158
990		0144	0129	0115	0101	0086	0072	0058	0043	0029	0014

APPENDIX 2 Information Radius for Normal Distributions

The information radius between two normal distributions with arbitrary weights is

$$K\begin{bmatrix} N(\beta_1, \sigma_1{}^2) & N(\beta_2, \sigma_2{}^2) \\ w_1 & w_2 \end{bmatrix}$$

Values of this are needed for dealing with continuously variable numerical attributes with approximately normal or lognormal distributions. Usually $w_1 = w_2$, since such attributes are in practice not usually conditionally defined, although they may be. If $\sigma_M = \max\{\sigma_1, \sigma_2\}$ and $\sigma_m = \min\{\sigma_1, \sigma_2\}$, then

$$K\begin{bmatrix} N(\beta_1, \sigma_1{}^2) & N(\beta_2, \sigma_2{}^2) \\ w_1 & w_2 \end{bmatrix} = K\begin{bmatrix} N(0,1) & N(\beta, \sigma^2) \\ t & 1-t \end{bmatrix} = k(\beta, \sigma, t), \text{ say.}$$

where

$$\sigma = \sigma_m/\sigma_M \leqslant 1$$

$$\beta = |\beta_1 - \beta_2|/\sigma_M \geqslant 0$$

$$t = w_M/(w_m + w_M) \in [0, 1]$$

Tables of $k(\beta, \sigma, t)$ for

$$t = 0.01, 0.02, 0.03, 0.05\ (0.05)\ 0.95, 0.98, 0.99$$

$$\sigma = 0.10\ (0.05)\ 1.00$$

$$\beta = 0.0\ (0.1)\ 1.0\ (0.2)\ 5.0\ (0.5)\ 7.0$$

have been prepared by C. J. van Rijsbergen using the Titan computer at the Cambridge University Mathematical Laboratory. The tables are given on the following pages. Tables are given for different values of t on separate pages; values in the same column are for fixed σ; and those in the same row, for fixed β.

WEIGHT= .010

	.100	.150	.200	.250	.300	.350	.400	.450	.500	.550	.600	.650	.700	.750	.800	.850	.900	.950	1.000
.0*	.056	.047	.039	.032	.026	.021	.016	.013	.010	.007	.005	.003	.002	.001	.001	.000	.000	.000	.000
.1*	.056	.047	.040	.033	.026	.021	.017	.013	.010	.007	.005	.004	.002	.002	.000	.000	.000	.000	.000
.2*	.057	.048	.040	.033	.027	.022	.017	.013	.010	.008	.006	.004	.003	.002	.001	.000	.000	.000	.000
.3*	.057	.048	.041	.034	.028	.023	.018	.014	.011	.009	.006	.005	.004	.003	.002	.001	.001	.001	.000
.4*	.058	.049	.042	.035	.029	.024	.020	.016	.013	.009	.007	.006	.001	.001	.000	.000	.000	.000	.000
.5*	.059	.050	.043	.037	.031	.026	.021	.017	.014	.011	.009	.007	.005	.001	.001	.001	.001	.001	.001
.6*	.060	.052	.045	.038	.033	.028	.023	.019	.016	.013	.010	.008	.007	.001	.001	.001	.001	.001	.001
.7*	.061	.053	.046	.040	.035	.030	.025	.022	.018	.015	.013	.011	.009	.001	.001	.001	.001	.001	.001
.8*	.062	.055	.048	.042	.037	.032	.028	.024	.021	.018	.015	.013	.011	.002	.002	.002	.002	.002	.002
.9*	.063	.056	.059	.045	.039	.035	.031	.027	.023	.020	.017	.015	.013	.002	.002	.002	.002	.002	.002
1.0*	.064	.058	.052	.047	.042	.038	.033	.029	.026	.023	.020	.017	.015	.003	.003	.003	.003	.003	.003
1.2*	.067	.062	.057	.052	.047	.043	.039	.035	.032	.029	.025	.023	.020	.018	.004	.004	.004	.004	.004
1.4*	.070	.065	.061	.057	.053	.049	.045	.041	.038	.035	.031	.028	.026	.023	.021	.019	.017	.015	.014
1.6*	.072	.069	.065	.061	.058	.054	.051	.047	.044	.041	.037	.034	.031	.029	.026	.024	.021	.019	.018
1.8*	.074	.071	.069	.066	.063	.060	.056	.053	.050	.047	.044	.040	.037	.034	.031	.029	.026	.024	.022
2.0*	.076	.074	.072	.069	.067	.064	.061	.058	.055	.052	.049	.046	.043	.040	.037	.034	.032	.029	.027
2.2*	.077	.076	.074	.072	.070	.068	.065	.063	.060	.057	.054	.051	.048	.046	.043	.040	.037	.034	.032
2.4*	.078	.077	.076	.075	.073	.071	.069	.067	.064	.062	.059	.056	.053	.051	.048	.045	.042	.039	.037
2.6*	.079	.079	.078	.077	.075	.073	.072	.070	.068	.066	.063	.061	.058	.055	.053	.050	.047	.044	.042
2.8*	.080	.079	.079	.078	.077	.075	.074	.073	.071	.069	.067	.065	.062	.060	.057	.054	.052	.049	.047
3.0*	.080	.080	.079	.079	.078	.077	.076	.075	.073	.072	.070	.068	.066	.063	.061	.059	.056	.054	.051
3.2*	.080	.080	.080	.079	.079	.078	.077	.077	.075	.074	.072	.071	.069	.067	.065	.063	.060	.058	.055
3.4*	.081	.080	.080	.080	.080	.079	.079	.078	.077	.076	.074	.073	.072	.070	.068	.066	.064	.062	.059
3.6*	.081	.081	.080	.080	.080	.080	.079	.079	.078	.077	.076	.075	.074	.072	.071	.069	.067	.065	.063
3.8*	.081	.081	.081	.080	.080	.080	.080	.079	.079	.078	.077	.076	.075	.074	.073	.071	.070	.068	.066
4.0*	.081	.081	.081	.081	.080	.080	.080	.080	.079	.079	.078	.078	.077	.076	.075	.073	.072	.070	.069
4.2*	.081	.081	.081	.081	.081	.080	.080	.080	.080	.079	.079	.078	.078	.077	.076	.075	.074	.072	.071
4.4*	.081	.081	.081	.081	.081	.081	.081	.080	.080	.080	.080	.079	.078	.078	.077	.076	.075	.074	.073
4.6*	.081	.081	.081	.081	.081	.081	.081	.081	.080	.080	.080	.080	.079	.079	.078	.077	.077	.076	.075
4.8*	.081	.081	.081	.081	.081	.081	.081	.081	.081	.080	.080	.080	.080	.079	.079	.078	.078	.077	.076
5.0*	.081	.081	.081	.081	.081	.081	.081	.081	.081	.081	.080	.080	.080	.080	.079	.079	.078	.078	.077
5.5*	.081	.081	.081	.081	.081	.081	.081	.081	.081	.081	.081	.081	.081	.080	.080	.080	.080	.079	.079
6.0*	.081	.081	.081	.081	.081	.081	.081	.081	.081	.081	.081	.081	.081	.081	.081	.080	.080	.080	.080
6.5*	.081	.081	.081	.081	.081	.081	.081	.081	.081	.081	.081	.081	.081	.081	.081	.081	.081	.081	.080
7.0*	.081	.081	.081	.081	.081	.081	.081	.081	.081	.081	.081	.081	.081	.081	.081	.081	.081	.081	.081

WEIGHT= .020

	.100	.150	.200	.250	.300	.350	.400	.450	.500	.550	.600	.650	.700	.750	.800	.850	.900	.950	1.000
.0*	.099	.083	.070	.058	.047	.038	.030	.023	.018	.013	.010	.007	.004	.003	.002	.001	.000	.000	.000
.1*	.099	.083	.070	.058	.047	.038	.030	.024	.018	.014	.010	.007	.005	.003	.002	.000	.000	.000	.000
.2*	.099	.084	.071	.059	.048	.039	.031	.025	.019	.015	.011	.008	.005	.004	.002	.002	.001	.001	.000
.3*	.100	.085	.072	.060	.050	.041	.033	.026	.021	.016	.012	.009	.007	.005	.004	.003	.002	.002	.001
.4*	.101	.087	.074	.062	.052	.043	.035	.029	.023	.018	.014	.011	.008	.006	.005	.001	.001	.001	.001
.5*	.103	.088	.076	.065	.055	.046	.038	.032	.026	.021	.017	.014	.010	.009	.007	.001	.001	.001	.002
.6*	.104	.091	.079	.068	.058	.050	.042	.035	.029	.024	.020	.017	.014	.011	.009	.008	.002	.002	.002
.7*	.106	.093	.081	.071	.062	.053	.046	.039	.033	.028	.024	.020	.017	.014	.012	.010	.003	.003	.003
.8*	.108	.096	.085	.075	.066	.058	.050	.043	.038	.032	.028	.024	.020	.017	.015	.013	.003	.004	.004
.9*	.110	.099	.088	.079	.070	.062	.055	.048	.042	.037	.032	.028	.024	.021	.018	.016	.004	.004	.065
1.0*	.113	.102	.092	.083	.074	.067	.060	.053	.047	.042	.037	.032	.028	.025	.022	.019	.017	.005	.005
1.2*	.117	.107	.099	.091	.084	.077	.070	.064	.057	.052	.047	.042	.038	.034	.030	.027	.024	.022	.020
1.4*	.122	.114	.107	.100	.093	.086	.080	.074	.068	.063	.057	.052	.048	.043	.039	.036	.032	.029	.027
1.6*	.126	.120	.114	.108	.102	.096	.090	.084	.079	.073	.068	.063	.058	.053	.049	.045	.041	.038	.034
1.8*	.130	.125	.120	.115	.110	.105	.100	.094	.089	.084	.078	.073	.068	.063	.059	.054	.050	.046	.043
2.0*	.133	.129	.125	.121	.117	.113	.108	.103	.098	.093	.088	.083	.078	.073	.068	.064	.059	.055	.051
2.2*	.135	.133	.130	.126	.123	.119	.115	.111	.107	.102	.097	.092	.088	.083	.078	.074	.069	.064	.060
2.4*	.137	.135	.133	.131	.128	.125	.121	.118	.114	.110	.105	.101	.096	.092	.087	.082	.078	.073	.069
2.6*	.139	.137	.135	.134	.131	.129	.126	.123	.120	.116	.112	.108	.104	.100	.095	.091	.086	.082	.078
2.8*	.140	.139	.138	.136	.134	.132	.130	.128	.125	.122	.118	.115	.111	.107	.103	.099	.094	.090	.086
3.0*	.140	.140	.139	.138	.137	.135	.133	.131	.129	.126	.123	.120	.117	.113	.110	.106	.102	.098	.094
3.2*	.141	.140	.140	.139	.138	.137	.136	.134	.132	.130	.128	.125	.122	.119	.116	.112	.109	.105	.101
3.4*	.141	.141	.140	.140	.139	.139	.138	.136	.135	.133	.131	.129	.126	.124	.121	.118	.114	.111	.108
3.6*	.141	.141	.141	.141	.140	.140	.139	.138	.137	.135	.134	.132	.130	.128	.125	.122	.119	.116	.113
3.8*	.141	.141	.141	.141	.141	.140	.140	.139	.138	.137	.136	.134	.133	.131	.129	.126	.124	.121	.118
4.0*	.141	.141	.141	.141	.141	.141	.140	.140	.139	.138	.137	.136	.135	.133	.132	.130	.127	.125	.123
4.2*	.141	.141	.141	.141	.141	.141	.141	.140	.140	.139	.139	.138	.137	.135	.134	.132	.130	.128	.126
4.4*	.141	.141	.141	.141	.141	.141	.141	.141	.140	.140	.139	.139	.138	.137	.136	.134	.133	.131	.129
4.6*	.141	.141	.141	.141	.141	.141	.141	.141	.141	.140	.140	.140	.139	.138	.137	.136	.135	.133	.132
4.8*	.141	.141	.141	.141	.141	.141	.141	.141	.141	.141	.141	.140	.140	.139	.138	.138	.137	.135	.134
5.0*	.141	.141	.141	.141	.141	.141	.141	.141	.141	.141	.141	.141	.140	.140	.139	.139	.138	.137	.136
5.5*	.141	.141	.141	.141	.141	.141	.141	.141	.141	.141	.141	.141	.141	.141	.141	.140	.140	.139	.139
6.0*	.141	.141	.141	.141	.141	.141	.141	.141	.141	.141	.141	.141	.141	.141	.141	.141	.141	.140	.140
6.5*	.141	.141	.141	.141	.141	.141	.141	.141	.141	.141	.141	.141	.141	.141	.141	.141	.141	.141	.141
7.0*	.141	.141	.141	.141	.141	.141	.141	.141	.141	.141	.141	.141	.141	.141	.141	.141	.141	.141	.141

WEIGHT= .030

	.100	.150	.200	.250	.300	.350	.400	.450	.500	.550	.600	.650	.700	.750	.800	.850	.900	.950	1.000
.0*	.136	.114	.096	.080	.065	.053	.042	.033	.025	.019	.014	.010	.006	.004	.002	.001	.000	.000	.000
.1*	.136	.115	.096	.080	.066	.054	.043	.034	.026	.019	.014	.010	.007	.004	.003	.001	.000	.000	.000
.2*	.136	.116	.097	.081	.067	.055	.044	.035	.027	.021	.016	.011	.008	.005	.004	.002	.002	.001	.001
.3*	.138	.117	.099	.083	.069	.057	.047	.037	.030	.023	.018	.013	.010	.007	.005	.004	.003	.002	.002
.4*	.139	.119	.102	.086	.072	.060	.050	.041	.033	.026	.021	.016	.012	.009	.007	.006	.001	.001	.002
.5*	.141	.122	.105	.089	.076	.064	.054	.045	.037	.030	.024	.020	.016	.013	.010	.008	.002	.002	.002
.6*	.143	.125	.108	.093	.080	.069	.058	.049	.041	.035	.029	.024	.020	.016	.013	.011	.010	.003	.003
.7*	.146	.128	.112	.098	.085	.074	.064	.055	.047	.040	.034	.029	.024	.020	.017	.015	.013	.004	.004
.8*	.148	.132	.117	.103	.091	.080	.070	.061	.053	.046	.039	.034	.029	.025	.022	.019	.017	.015	.005
.9*	.151	.136	.121	.108	.097	.086	.076	.067	.059	.052	.045	.040	.035	.030	.027	.023	.021	.019	.007
1.0*	.155	.140	.126	.114	.103	.092	.083	.074	.066	.059	.052	.046	.041	.036	.032	.028	.025	.023	.021
1.2*	.161	.148	.136	.125	.115	.106	.097	.088	.080	.073	.066	.060	.054	.048	.043	.039	.035	.032	.029
1.4*	.167	.156	.147	.137	.128	.119	.111	.103	.095	.088	.081	.074	.068	.062	.056	.051	.047	.043	.039
1.6*	.173	.164	.156	.148	.140	.132	.125	.117	.110	.102	.095	.088	.082	.076	.070	.064	.059	.055	.050
1.8*	.178	.171	.165	.158	.151	.144	.138	.130	.123	.116	.110	.103	.096	.090	.084	.078	.072	.067	.062
2.0*	.182	.177	.172	.166	.161	.155	.149	.143	.136	.130	.123	.116	.110	.103	.097	.091	.085	.079	.074
2.2*	.186	.182	.178	.174	.169	.164	.159	.153	.147	.141	.135	.129	.123	.116	.110	.104	.098	.092	.087
2.4*	.189	.186	.183	.179	.176	.172	.167	.162	.157	.152	.146	.140	.134	.128	.122	.116	.110	.104	.099
2.6*	.191	.189	.186	.184	.181	.178	.174	.170	.165	.161	.156	.151	.145	.139	.134	.128	.122	.116	.111
2.8*	.192	.191	.189	.187	.185	.182	.179	.176	.172	.168	.164	.159	.154	.149	.144	.138	.133	.127	.122
3.0*	.193	.192	.191	.190	.188	.186	.183	.181	.178	.174	.170	.166	.162	.158	.153	.148	.143	.137	.132
3.2*	.193	.193	.192	.191	.190	.189	.187	.184	.182	.179	.176	.173	.169	.165	.161	.156	.152	.147	.142
3.4*	.194	.194	.193	.192	.192	.190	.189	.187	.185	.183	.181	.178	.175	.171	.168	.164	.159	.155	.150
3.6*	.194	.194	.194	.193	.193	.192	.191	.190	.188	.186	.184	.182	.179	.176	.173	.170	.166	.162	.158
3.8*	.194	.194	.194	.194	.193	.193	.192	.191	.190	.189	.187	.185	.183	.181	.178	.175	.172	.168	.164
4.0*	.194	.194	.194	.194	.194	.193	.193	.192	.191	.190	.189	.188	.186	.184	.182	.179	.176	.173	.170
4.2*	.194	.194	.194	.194	.194	.194	.193	.193	.192	.192	.191	.189	.188	.187	.185	.183	.180	.178	.175
4.4*	.194	.194	.194	.194	.194	.194	.194	.193	.193	.192	.192	.191	.190	.189	.187	.185	.184	.181	.179
4.6*	.194	.194	.194	.194	.194	.194	.194	.194	.193	.193	.193	.192	.191	.190	.189	.188	.186	.184	.182
4.8*	.194	.194	.194	.194	.194	.194	.194	.194	.194	.193	.193	.193	.192	.191	.190	.189	.188	.187	.185
5.0*	.194	.194	.194	.194	.194	.194	.194	.194	.194	.194	.194	.193	.193	.192	.191	.191	.190	.188	.187
5.5*	.194	.194	.194	.194	.194	.194	.194	.194	.194	.194	.194	.194	.194	.194	.193	.193	.192	.192	.191
6.0*	.194	.194	.194	.194	.194	.194	.194	.194	.194	.194	.194	.194	.194	.194	.194	.194	.193	.193	.193
6.5*	.194	.194	.194	.194	.194	.194	.194	.194	.194	.194	.194	.194	.194	.194	.194	.194	.194	.194	.194
7.0*	.194	.194	.194	.194	.194	.194	.194	.194	.194	.194	.194	.194	.194	.194	.194	.194	.194	.194	.194

WEIGHT= .050

	.100	.150	.200	.250	.300	.350	.400	.450	.500	.550	.600	.650	.700	.750	.800	.850	.900	.950	1.000
.0*	.199	.168	.142	.118	.098	.080	.064	.050	.039	.029	.022	.015	.010	.007	.004	.002	.001	.000	.000
.1*	.199	.169	.142	.119	.098	.080	.065	.051	.040	.030	.022	.016	.011	.007	.004	.002	.001	.000	.000
.2*	.200	.170	.144	.120	.100	.082	.067	.053	.042	.032	.024	.018	.013	.009	.006	.004	.002	.002	.001
.3*	.202	.172	.145	.123	.103	.086	.070	.057	.045	.036	.027	.021	.016	.011	.008	.006	.005	.004	.003
.4*	.204	.175	.150	.127	.107	.090	.075	.061	.050	.040	.032	.025	.020	.015	.012	.009	.007	.002	.002
.5*	.207	.179	.154	.132	.113	.096	.081	.067	.056	.046	.038	.030	.025	.020	.016	.013	.011	.009	.004
.6*	.210	.183	.159	.138	.119	.102	.087	.074	.063	.053	.044	.037	.031	.026	.021	.018	.016	.014	.005
.7*	.214	.188	.165	.145	.126	.110	.095	.082	.071	.061	.052	.044	.038	.032	.028	.024	.021	.018	.017
.8*	.218	.193	.171	.152	.134	.118	.104	.091	.080	.069	.060	.052	.045	.039	.034	.030	.027	.024	.022
.9*	.222	.199	.178	.160	.143	.127	.113	.101	.089	.079	.069	.061	.054	.047	.042	.037	.033	.030	.027
1.0*	.227	.205	.185	.168	.151	.137	.123	.111	.099	.089	.079	.071	.063	.056	.050	.045	.040	.036	.033
1.2*	.236	.217	.209	.185	.170	.156	.144	.132	.120	.110	.100	.091	.083	.075	.068	.062	.056	.051	.047
1.4*	.246	.230	.215	.202	.189	.176	.164	.153	.142	.132	.122	.112	.103	.095	.087	.080	.074	.068	.063
1.6*	.254	.241	.229	.218	.206	.195	.184	.174	.163	.153	.143	.134	.125	.116	.108	.100	.093	.086	.080
1.8*	.262	.252	.242	.232	.223	.213	.203	.193	.184	.174	.164	.155	.146	.137	.128	.120	.112	.104	.098
2.0*	.268	.260	.253	.245	.237	.229	.220	.211	.202	.193	.184	.175	.166	.157	.148	.140	.131	.124	.116
2.2*	.273	.268	.262	.255	.249	.242	.234	.227	.219	.210	.202	.193	.184	.176	.167	.159	.150	.142	.134
2.4*	.277	.273	.269	.264	.259	.253	.247	.240	.233	.225	.218	.210	.202	.193	.185	.177	.168	.160	.152
2.6*	.280	.278	.274	.271	.266	.262	.257	.251	.245	.238	.231	.224	.217	.209	.201	.193	.185	.177	.170
2.8*	.282	.281	.279	.275	.272	.268	.264	.260	.254	.249	.243	.237	.230	.223	.216	.208	.201	.193	.186
3.0*	.284	.283	.281	.279	.277	.274	.271	.267	.263	.258	.253	.247	.241	.235	.229	.222	.215	.208	.201
3.2*	.285	.284	.283	.282	.280	.278	.275	.272	.269	.265	.261	.256	.251	.246	.240	.234	.227	.221	.214
3.4*	.286	.285	.284	.283	.282	.281	.279	.276	.274	.271	.267	.263	.259	.254	.249	.244	.238	.232	.226
3.6*	.286	.286	.285	.285	.284	.283	.281	.279	.277	.275	.272	.269	.265	.262	.257	.252	.247	.242	.237
3.8*	.286	.286	.285	.285	.285	.284	.283	.282	.280	.278	.276	.273	.271	.267	.264	.260	.255	.251	.246
4.0*	.286	.286	.286	.286	.285	.285	.284	.283	.282	.281	.279	.277	.275	.272	.269	.266	.262	.258	.253
4.2*	.286	.286	.286	.286	.286	.285	.285	.284	.283	.282	.281	.279	.278	.276	.273	.270	.267	.264	.260
4.4*	.286	.286	.285	.286	.286	.286	.285	.285	.284	.284	.283	.282	.280	.278	.276	.274	.272	.269	.266
4.6*	.286	.286	.286	.286	.286	.286	.286	.286	.285	.284	.284	.283	.282	.281	.279	.277	.275	.273	.270
4.8*	.286	.286	.285	.286	.286	.286	.286	.286	.286	.285	.285	.284	.283	.282	.281	.280	.278	.276	.274
5.0*	.286	.286	.285	.286	.286	.286	.286	.286	.286	.286	.285	.285	.284	.283	.282	.281	.280	.279	.277
5.5*	.286	.286	.286	.286	.286	.286	.286	.286	.286	.286	.286	.286	.286	.285	.285	.284	.283	.283	.282
6.0*	.286	.286	.286	.286	.286	.286	.286	.286	.286	.286	.286	.286	.286	.286	.286	.285	.285	.285	.284
6.5*	.286	.286	.285	.286	.286	.286	.286	.286	.286	.286	.286	.286	.286	.286	.286	.286	.286	.286	.285
7.0*	.286	.286	.286	.286	.286	.286	.286	.286	.286	.286	.286	.286	.286	.286	.286	.286	.286	.286	.286

WEIGHT= .100

	.100	.150	.200	.250	.300	.350	.400	.450	.500	.550	.600	.650	.700	.750	.800	.850	.900	.950	1.000
.0*	.323	.273	.231	.193	.161	.132	.107	.085	.067	.051	.038	.027	.019	.012	.007	.004	.002	.000	.000
.1*	.323	.274	.231	.194	.162	.133	.108	.086	.068	.052	.039	.028	.020	.013	.008	.005	.002	.001	.001
.2*	.325	.276	.234	.197	.165	.136	.111	.090	.071	.055	.042	.031	.023	.016	.011	.007	.005	.003	.003
.3*	.327	.279	.238	.202	.170	.141	.117	.095	.077	.061	.048	.037	.028	.021	.015	.011	.008	.007	.006
.4*	.331	.284	.243	.208	.176	.149	.124	.103	.085	.069	.055	.044	.035	.027	.021	.017	.014	.012	.010
.5*	.336	.290	.250	.216	.185	.158	.134	.113	.094	.078	.064	.053	.044	.036	.029	.024	.021	.018	.016
.6*	.341	.297	.259	.225	.195	.168	.145	.124	.106	.090	.076	.064	.054	.046	.039	.033	.029	.025	.023
.7*	.347	.305	.268	.235	.206	.180	.157	.137	.119	.103	.088	.076	.066	.057	.049	.043	.038	.034	.031
.8*	.354	.314	.278	.247	.219	.194	.171	.151	.133	.117	.103	.090	.079	.070	.062	.055	.049	.044	.040
.9*	.361	.323	.289	.260	.233	.208	.186	.166	.149	.133	.118	.105	.094	.084	.075	.067	.061	.055	.050
1.0*	.368	.333	.301	.273	.247	.224	.202	.183	.165	.149	.134	.121	.109	.098	.089	.081	.073	.067	.062
1.2*	.384	.353	.325	.300	.277	.255	.235	.217	.200	.184	.168	.155	.142	.130	.120	.110	.101	.093	.087
1.4*	.399	.373	.350	.328	.307	.288	.269	.252	.235	.220	.205	.190	.177	.164	.153	.142	.132	.123	.114
1.6*	.414	.392	.373	.354	.336	.319	.302	.286	.270	.255	.240	.226	.212	.199	.186	.175	.164	.153	.144
1.8*	.426	.410	.394	.378	.363	.348	.333	.318	.303	.289	.274	.260	.246	.233	.220	.208	.196	.185	.174
2.0*	.437	.424	.412	.399	.386	.373	.360	.347	.333	.320	.306	.292	.279	.266	.253	.240	.228	.216	.205
2.2*	.446	.437	.427	.417	.406	.395	.384	.372	.360	.347	.335	.322	.309	.296	.284	.271	.259	.247	.235
2.4*	.453	.446	.439	.431	.423	.414	.404	.394	.383	.372	.360	.348	.336	.324	.312	.300	.288	.276	.264
2.6*	.458	.453	.448	.442	.436	.428	.420	.412	.402	.393	.382	.372	.361	.349	.338	.326	.315	.303	.292
2.8*	.462	.459	.455	.450	.445	.440	.433	.426	.418	.410	.401	.392	.382	.371	.361	.350	.339	.328	.317
3.0*	.465	.462	.460	.457	.453	.448	.443	.437	.431	.424	.416	.408	.400	.390	.381	.371	.361	.351	.340
3.2*	.466	.465	.463	.461	.458	.455	.451	.446	.441	.435	.429	.422	.415	.407	.398	.389	.380	.371	.361
3.4*	.468	.467	.465	.464	.462	.459	.456	.453	.449	.444	.439	.433	.427	.420	.413	.405	.397	.388	.379
3.6*	.468	.468	.467	.466	.464	.463	.461	.458	.455	.451	.447	.442	.437	.431	.425	.418	.411	.403	.395
3.8*	.469	.468	.468	.467	.466	.465	.463	.461	.459	.456	.453	.449	.445	.440	.435	.429	.423	.416	.409
4.0*	.469	.469	.468	.468	.467	.466	.465	.464	.462	.460	.458	.455	.451	.447	.443	.438	.433	.427	.421
4.2*	.469	.469	.469	.468	.468	.467	.467	.466	.465	.463	.461	.459	.456	.453	.449	.445	.441	.436	.431
4.4*	.469	.469	.469	.469	.468	.468	.468	.467	.466	.465	.463	.462	.460	.457	.454	.451	.447	.443	.439
4.6*	.469	.469	.469	.469	.469	.468	.468	.468	.467	.466	.465	.464	.462	.460	.458	.455	.452	.449	.445
4.8*	.469	.469	.469	.469	.469	.469	.468	.468	.468	.467	.466	.466	.464	.463	.461	.459	.457	.454	.451
5.0*	.469	.469	.469	.469	.469	.469	.469	.468	.468	.468	.467	.467	.466	.465	.463	.462	.460	.457	.455
5.5*	.469	.469	.469	.469	.469	.469	.469	.469	.469	.469	.468	.468	.468	.467	.467	.466	.465	.464	.462
6.0*	.469	.469	.469	.469	.469	.469	.469	.469	.469	.469	.469	.469	.469	.468	.468	.468	.467	.467	.466
6.5*	.469	.469	.469	.469	.469	.469	.469	.469	.469	.469	.469	.469	.469	.469	.469	.469	.468	.468	.468
7.0*	.469	.469	.469	.469	.469	.469	.469	.469	.469	.469	.469	.469	.469	.469	.469	.469	.469	.469	.468

WEIGHT= .150

	.100	.150	.200	.250	.300	.350	.400	.450	.500	.550	.600	.650	.700	.750	.800	.850	.900	.950	1.000
.0*	.415	.351	.297	.249	.208	.171	.140	.112	.088	.068	.051	.037	.025	.016	.010	.005	.002	.000	.000
.1*	.416	.352	.299	.251	.209	.173	.141	.113	.089	.069	.052	.038	.027	.018	.011	.006	.003	.001	.001
.2*	.418	.355	.301	.254	.213	.177	.145	.118	.094	.074	.056	.042	.031	.022	.015	.010	.006	.004	.004
.3*	.421	.359	.306	.260	.219	.183	.152	.125	.101	.081	.064	.049	.038	.028	.021	.016	.012	.009	.008
.4*	.426	.366	.313	.268	.228	.193	.162	.134	.111	.091	.073	.059	.047	.037	.029	.023	.019	.016	.015
.5*	.432	.373	.322	.278	.239	.204	.174	.147	.123	.103	.086	.071	.059	.048	.040	.034	.029	.025	.023
.6*	.439	.382	.333	.290	.251	.218	.188	.161	.138	.118	.100	.085	.072	.062	.053	.045	.040	.036	.032
.7*	.447	.392	.345	.303	.266	.233	.204	.178	.155	.135	.117	.102	.088	.077	.067	.059	.053	.048	.044
.8*	.456	.403	.359	.318	.282	.250	.222	.196	.174	.153	.135	.120	.106	.094	.084	.075	.067	.061	.056
.9*	.465	.416	.372	.334	.300	.269	.241	.216	.194	.173	.155	.139	.125	.112	.101	.092	.083	.076	.071
1.0*	.475	.428	.387	.351	.318	.289	.262	.237	.215	.195	.177	.160	.145	.132	.120	.110	.101	.093	.086
1.2*	.495	.455	.419	.387	.357	.330	.305	.281	.260	.240	.221	.205	.189	.174	.161	.149	.138	.128	.120
1.4*	.515	.481	.451	.422	.396	.372	.349	.327	.306	.287	.268	.251	.234	.219	.205	.191	.179	.167	.157
1.6*	.535	.507	.481	.457	.434	.412	.391	.371	.352	.333	.314	.297	.280	.264	.249	.234	.221	.208	.196
1.8*	.552	.530	.509	.489	.469	.450	.431	.413	.394	.376	.359	.342	.325	.309	.293	.278	.263	.250	.237
2.0*	.567	.549	.533	.516	.500	.484	.467	.450	.434	.417	.400	.383	.367	.351	.335	.320	.305	.291	.277
2.2*	.579	.566	.553	.540	.526	.512	.498	.483	.468	.453	.437	.422	.406	.390	.375	.360	.345	.330	.316
2.4*	.588	.579	.569	.559	.548	.536	.524	.511	.498	.484	.470	.456	.441	.426	.411	.397	.382	.368	.353
2.6*	.595	.589	.581	.574	.565	.556	.546	.535	.523	.511	.499	.486	.472	.458	.444	.430	.416	.402	.388
2.8*	.600	.596	.591	.585	.578	.571	.563	.554	.544	.534	.523	.511	.499	.487	.474	.461	.447	.434	.421
3.0*	.604	.601	.597	.593	.588	.582	.576	.569	.561	.552	.543	.532	.522	.511	.499	.487	.475	.463	.450
3.2*	.606	.604	.602	.599	.595	.591	.586	.580	.574	.567	.559	.550	.541	.531	.521	.510	.499	.488	.476
3.4*	.608	.607	.605	.603	.600	.597	.593	.589	.584	.578	.572	.564	.557	.548	.539	.530	.520	.510	.499
3.6*	.609	.608	.607	.606	.604	.601	.599	.595	.591	.587	.582	.576	.569	.562	.555	.546	.538	.529	.519
3.8*	.609	.609	.608	.607	.606	.604	.602	.600	.597	.593	.589	.585	.579	.574	.567	.560	.553	.545	.536
4.0*	.610	.609	.609	.608	.608	.606	.605	.603	.601	.598	.595	.591	.587	.583	.577	.571	.565	.558	.551
4.2*	.610	.610	.609	.609	.608	.608	.607	.606	.604	.602	.599	.597	.593	.589	.585	.580	.575	.569	.563
4.4*	.610	.610	.610	.609	.609	.609	.608	.607	.606	.604	.603	.600	.598	.595	.591	.587	.583	.578	.573
4.6*	.610	.610	.610	.610	.609	.609	.609	.608	.607	.606	.605	.603	.601	.599	.596	.593	.589	.585	.581
4.8*	.610	.610	.610	.610	.610	.609	.609	.609	.608	.608	.607	.605	.604	.602	.600	.597	.595	.591	.588
5.0*	.610	.610	.610	.610	.610	.610	.610	.610	.609	.608	.608	.607	.606	.604	.603	.601	.598	.596	.593
5.5*	.610	.610	.610	.610	.610	.610	.610	.610	.610	.609	.609	.609	.608	.608	.607	.606	.605	.603	.601
6.0*	.610	.610	.610	.610	.610	.610	.610	.610	.610	.610	.610	.610	.609	.609	.609	.608	.608	.607	.606
6.5*	.610	.610	.610	.610	.610	.610	.610	.610	.610	.610	.610	.610	.610	.610	.609	.609	.609	.609	.608
7.0*	.610	.610	.610	.610	.610	.610	.610	.610	.610	.610	.610	.610	.610	.610	.610	.610	.610	.609	.609

WEIGHT= .200

	.100	.150	.200	.250	.300	.350	.400	.450	.500	.550	.600	.650	.700	.750	.800	.850	.900	.950	1.000
.0*	.486	.411	.347	.292	.244	.201	.164	.132	.105	.081	.061	.044	.031	.020	.012	.006	.003	.001	.000
.1*	.487	.412	.349	.293	.245	.203	.166	.134	.106	.083	.062	.046	.032	.022	.014	.008	.004	.002	.001
.2*	.490	.415	.352	.297	.250	.208	.171	.139	.111	.088	.068	.051	.037	.026	.018	.012	.008	.006	.005
.3*	.494	.421	.358	.304	.257	.215	.179	.147	.120	.096	.076	.059	.045	.034	.026	.019	.015	.012	.010
.4*	.499	.428	.367	.314	.267	.226	.190	.159	.131	.108	.088	.071	.056	.045	.036	.029	.024	.020	.018
.5*	.506	.437	.377	.325	.280	.239	.204	.173	.146	.122	.102	.085	.070	.058	.049	.041	.035	.031	.028
.6*	.515	.447	.389	.339	.295	.255	.221	.190	.163	.140	.119	.102	.087	.074	.064	.056	.049	.044	.040
.7*	.524	.459	.404	.355	.312	.273	.239	.209	.183	.159	.139	.121	.106	.093	.082	.073	.065	.059	.054
.8*	.535	.473	.419	.372	.331	.293	.260	.231	.205	.181	.161	.142	.127	.113	.101	.091	.083	.076	.070
.9*	.546	.487	.436	.391	.351	.315	.283	.254	.228	.205	.184	.166	.150	.135	.123	.112	.102	.094	.087
1.0*	.557	.502	.454	.411	.373	.338	.307	.279	.253	.230	.210	.191	.174	.159	.145	.134	.123	.114	.106
1.2*	.582	.534	.491	.453	.418	.387	.358	.331	.306	.284	.263	.243	.226	.209	.194	.181	.169	.158	.148
1.4*	.606	.565	.529	.496	.465	.436	.410	.385	.361	.339	.318	.298	.280	.263	.246	.231	.217	.204	.193
1.6*	.629	.596	.565	.537	.510	.485	.460	.437	.415	.393	.373	.353	.334	.316	.299	.283	.268	.253	.240
1.8*	.650	.623	.598	.575	.552	.529	.508	.486	.466	.445	.425	.406	.387	.369	.351	.334	.318	.303	.288
2.0*	.668	.648	.628	.608	.589	.570	.550	.531	.512	.493	.474	.455	.437	.419	.401	.384	.367	.351	.336
2.2*	.683	.667	.652	.636	.620	.604	.588	.571	.553	.536	.518	.500	.483	.465	.448	.431	.414	.398	.382
2.4*	.695	.683	.672	.659	.647	.633	.619	.604	.589	.573	.557	.541	.524	.508	.491	.474	.458	.442	.426
2.6*	.704	.696	.687	.678	.667	.656	.645	.632	.619	.605	.591	.576	.561	.545	.530	.514	.498	.482	.467
2.8*	.710	.704	.698	.691	.683	.675	.665	.655	.644	.632	.619	.606	.592	.578	.564	.549	.534	.519	.504
3.0*	.715	.711	.706	.701	.695	.689	.681	.673	.663	.653	.643	.631	.619	.607	.594	.580	.566	.552	.538
3.2*	.717	.715	.712	.708	.704	.699	.693	.686	.679	.671	.662	.652	.642	.630	.619	.607	.594	.582	.569
3.4*	.719	.718	.716	.713	.710	.706	.702	.697	.691	.684	.677	.669	.660	.650	.640	.630	.619	.607	.595
3.6*	.720	.720	.718	.717	.714	.712	.708	.705	.700	.695	.689	.682	.675	.667	.658	.649	.639	.629	.618
3.8*	.721	.721	.720	.719	.717	.715	.713	.710	.707	.703	.698	.692	.686	.680	.672	.665	.656	.647	.638
4.0*	.722	.721	.721	.720	.719	.718	.716	.714	.711	.708	.705	.700	.696	.690	.684	.678	.670	.663	.654
4.2*	.722	.722	.721	.721	.720	.719	.718	.717	.715	.713	.710	.707	.703	.698	.693	.688	.682	.675	.668
4.4*	.722	.722	.722	.721	.721	.720	.720	.719	.717	.716	.714	.711	.708	.705	.701	.696	.691	.686	.680
4.6*	.722	.722	.722	.722	.721	.721	.721	.720	.719	.718	.716	.714	.712	.709	.706	.703	.699	.694	.689
4.8*	.722	.722	.722	.722	.722	.721	.721	.721	.720	.719	.718	.717	.715	.713	.711	.708	.704	.701	.697
5.0*	.722	.722	.722	.722	.722	.722	.721	.721	.721	.720	.719	.718	.717	.716	.714	.712	.709	.706	.703
5.5*	.722	.722	.722	.722	.722	.722	.722	.722	.722	.721	.721	.721	.720	.719	.719	.717	.716	.714	.712
6.0*	.722	.722	.722	.722	.722	.722	.722	.722	.722	.722	.722	.722	.721	.721	.721	.720	.719	.719	.718
6.5*	.722	.722	.722	.722	.722	.722	.722	.722	.722	.722	.722	.722	.722	.722	.721	.721	.721	.721	.720
7.0*	.722	.722	.722	.722	.722	.722	.722	.722	.722	.722	.722	.722	.722	.722	.722	.722	.722	.721	.721

WEIGHT= .250

	.100	.150	.200	.250	.300	.350	.400	.450	.500	.550	.600	.650	.700	.750	.800	.850	.900	.950	1.000
.0*	.541	.456	.385	.324	.270	.224	.183	.148	.117	.091	.069	.050	.035	.023	.014	.007	.003	.001	.000
.1*	.542	.458	.387	.325	.272	.226	.185	.150	.119	.093	.070	.052	.037	.025	.016	.009	.005	.002	.001
.2*	.545	.461	.391	.330	.277	.231	.190	.155	.125	.098	.076	.058	.042	.030	.021	.014	.009	.007	.005
.3*	.549	.467	.397	.337	.285	.239	.199	.164	.134	.108	.085	.067	.051	.039	.029	.022	.017	.014	.012
.4*	.555	.475	.407	.348	.296	.251	.212	.177	.147	.121	.098	.080	.064	.051	.041	.033	.027	.024	.021
.5*	.563	.485	.418	.361	.310	.266	.227	.193	.163	.137	.115	.096	.080	.066	.056	.047	.041	.036	.033
.6*	.573	.497	.432	.376	.327	.283	.245	.211	.182	.156	.134	.114	.098	.085	.073	.064	.057	.051	.047
.7*	.583	.510	.448	.394	.346	.304	.266	.233	.204	.178	.156	.136	.120	.105	.093	.083	.075	.068	.063
.8*	.595	.525	.465	.413	.367	.326	.289	.257	.228	.203	.180	.160	.143	.128	.115	.105	.095	.088	.082
.9*	.608	.541	.484	.434	.390	.350	.315	.283	.254	.229	.207	.186	.169	.153	.140	.128	.118	.109	.102
1.0*	.621	.558	.504	.457	.414	.376	.341	.310	.282	.257	.235	.214	.196	.180	.165	.153	.141	.132	.123
1.2*	.649	.594	.546	.504	.465	.430	.398	.369	.342	.317	.295	.274	.254	.237	.221	.206	.193	.181	.171
1.4*	.677	.630	.589	.552	.517	.486	.456	.429	.403	.379	.356	.335	.315	.297	.279	.263	.248	.234	.222
1.6*	.704	.665	.630	.598	.568	.540	.513	.488	.464	.440	.418	.397	.377	.357	.339	.321	.305	.290	.275
1.8*	.728	.697	.663	.642	.616	.591	.567	.544	.521	.499	.477	.456	.436	.416	.397	.379	.362	.345	.330
2.0*	.749	.724	.702	.680	.658	.637	.615	.594	.573	.553	.532	.512	.492	.472	.453	.435	.417	.400	.383
2.2*	.766	.748	.730	.712	.694	.676	.658	.639	.620	.601	.582	.563	.544	.525	.506	.488	.470	.452	.435
2.4*	.779	.766	.753	.739	.724	.709	.694	.677	.660	.643	.626	.608	.590	.572	.554	.536	.518	.501	.484
2.6*	.790	.780	.770	.760	.748	.736	.723	.709	.695	.679	.664	.648	.631	.614	.597	.580	.563	.546	.530
2.8*	.797	.791	.783	.776	.767	.757	.746	.735	.722	.709	.696	.681	.666	.651	.636	.620	.604	.588	.572
3.0*	.803	.798	.793	.787	.781	.773	.764	.755	.745	.734	.722	.710	.696	.683	.669	.654	.639	.624	.609
3.2*	.806	.803	.800	.796	.791	.785	.778	.771	.763	.753	.743	.733	.721	.710	.697	.684	.671	.657	.643
3.4*	.808	.806	.804	.801	.798	.793	.788	.783	.776	.769	.761	.752	.742	.732	.721	.709	.697	.685	.672
3.6*	.810	.808	.807	.805	.803	.800	.796	.791	.786	.781	.774	.767	.759	.750	.740	.730	.720	.709	.697
3.8*	.810	.810	.809	.807	.806	.804	.801	.798	.794	.789	.784	.778	.772	.764	.757	.748	.739	.729	.719
4.0*	.811	.810	.810	.809	.808	.806	.805	.802	.799	.796	.792	.787	.782	.776	.769	.762	.755	.746	.737
4.2*	.811	.811	.810	.810	.809	.808	.807	.805	.803	.801	.798	.794	.790	.785	.780	.774	.767	.760	.753
4.4*	.811	.811	.811	.811	.810	.810	.809	.808	.806	.804	.802	.799	.796	.792	.788	.783	.778	.772	.765
4.6*	.811	.811	.811	.811	.811	.810	.810	.809	.808	.807	.805	.803	.800	.797	.794	.790	.786	.781	.775
4.8*	.811	.811	.811	.811	.811	.811	.810	.810	.809	.808	.807	.805	.804	.801	.799	.796	.792	.788	.784
5.0*	.811	.811	.811	.811	.811	.811	.811	.810	.810	.809	.808	.807	.806	.804	.802	.800	.797	.794	.790
5.5*	.811	.811	.811	.811	.811	.811	.811	.810	.810	.809	.808	.807	.806	.804	.802	.800	.797	.794	.790
6.0*	.811	.811	.811	.811	.811	.811	.811	.811	.811	.811	.810	.810	.809	.809	.808	.806	.805	.803	.801
6.5*	.811	.811	.811	.811	.811	.811	.811	.811	.811	.811	.811	.811	.811	.810	.810	.809	.809	.808	.807
7.0*	.811	.811	.811	.811	.811	.811	.811	.811	.811	.811	.811	.811	.811	.811	.811	.811	.810	.810	.809

WEIGHT= .300

	.100	.150	.200	.250	.300	.350	.400	.450	.500	.550	.600	.650	.700	.750	.800	.850	.900	.950	1.000
.0*	.581	.489	.413	.347	.290	.240	.196	.159	.126	.098	.074	.054	.038	.025	.015	.008	.003	.001	.000
.1*	.582	.491	.414	.348	.291	.242	.198	.161	.128	.100	.076	.056	.040	.027	.017	.010	.005	.002	.002
.2*	.585	.495	.419	.353	.297	.247	.204	.167	.134	.106	.082	.062	.046	.033	.023	.015	.010	.007	.006
.3*	.590	.501	.426	.361	.305	.256	.214	.176	.144	.116	.092	.072	.056	.043	.032	.024	.019	.015	.013
.4*	.597	.509	.436	.372	.317	.269	.227	.190	.158	.130	.106	.086	.069	.056	.045	.036	.030	.026	.024
.5*	.606	.520	.448	.386	.332	.285	.243	.207	.175	.147	.124	.103	.086	.072	.061	.052	.045	.040	.037
.6*	.616	.533	.463	.403	.350	.304	.263	.227	.195	.168	.144	.124	.107	.092	.080	.070	.063	.057	.053
.7*	.627	.548	.480	.422	.370	.325	.285	.250	.219	.192	.168	.147	.130	.115	.102	.091	.083	.076	.071
.8*	.640	.564	.499	.443	.393	.349	.310	.276	.245	.218	.194	.173	.155	.140	.126	.115	.105	.097	.091
.9*	.654	.582	.520	.466	.418	.375	.337	.303	.273	.247	.223	.201	.183	.167	.153	.140	.130	.120	.113
1.0*	.669	.600	.541	.490	.444	.403	.366	.333	.304	.277	.253	.231	.213	.196	.181	.167	.156	.146	.137
1.2*	.699	.639	.587	.541	.499	.462	.427	.396	.368	.342	.318	.296	.276	.258	.241	.226	.212	.200	.189
1.4*	.730	.679	.634	.593	.556	.522	.491	.461	.434	.409	.385	.363	.342	.323	.305	.288	.273	.258	.245
1.6*	.760	.717	.679	.644	.612	.582	.553	.526	.500	.476	.452	.430	.409	.389	.370	.352	.335	.319	.304
1.8*	.787	.753	.721	.692	.664	.637	.612	.587	.563	.539	.517	.495	.473	.453	.433	.414	.396	.379	.363
2.0*	.810	.783	.758	.734	.711	.688	.665	.642	.620	.598	.577	.555	.534	.514	.494	.475	.456	.438	.421
2.2*	.830	.809	.790	.770	.751	.731	.711	.691	.672	.651	.631	.611	.590	.571	.551	.532	.513	.495	.477
2.4*	.845	.830	.815	.800	.784	.768	.751	.733	.715	.697	.679	.660	.641	.622	.603	.584	.566	.548	.530
2.6*	.857	.846	.835	.823	.811	.797	.783	.768	.753	.737	.720	.703	.685	.668	.650	.632	.614	.597	.579
2.8*	.865	.858	.850	.841	.831	.821	.809	.797	.784	.770	.755	.740	.724	.708	.691	.675	.658	.641	.624
3.0*	.871	.866	.861	.854	.847	.838	.829	.819	.808	.796	.784	.770	.757	.742	.727	.712	.697	.681	.665
3.2*	.875	.872	.868	.863	.858	.852	.845	.837	.828	.818	.807	.796	.784	.771	.758	.744	.730	.716	.701
3.4*	.878	.876	.873	.870	.866	.861	.856	.850	.843	.835	.826	.816	.806	.795	.784	.772	.759	.746	.732
3.6*	.879	.878	.876	.874	.872	.868	.864	.859	.854	.848	.840	.833	.824	.815	.805	.794	.783	.772	.760
3.8*	.880	.879	.878	.877	.875	.873	.870	.866	.862	.857	.852	.845	.838	.831	.822	.813	.804	.793	.783
4.0*	.881	.880	.880	.879	.878	.876	.874	.871	.868	.864	.860	.855	.849	.843	.836	.829	.820	.812	.802
4.2*	.881	.881	.880	.880	.879	.878	.877	.875	.872	.870	.866	.863	.858	.853	.847	.841	.834	.827	.819
4.4*	.881	.881	.881	.880	.880	.879	.878	.877	.875	.873	.871	.868	.865	.861	.856	.851	.845	.839	.832
4.6*	.881	.881	.881	.881	.881	.880	.880	.879	.878	.876	.874	.872	.869	.866	.863	.858	.854	.849	.843
4.8*	.881	.881	.881	.881	.881	.881	.880	.880	.879	.878	.877	.875	.873	.871	.868	.864	.861	.856	.852
5.0*	.881	.881	.881	.881	.881	.881	.881	.880	.880	.879	.878	.877	.875	.874	.872	.869	.866	.863	.859
5.5*	.881	.881	.881	.881	.881	.881	.881	.881	.881	.881	.880	.880	.879	.878	.877	.876	.874	.873	.870
6.0*	.881	.881	.881	.881	.881	.881	.881	.881	.881	.881	.881	.881	.881	.880	.880	.879	.878	.877	.876
6.5*	.881	.881	.881	.881	.881	.881	.881	.881	.881	.881	.881	.881	.881	.881	.881	.880	.880	.880	.879
7.0*	.881	.881	.881	.881	.881	.881	.881	.881	.881	.881	.881	.881	.881	.881	.881	.881	.881	.881	.880

WEIGHT= .350

	.100	.150	.200	.250	.300	.350	.400	.450	.500	.550	.600	.650	.700	.750	.800	.850	.900	.950	1.000
.0*	.608	.512	.431	.362	.302	.250	.205	.166	.132	.103	.078	.057	.040	.027	.016	.009	.004	.001	.000
.1*	.610	.513	.432	.364	.304	.252	.207	.168	.134	.105	.080	.060	.042	.029	.018	.011	.006	.003	.002
.2*	.613	.517	.437	.369	.309	.258	.213	.174	.140	.111	.087	.066	.049	.035	.024	.016	.011	.008	.007
.3*	.618	.524	.445	.377	.318	.268	.223	.184	.151	.122	.097	.076	.059	.045	.034	.026	.020	.016	.015
.4*	.625	.533	.455	.389	.331	.281	.237	.198	.165	.136	.111	.091	.073	.059	.048	.039	.032	.028	.026
.5*	.635	.544	.468	.403	.346	.297	.254	.216	.183	.154	.130	.109	.091	.076	.065	.055	.048	.043	.040
.6*	.645	.558	.484	.421	.365	.317	.274	.237	.204	.176	.151	.130	.112	.097	.085	.075	.067	.061	.057
.7*	.658	.573	.502	.440	.387	.339	.298	.261	.229	.201	.176	.155	.137	.121	.108	.097	.088	.081	.076
.8*	.672	.591	.522	.462	.410	.364	.324	.288	.256	.228	.204	.182	.164	.148	.134	.122	.112	.104	.098
.9*	.687	.609	.544	.486	.436	.392	.352	.317	.286	.258	.234	.212	.193	.176	.162	.149	.138	.129	.122
1.0*	.702	.629	.567	.512	.464	.421	.383	.348	.318	.290	.266	.244	.224	.207	.192	.178	.166	.156	.147
1.2*	.735	.671	.615	.566	.522	.483	.447	.415	.385	.359	.334	.312	.291	.273	.256	.241	.227	.214	.203
1.4*	.769	.714	.665	.622	.583	.547	.514	.484	.456	.430	.405	.383	.361	.342	.324	.307	.291	.276	.263
1.6*	.801	.755	.714	.677	.642	.610	.580	.552	.526	.500	.476	.454	.432	.412	.392	.374	.357	.340	.325
1.8*	.830	.793	.759	.728	.698	.670	.643	.617	.592	.568	.545	.522	.501	.480	.460	.440	.422	.405	.388
2.0*	.856	.826	.799	.773	.748	.724	.700	.677	.654	.631	.609	.587	.565	.544	.524	.505	.486	.467	.450
2.2*	.877	.855	.833	.813	.792	.771	.750	.729	.708	.687	.667	.646	.625	.605	.585	.565	.546	.527	.509
2.4*	.894	.877	.861	.845	.828	.811	.793	.774	.756	.737	.717	.698	.679	.659	.640	.621	.602	.583	.565
2.6*	.907	.895	.883	.870	.857	.843	.828	.812	.796	.779	.762	.744	.726	.708	.689	.671	.653	.635	.617
2.8*	.916	.908	.899	.890	.879	.868	.856	.843	.829	.814	.799	.783	.767	.750	.733	.716	.699	.682	.665
3.0*	.923	.917	.911	.904	.896	.887	.878	.867	.855	.843	.830	.816	.802	.787	.771	.755	.740	.723	.707
3.2*	.927	.924	.919	.914	.909	.902	.894	.886	.876	.866	.855	.843	.831	.817	.804	.789	.775	.760	.745
3.4*	.930	.928	.925	.921	.917	.912	.906	.900	.892	.884	.875	.865	.854	.843	.831	.818	.805	.792	.778
3.6*	.932	.930	.929	.926	.923	.920	.915	.910	.904	.898	.890	.882	.873	.864	.853	.842	.831	.819	.807
3.8*	.933	.932	.931	.929	.927	.925	.922	.918	.913	.908	.902	.896	.888	.880	.872	.862	.852	.842	.831
4.0*	.933	.933	.932	.931	.930	.928	.926	.923	.920	.916	.911	.906	.900	.894	.886	.879	.870	.861	.851
4.2*	.934	.933	.933	.932	.932	.930	.929	.927	.925	.922	.918	.914	.909	.904	.898	.892	.884	.877	.868
4.4*	.934	.934	.934	.933	.933	.932	.931	.929	.928	.926	.923	.920	.916	.912	.907	.902	.896	.890	.883
4.6*	.934	.934	.934	.934	.933	.933	.932	.931	.930	.928	.927	.924	.921	.918	.914	.910	.905	.900	.894
4.8*	.934	.934	.934	.934	.934	.933	.933	.932	.931	.930	.929	.927	.925	.923	.920	.916	.912	.908	.903
5.0*	.934	.934	.934	.934	.934	.934	.933	.933	.932	.932	.931	.929	.928	.926	.924	.921	.918	.914	.910
5.5*	.934	.934	.934	.934	.934	.934	.934	.934	.934	.933	.933	.932	.932	.931	.930	.928	.927	.925	.923
6.0*	.934	.934	.934	.934	.934	.934	.934	.934	.934	.934	.934	.934	.933	.933	.932	.932	.931	.930	.929
6.5*	.934	.934	.934	.934	.934	.934	.934	.934	.934	.934	.934	.934	.934	.934	.933	.933	.933	.932	.932
7.0*	.934	.934	.934	.934	.934	.934	.934	.934	.934	.934	.934	.934	.934	.934	.934	.934	.934	.933	.933

WEIGHT= .400

	.100	.150	.200	.250	.300	.350	.400	.450	.500	.550	.600	.650	.700	.750	.800	.850	.900	.950	1.000
.0*	.625	.524	.440	.370	.309	.256	.210	.170	.135	.105	.080	.059	.042	.028	.017	.009	.004	.001	.000
.1*	.626	.525	.442	.371	.310	.258	.212	.172	.137	.108	.082	.061	.044	.030	.019	.011	.006	.003	.002
.2*	.629	.529	.447	.377	.316	.263	.218	.178	.144	.114	.089	.068	.050	.036	.025	.017	.011	.008	.007
.3*	.635	.536	.455	.385	.325	.273	.228	.188	.154	.125	.100	.078	.061	.047	.035	.027	.021	.017	.015
.4*	.642	.546	.466	.397	.338	.287	.242	.203	.168	.139	.114	.093	.075	.061	.049	.040	.034	.030	.027
.5*	.652	.558	.479	.412	.354	.304	.259	.221	.187	.158	.133	.112	.094	.079	.067	.057	.050	.045	.042
.6*	.663	.572	.495	.430	.373	.324	.280	.242	.209	.180	.155	.134	.115	.100	.088	.078	.070	.064	.060
.7*	.676	.588	.514	.450	.395	.347	.304	.267	.234	.205	.180	.159	.140	.125	.112	.101	.092	.085	.080
.8*	.691	.606	.535	.473	.420	.373	.331	.294	.262	.234	.209	.187	.168	.152	.139	.127	.117	.109	.103
.9*	.706	.625	.557	.498	.446	.401	.360	.324	.293	.265	.240	.218	.199	.182	.168	.155	.144	.135	.128
1.0*	.723	.646	.581	.525	.475	.431	.391	.357	.325	.298	.273	.251	.231	.214	.199	.185	.174	.163	.155
1.2*	.758	.690	.632	.581	.536	.495	.458	.425	.395	.368	.344	.321	.301	.282	.265	.250	.236	.224	.213
1.4*	.793	.735	.684	.639	.599	.562	.528	.497	.468	.442	.417	.395	.373	.354	.336	.319	.303	.289	.276
1.6*	.827	.779	.736	.697	.661	.628	.597	.568	.541	.516	.491	.468	.447	.426	.407	.389	.372	.356	.341
1.8*	.858	.819	.783	.750	.720	.691	.663	.636	.611	.586	.563	.540	.518	.497	.477	.458	.440	.422	.406
2.0*	.886	.855	.826	.799	.773	.747	.723	.699	.675	.652	.629	.607	.586	.565	.544	.525	.506	.487	.470
2.2*	.909	.885	.862	.840	.819	.797	.776	.754	.733	.711	.690	.669	.648	.627	.607	.587	.568	.550	.532
2.4*	.927	.909	.892	.875	.857	.839	.821	.802	.782	.763	.743	.724	.704	.684	.665	.645	.626	.608	.590
2.6*	.941	.928	.916	.902	.888	.873	.858	.842	.825	.807	.790	.772	.753	.735	.716	.698	.680	.661	.643
2.8*	.951	.943	.933	.923	.912	.900	.887	.874	.860	.845	.829	.813	.796	.779	.762	.745	.727	.710	.693
3.0*	.959	.953	.945	.939	.930	.921	.911	.900	.888	.875	.861	.847	.832	.817	.801	.786	.769	.753	.737
3.2*	.963	.959	.955	.950	.943	.936	.928	.919	.910	.899	.888	.875	.863	.849	.835	.821	.806	.791	.776
3.4*	.967	.964	.961	.957	.953	.948	.941	.935	.927	.918	.909	.898	.887	.876	.864	.851	.838	.824	.810
3.6*	.968	.967	.965	.962	.959	.956	.951	.946	.940	.933	.925	.917	.907	.897	.887	.876	.864	.852	.839
3.8*	.970	.969	.967	.966	.964	.961	.958	.954	.949	.944	.938	.931	.923	.915	.906	.897	.886	.876	.865
4.0*	.970	.970	.969	.968	.967	.965	.962	.959	.956	.952	.947	.942	.936	.929	.921	.913	.905	.896	.886
4.2*	.971	.970	.970	.969	.968	.967	.965	.963	.961	.958	.954	.950	.945	.940	.934	.927	.920	.912	.903
4.4*	.971	.971	.970	.970	.969	.969	.968	.966	.964	.962	.959	.956	.952	.948	.943	.938	.932	.925	.918
4.6*	.971	.971	.971	.970	.970	.970	.969	.968	.967	.965	.963	.961	.958	.954	.950	.946	.941	.936	.930
4.8*	.971	.971	.971	.971	.970	.970	.970	.969	.968	.967	.966	.964	.962	.959	.956	.953	.949	.944	.939
5.0*	.971	.971	.971	.971	.971	.971	.970	.970	.969	.968	.967	.966	.965	.963	.960	.958	.954	.951	.947
5.5*	.971	.971	.971	.971	.971	.971	.971	.971	.970	.970	.970	.969	.969	.968	.967	.965	.963	.961	.959
6.0*	.971	.971	.971	.971	.971	.971	.971	.971	.971	.971	.971	.970	.970	.970	.969	.969	.968	.967	.966
6.5*	.971	.971	.971	.971	.971	.971	.971	.971	.971	.971	.971	.971	.971	.971	.970	.970	.970	.969	.969
7.0*	.971	.971	.971	.971	.971	.971	.971	.971	.971	.971	.971	.971	.971	.971	.971	.971	.971	.970	.970

WEIGHT= .450

	.100	.150	.200	.250	.300	.350	.400	.450	.500	.550	.600	.650	.700	.750	.800	.850	.900	.950	1.000
.0*	.630	.527	.442	.371	.309	.256	.210	.170	.136	.106	.081	.060	.042	.028	.017	.009	.004	.001	.000
.1*	.631	.528	.444	.373	.311	.258	.212	.172	.138	.108	.083	.062	.044	.030	.019	.011	.006	.003	.002
.2*	.635	.533	.449	.378	.317	.264	.218	.179	.144	.115	.089	.068	.051	.037	.026	.017	.012	.008	.007
.3*	.640	.540	.457	.386	.326	.274	.228	.189	.155	.125	.100	.079	.061	.047	.036	.027	.021	.018	.016
.4*	.648	.549	.468	.399	.339	.287	.242	.203	.169	.140	.115	.094	.076	.062	.050	.041	.035	.030	.028
.5*	.658	.561	.482	.414	.355	.304	.260	.221	.188	.159	.134	.112	.094	.080	.068	.058	.051	.046	.043
.6*	.670	.576	.499	.431	.374	.324	.281	.243	.210	.181	.156	.135	.116	.101	.089	.079	.071	.066	.062
.7*	.683	.592	.517	.452	.397	.348	.305	.268	.235	.206	.181	.160	.142	.126	.113	.103	.094	.088	.083
.8*	.698	.611	.533	.476	.421	.374	.332	.295	.263	.235	.210	.189	.170	.154	.141	.129	.120	.112	.106
.9*	.714	.631	.561	.501	.448	.402	.362	.326	.294	.266	.241	.220	.201	.184	.170	.158	.148	.139	.131
1.0*	.731	.652	.585	.528	.477	.433	.393	.358	.327	.299	.275	.253	.233	.217	.202	.189	.177	.168	.159
1.2*	.767	.697	.638	.586	.539	.498	.461	.428	.398	.371	.347	.324	.304	.286	.270	.255	.242	.230	.219
1.4*	.804	.744	.692	.645	.604	.566	.532	.501	.473	.446	.422	.399	.379	.359	.342	.325	.310	.296	.283
1.6*	.840	.789	.745	.705	.668	.634	.603	.574	.547	.522	.498	.475	.454	.434	.415	.397	.380	.364	.350
1.8*	.873	.832	.795	.761	.729	.699	.671	.644	.619	.594	.571	.548	.527	.506	.486	.468	.450	.433	.416
2.0*	.902	.869	.839	.811	.784	.758	.733	.709	.685	.662	.639	.617	.596	.575	.555	.536	.517	.499	.482
2.2*	.926	.901	.877	.855	.832	.810	.788	.766	.745	.723	.702	.681	.660	.640	.620	.600	.581	.563	.545
2.4*	.946	.927	.909	.891	.872	.854	.835	.816	.796	.777	.757	.737	.718	.698	.679	.659	.641	.622	.604
2.6*	.961	.947	.934	.920	.905	.890	.874	.857	.840	.823	.805	.787	.768	.750	.732	.713	.695	.677	.659
2.8*	.972	.962	.952	.942	.930	.918	.905	.891	.876	.861	.845	.829	.812	.795	.778	.761	.744	.726	.709
3.0*	.979	.973	.967	.958	.950	.940	.929	.918	.906	.893	.879	.865	.850	.834	.819	.803	.787	.770	.754
3.2*	.985	.980	.976	.970	.964	.956	.948	.939	.929	.918	.906	.894	.881	.867	.853	.839	.824	.809	.794
3.4*	.988	.985	.982	.978	.974	.968	.962	.955	.947	.938	.928	.918	.907	.895	.883	.870	.856	.843	.829
3.6*	.990	.988	.986	.984	.980	.976	.972	.966	.960	.953	.945	.936	.927	.917	.907	.895	.884	.871	.859
3.8*	.991	.990	.989	.987	.985	.982	.979	.975	.970	.964	.958	.951	.943	.935	.926	.917	.906	.896	.885
4.0*	.992	.991	.991	.989	.988	.986	.984	.981	.977	.973	.968	.962	.956	.949	.942	.934	.925	.916	.906
4.2*	.992	.992	.992	.991	.990	.989	.987	.985	.982	.979	.975	.971	.966	.961	.954	.948	.940	.932	.924
4.4*	.993	.992	.992	.992	.991	.990	.989	.988	.986	.983	.981	.977	.974	.969	.964	.959	.953	.946	.939
4.6*	.993	.993	.992	.992	.992	.991	.991	.990	.988	.987	.985	.982	.979	.976	.972	.967	.962	.957	.951
4.8*	.993	.993	.993	.992	.992	.992	.992	.991	.990	.989	.987	.985	.983	.981	.977	.974	.970	.965	.960
5.0*	.993	.993	.993	.993	.993	.992	.992	.992	.991	.990	.989	.988	.986	.984	.982	.979	.976	.972	.968
5.5*	.993	.993	.993	.993	.993	.993	.993	.992	.992	.992	.992	.991	.990	.989	.988	.987	.985	.983	.981
6.0*	.993	.993	.993	.993	.993	.993	.993	.993	.993	.992	.992	.992	.992	.992	.991	.990	.990	.989	.987
6.5*	.993	.993	.993	.993	.993	.993	.993	.993	.993	.993	.993	.993	.993	.992	.992	.992	.991	.991	.990
7.0*	.993	.993	.993	.993	.993	.993	.993	.993	.993	.993	.993	.993	.993	.993	.993	.992	.992	.992	.992

WEIGHT= .500

	.100	.150	.200	.250	.300	.350	.400	.450	.500	.550	.600	.650	.700	.750	.800	.850	.900	.950	1.000
.0*	.625	.521	.437	.366	.305	.252	.207	.168	.134	.105	.080	.059	.042	.028	.017	.009	.004	.001	.000
.1*	.626	.523	.438	.367	.307	.254	.209	.170	.136	.107	.082	.061	.044	.030	.019	.011	.006	.003	.002
.2*	.630	.527	.443	.373	.312	.260	.215	.176	.142	.113	.088	.068	.050	.036	.025	.017	.012	.008	.007
.3*	.636	.534	.451	.381	.321	.270	.225	.186	.152	.124	.099	.078	.061	.047	.036	.027	.021	.018	.016
.4*	.644	.544	.462	.393	.334	.283	.239	.200	.167	.138	.113	.093	.075	.061	.050	.041	.035	.031	.028
.5*	.654	.556	.476	.408	.350	.300	.256	.218	.185	.156	.132	.111	.093	.079	.067	.058	.052	.047	.044
.6*	.665	.571	.492	.426	.369	.320	.277	.239	.207	.178	.154	.133	.115	.101	.089	.079	.071	.066	.062
.7*	.679	.587	.511	.447	.391	.343	.301	.264	.232	.204	.179	.158	.140	.125	.113	.103	.094	.088	.083
.8*	.694	.606	.533	.470	.416	.369	.328	.291	.260	.232	.208	.187	.168	.153	.140	.129	.120	.113	.107
.9*	.711	.626	.556	.495	.443	.397	.357	.322	.290	.263	.239	.218	.199	.183	.170	.158	.148	.140	.133
1.0*	.728	.648	.580	.522	.472	.428	.389	.354	.323	.296	.272	.251	.232	.216	.201	.189	.178	.169	.161
1.2*	.765	.694	.633	.581	.534	.493	.457	.424	.394	.368	.344	.322	.303	.285	.269	.255	.243	.231	.221
1.4*	.803	.741	.688	.641	.600	.562	.528	.497	.469	.443	.419	.397	.377	.359	.342	.326	.311	.298	.286
1.6*	.840	.788	.742	.702	.665	.631	.600	.571	.544	.519	.496	.474	.453	.433	.415	.398	.382	.367	.353
1.8*	.874	.831	.793	.759	.727	.697	.669	.642	.617	.593	.570	.548	.527	.507	.487	.469	.452	.436	.420
2.0*	.904	.870	.839	.811	.783	.757	.732	.708	.684	.661	.639	.618	.597	.577	.557	.538	.520	.503	.486
2.2*	.930	.903	.879	.856	.833	.810	.788	.767	.745	.724	.703	.682	.661	.641	.622	.603	.584	.567	.549
2.4*	.950	.931	.912	.893	.874	.856	.837	.817	.798	.778	.759	.739	.720	.701	.682	.663	.644	.627	.609
2.6*	.966	.952	.938	.923	.908	.893	.877	.860	.843	.825	.808	.790	.772	.753	.735	.717	.699	.682	.664
2.8*	.978	.968	.957	.946	.935	.922	.909	.895	.880	.865	.849	.833	.816	.800	.783	.766	.749	.732	.715
3.0*	.986	.979	.972	.964	.955	.945	.934	.923	.910	.897	.883	.869	.854	.839	.824	.808	.792	.776	.760
3.2*	.991	.987	.982	.976	.969	.962	.953	.944	.934	.923	.911	.899	.886	.873	.859	.844	.830	.815	.800
3.4*	.995	.992	.989	.985	.980	.974	.968	.960	.952	.943	.934	.923	.912	.900	.888	.875	.862	.849	.835
3.6*	.997	.995	.993	.990	.987	.983	.978	.972	.966	.959	.951	.942	.933	.923	.912	.901	.890	.878	.865
3.8*	.998	.997	.996	.994	.992	.989	.985	.981	.976	.971	.964	.957	.950	.941	.932	.923	.913	.902	.891
4.0*	.999	.999	.998	.997	.995	.993	.990	.987	.984	.979	.975	.969	.963	.956	.948	.940	.932	.922	.913
4.2*	1.000	.999	.999	.998	.997	.996	.994	.992	.989	.986	.982	.978	.973	.967	.961	.954	.947	.939	.931
4.4*	1.000	1.000	.999	.999	.998	.997	.996	.995	.993	.990	.988	.984	.980	.976	.971	.965	.959	.953	.946
4.6*	1.000	1.000	1.000	.999	.999	.998	.998	.997	.995	.994	.992	.989	.986	.983	.979	.974	.969	.964	.958
4.8*	1.000	1.000	1.000	1.000	.999	.999	.999	.998	.997	.996	.994	.992	.990	.988	.984	.981	.977	.972	.967
5.0*	1.000	1.000	1.000	1.000	1.000	1.000	.999	.999	.998	.997	.996	.995	.993	.991	.989	.986	.983	.979	.975
5.5*	1.000	1.000	1.000	1.000	1.000	1.000	1.000	1.000	.999	.999	.999	.998	.997	.997	.995	.994	.992	.990	.988
6.0*	1.000	1.000	1.000	1.000	1.000	1.000	1.000	1.000	1.000	1.000	1.000	.999	.999	.999	.998	.998	.997	.996	.994
6.5*	1.000	1.000	1.000	1.000	1.000	1.000	1.000	1.000	1.000	1.000	1.000	1.000	1.000	1.000	.999	.999	.999	.998	.998
7.0*	1.000	1.000	1.000	1.000	1.000	1.000	1.000	1.000	1.000	1.000	1.000	1.000	1.000	1.000	1.000	1.000	1.000	.999	.999

WEIGHT= .550

	.100	.150	.200	.250	.300	.350	.400	.450	.500	.550	.600	.650	.700	.750	.800	.850	.900	.950	1.000
.0*	.610	.508	.424	.355	.295	.244	.200	.162	.130	.101	.077	.057	.041	.027	.017	.009	.004	.001	.000
.1*	.612	.509	.426	.356	.297	.246	.202	.164	.132	.103	.080	.059	.043	.029	.019	.011	.006	.003	.002
.2*	.615	.513	.431	.361	.303	.252	.208	.170	.138	.110	.086	.066	.049	.035	.025	.017	.012	.008	.007
.3*	.621	.520	.438	.370	.311	.261	.218	.180	.148	.120	.096	.076	.059	.045	.035	.027	.021	.017	.016
.4*	.629	.530	.449	.382	.324	.274	.231	.194	.162	.134	.110	.090	.073	.059	.048	.040	.034	.030	.028
.5*	.639	.542	.463	.396	.339	.291	.248	.211	.179	.151	.128	.108	.091	.077	.066	.057	.051	.046	.043
.6*	.651	.556	.479	.414	.358	.310	.268	.232	.200	.173	.149	.129	.112	.098	.087	.077	.070	.065	.062
.7*	.664	.573	.499	.434	.380	.333	.292	.256	.224	.197	.174	.154	.136	.122	.110	.101	.093	.087	.083
.8*	.680	.591	.519	.457	.404	.358	.318	.283	.252	.225	.202	.181	.164	.149	.137	.126	.118	.111	.106
.9*	.696	.611	.541	.482	.431	.386	.346	.312	.282	.255	.232	.212	.194	.179	.166	.155	.146	.138	.131
1.0*	.714	.633	.566	.509	.459	.416	.377	.344	.314	.288	.265	.244	.226	.211	.197	.185	.175	.167	.159
1.2*	.751	.679	.613	.566	.521	.480	.444	.412	.384	.358	.335	.314	.296	.279	.264	.251	.239	.228	.219
1.4*	.789	.727	.673	.627	.586	.549	.515	.485	.458	.433	.410	.389	.370	.352	.336	.321	.307	.295	.283
1.6*	.827	.774	.728	.687	.651	.617	.587	.558	.532	.508	.486	.464	.445	.426	.409	.392	.377	.363	.350
1.8*	.862	.818	.780	.745	.713	.683	.656	.630	.605	.582	.559	.538	.518	.499	.481	.463	.447	.431	.416
2.0*	.893	.858	.827	.798	.770	.744	.720	.696	.673	.650	.629	.608	.588	.568	.550	.532	.514	.498	.482
2.2*	.919	.892	.867	.844	.821	.798	.776	.755	.734	.713	.693	.672	.653	.633	.615	.596	.579	.561	.545
2.4*	.940	.920	.901	.882	.863	.844	.825	.806	.787	.768	.749	.730	.711	.693	.674	.656	.638	.621	.604
2.6*	.957	.942	.928	.913	.898	.882	.866	.850	.833	.816	.798	.781	.763	.745	.728	.710	.693	.676	.659
2.8*	.969	.959	.948	.937	.925	.913	.899	.885	.871	.856	.840	.824	.808	.791	.775	.758	.742	.725	.709
3.0*	.978	.971	.964	.955	.946	.936	.925	.913	.901	.888	.875	.861	.846	.831	.816	.801	.785	.770	.754
3.2*	.984	.979	.974	.968	.961	.953	.945	.935	.925	.914	.903	.891	.878	.865	.851	.837	.823	.809	.794
3.4*	.987	.984	.981	.977	.972	.966	.959	.952	.944	.935	.925	.915	.904	.893	.881	.868	.855	.842	.829
3.6*	.990	.988	.986	.983	.979	.975	.970	.964	.958	.951	.943	.934	.925	.915	.905	.894	.883	.871	.859
3.8*	.991	.990	.989	.987	.984	.981	.978	.973	.969	.963	.957	.950	.942	.934	.925	.916	.906	.895	.885
4.0*	.992	.991	.990	.989	.988	.986	.983	.980	.976	.972	.967	.961	.955	.948	.941	.933	.925	.916	.906
4.2*	.992	.992	.991	.991	.990	.988	.986	.984	.982	.978	.975	.970	.965	.960	.954	.947	.940	.932	.924
4.4*	.993	.992	.992	.992	.991	.990	.989	.987	.985	.983	.980	.977	.973	.969	.964	.958	.952	.946	.939
4.6*	.993	.993	.992	.992	.992	.991	.990	.989	.988	.986	.984	.982	.979	.975	.971	.967	.962	.957	.951
4.8*	.993	.993	.993	.992	.992	.992	.991	.991	.990	.988	.987	.985	.983	.980	.977	.974	.970	.965	.960
5.0*	.993	.993	.993	.993	.992	.992	.992	.991	.991	.990	.989	.988	.986	.984	.982	.979	.976	.972	.968
5.5*	.993	.993	.993	.993	.993	.993	.993	.992	.992	.992	.992	.991	.990	.989	.988	.987	.985	.983	.981
6.0*	.993	.993	.993	.993	.993	.993	.993	.993	.993	.992	.992	.992	.992	.991	.991	.990	.989	.988	.987
6.5*	.993	.993	.993	.993	.993	.993	.993	.993	.993	.993	.993	.993	.993	.992	.992	.992	.991	.991	.990
7.0*	.993	.993	.993	.993	.993	.993	.993	.993	.993	.993	.993	.993	.993	.993	.993	.992	.992	.992	.992

WEIGHT= .600

	.100	.150	.200	.250	.300	.350	.400	.450	.500	.550	.600	.650	.700	.750	.800	.850	.900	.950	1.000
.0*	.586	.486	.405	.338	.281	.232	.190	.154	.123	.096	.074	.055	.039	.026	.016	.009	.004	.001	.000
.1*	.587	.487	.406	.339	.283	.234	.192	.156	.125	.098	.076	.057	.041	.028	.018	.011	.006	.003	.002
.2*	.591	.491	.411	.344	.288	.239	.198	.162	.131	.104	.082	.062	.047	.034	.024	.016	.011	.008	.007
.3*	.596	.498	.418	.352	.296	.248	.207	.171	.140	.114	.091	.072	.056	.043	.033	.026	.020	.017	.015
.4*	.604	.507	.429	.364	.308	.261	.220	.184	.154	.127	.105	.086	.070	.057	.046	.038	.033	.029	.027
.5*	.614	.519	.442	.378	.323	.276	.236	.201	.170	.144	.122	.103	.087	.073	.063	.055	.049	.044	.042
.6*	.626	.533	.458	.395	.341	.295	.255	.221	.190	.164	.142	.123	.107	.093	.083	.074	.068	.063	.060
.7*	.639	.549	.476	.415	.362	.317	.277	.243	.214	.188	.165	.147	.130	.117	.106	.097	.089	.084	.080
.8*	.654	.567	.496	.437	.385	.341	.303	.269	.240	.214	.192	.173	.156	.143	.131	.122	.114	.108	.103
.9*	.671	.587	.519	.461	.411	.368	.330	.297	.268	.243	.221	.202	.185	.171	.159	.149	.141	.134	.128
1.0*	.688	.608	.543	.487	.439	.397	.360	.328	.299	.274	.253	.233	.216	.202	.189	.179	.169	.161	.155
1.2*	.725	.654	.594	.543	.499	.460	.425	.394	.367	.343	.321	.301	.283	.269	.255	.242	.231	.222	.213
1.4*	.764	.701	.648	.603	.562	.526	.494	.465	.439	.415	.393	.373	.356	.339	.324	.310	.298	.286	.276
1.6*	.801	.748	.703	.662	.626	.594	.564	.537	.512	.489	.467	.447	.429	.412	.395	.380	.366	.353	.341
1.8*	.837	.793	.754	.720	.688	.659	.632	.607	.583	.561	.540	.520	.501	.483	.466	.450	.434	.420	.406
2.0*	.868	.833	.801	.772	.745	.720	.696	.673	.650	.629	.609	.589	.570	.551	.534	.517	.500	.485	.470
2.2*	.895	.868	.842	.819	.796	.774	.753	.732	.711	.691	.671	.652	.633	.615	.597	.580	.563	.547	.532
2.4*	.917	.896	.877	.858	.839	.820	.802	.783	.765	.746	.728	.709	.691	.673	.656	.639	.622	.606	.590
2.6*	.934	.919	.904	.889	.874	.858	.843	.827	.810	.793	.776	.759	.743	.726	.709	.692	.676	.659	.643
2.8*	.946	.936	.925	.914	.902	.889	.876	.862	.848	.833	.818	.803	.787	.771	.755	.740	.724	.708	.693
3.0*	.955	.948	.940	.932	.923	.913	.902	.891	.879	.866	.853	.839	.825	.810	.796	.781	.766	.752	.737
3.2*	.961	.956	.951	.945	.938	.931	.922	.913	.903	.892	.881	.869	.857	.844	.831	.817	.804	.790	.776
3.4*	.965	.962	.959	.954	.949	.944	.937	.930	.922	.913	.903	.893	.883	.872	.860	.848	.835	.823	.810
3.6*	.968	.966	.963	.961	.957	.953	.948	.942	.936	.929	.921	.913	.904	.894	.884	.873	.862	.851	.839
3.8*	.969	.968	.967	.965	.962	.959	.956	.951	.946	.941	.935	.928	.920	.912	.904	.895	.885	.875	.865
4.0*	.970	.969	.968	.967	.966	.963	.961	.958	.954	.950	.945	.940	.933	.927	.920	.912	.904	.895	.886
4.2*	.970	.970	.970	.969	.968	.966	.965	.962	.960	.956	.953	.948	.944	.938	.932	.926	.919	.911	.903
4.4*	.971	.971	.970	.970	.969	.968	.967	.965	.963	.961	.958	.955	.951	.947	.942	.937	.931	.925	.918
4.6*	.971	.971	.971	.970	.970	.969	.969	.967	.966	.964	.962	.960	.957	.954	.950	.945	.941	.935	.930
4.8*	.971	.971	.971	.971	.970	.970	.970	.969	.968	.967	.965	.963	.961	.958	.955	.952	.948	.944	.939
5.0*	.971	.971	.971	.971	.971	.970	.970	.970	.969	.968	.967	.966	.964	.962	.960	.957	.954	.951	.947
5.5*	.971	.971	.971	.971	.971	.971	.971	.971	.970	.970	.970	.969	.968	.967	.966	.965	.963	.961	.959
6.0*	.971	.971	.971	.971	.971	.971	.971	.971	.971	.971	.971	.970	.970	.970	.969	.969	.968	.967	.966
6.5*	.971	.971	.971	.971	.971	.971	.971	.971	.971	.971	.971	.971	.971	.971	.970	.970	.970	.969	.969
7.0*	.971	.971	.971	.971	.971	.971	.971	.971	.971	.971	.971	.971	.971	.971	.971	.971	.971	.970	.970

WEIGHT= .650

	.100	.150	.200	.250	.300	.350	.400	.450	.500	.550	.600	.650	.700	.750	.800	.850	.900	.950	1.000
.0*	.552	.455	.379	.315	.262	.216	.177	.144	.115	.090	.069	.051	.036	.024	.015	.008	.004	.001	.000
.1*	.553	.457	.380	.317	.264	.218	.179	.145	.116	.092	.071	.053	.038	.026	.017	.010	.005	.003	.002
.2*	.556	.461	.384	.322	.268	.223	.184	.151	.122	.097	.076	.058	.044	.032	.022	.015	.010	.008	.007
.3*	.562	.467	.392	.329	.276	.231	.193	.159	.131	.106	.085	.067	.053	.041	.031	.024	.019	.016	.015
.4*	.569	.476	.402	.340	.288	.243	.205	.171	.143	.118	.097	.080	.065	.053	.043	.036	.031	.027	.026
.5*	.579	.487	.414	.353	.301	.258	.220	.187	.158	.134	.113	.096	.081	.069	.059	.051	.046	.042	.040
.6*	.590	.501	.429	.369	.319	.275	.238	.206	.177	.153	.132	.115	.100	.087	.078	.070	.064	.059	.057
.7*	.603	.516	.445	.388	.338	.296	.259	.227	.199	.175	.154	.137	.122	.109	.099	.091	.084	.079	.076
.8*	.618	.534	.466	.409	.360	.319	.282	.251	.223	.200	.179	.162	.146	.134	.123	.115	.108	.102	.098
.9*	.634	.553	.487	.432	.385	.344	.308	.277	.251	.227	.207	.189	.173	.160	.150	.140	.133	.127	.122
1.0*	.651	.573	.510	.457	.411	.371	.337	.306	.280	.257	.236	.219	.203	.190	.178	.168	.160	.153	.147
1.2*	.687	.618	.560	.511	.468	.431	.398	.369	.344	.321	.301	.283	.267	.253	.240	.229	.219	.211	.203
1.4*	.725	.664	.612	.568	.529	.495	.465	.437	.413	.390	.370	.352	.336	.321	.307	.294	.283	.273	.263
1.6*	.763	.710	.665	.626	.592	.560	.532	.506	.483	.461	.441	.423	.406	.390	.375	.361	.348	.336	.325
1.8*	.798	.754	.716	.683	.652	.624	.598	.574	.552	.531	.511	.493	.475	.459	.443	.428	.414	.401	.388
2.0*	.829	.794	.763	.735	.708	.684	.660	.638	.617	.597	.578	.560	.542	.525	.508	.493	.478	.463	.450
2.2*	.856	.829	.804	.781	.758	.737	.716	.696	.677	.658	.639	.621	.604	.587	.570	.554	.538	.524	.509
2.4*	.878	.858	.838	.819	.801	.783	.765	.747	.729	.712	.694	.677	.660	.643	.627	.611	.595	.580	.565
2.6*	.896	.880	.866	.851	.836	.821	.806	.790	.774	.758	.742	.726	.710	.694	.678	.662	.647	.632	.617
2.8*	.909	.898	.887	.876	.864	.852	.839	.825	.812	.797	.783	.768	.753	.738	.724	.709	.694	.679	.665
3.0*	.918	.910	.903	.894	.885	.875	.865	.854	.842	.830	.817	.804	.790	.777	.763	.749	.735	.721	.707
3.2*	.924	.919	.914	.908	.901	.893	.885	.876	.866	.856	.845	.833	.822	.809	.797	.784	.771	.758	.745
3.4*	.928	.925	.921	.917	.912	.906	.900	.893	.885	.876	.867	.857	.847	.836	.825	.814	.802	.790	.778
3.6*	.931	.929	.926	.923	.920	.916	.911	.905	.899	.892	.885	.876	.868	.859	.849	.839	.828	.818	.807
3.8*	.932	.931	.930	.928	.925	.922	.918	.914	.910	.904	.898	.891	.884	.876	.868	.859	.850	.841	.831
4.0*	.933	.932	.931	.930	.929	.926	.924	.921	.917	.913	.908	.903	.897	.891	.884	.876	.868	.860	.851
4.2*	.934	.933	.933	.932	.931	.929	.928	.925	.923	.920	.916	.912	.907	.902	.896	.890	.883	.876	.868
4.4*	.934	.934	.933	.933	.932	.931	.930	.928	.927	.924	.921	.918	.915	.910	.906	.901	.895	.889	.883
4.6*	.934	.934	.934	.933	.933	.932	.932	.931	.929	.927	.925	.923	.920	.917	.913	.909	.904	.899	.894
4.8*	.934	.934	.934	.934	.933	.933	.933	.932	.931	.930	.928	.926	.924	.922	.919	.915	.912	.908	.903
5.0*	.934	.934	.934	.934	.934	.934	.933	.933	.932	.931	.930	.929	.927	.925	.923	.921	.918	.914	.911
5.5*	.934	.934	.934	.934	.934	.934	.934	.934	.934	.933	.933	.932	.931	.931	.930	.928	.927	.925	.923
6.0*	.934	.934	.934	.934	.934	.934	.934	.934	.934	.934	.934	.933	.933	.933	.932	.932	.931	.930	.929
6.5*	.934	.934	.934	.934	.934	.934	.934	.934	.934	.934	.934	.934	.934	.934	.933	.933	.933	.932	.932
7.0*	.934	.934	.934	.934	.934	.934	.934	.934	.934	.934	.934	.934	.934	.934	.934	.934	.934	.933	.933

WEIGHT= .700

	.100	.150	.200	.250	.300	.350	.400	.450	.500	.550	.600	.650	.700	.750	.800	.850	.900	.950	1.000
.0*	.508	.417	.346	.287	.238	.197	.161	.130	.104	.082	.062	.046	.033	.022	.014	.008	.003	.001	.000
.1*	.509	.418	.347	.289	.240	.198	.163	.132	.106	.083	.064	.048	.035	.024	.015	.009	.005	.002	.002
.2*	.512	.422	.351	.293	.244	.203	.167	.137	.111	.088	.069	.053	.040	.029	.020	.014	.010	.007	.006
.3*	.517	.428	.358	.300	.252	.210	.175	.145	.119	.096	.077	.061	.048	.037	.028	.022	.017	.015	.013
.4*	.524	.436	.367	.310	.262	.221	.186	.156	.130	.107	.088	.072	.059	.048	.040	.033	.028	.025	.024
.5*	.533	.447	.379	.322	.275	.234	.200	.170	.144	.122	.103	.087	.073	.063	.054	.047	.042	.039	.037
.6*	.544	.460	.393	.337	.290	.250	.216	.187	.161	.139	.120	.104	.091	.080	.071	.064	.058	.055	.053
.7*	.557	.474	.409	.354	.309	.269	.235	.206	.181	.159	.140	.124	.111	.100	.091	.083	.078	.073	.071
.8*	.571	.491	.427	.374	.329	.290	.257	.228	.203	.182	.163	.147	.134	.122	.113	.105	.099	.094	.091
.9*	.586	.509	.447	.395	.352	.314	.281	.253	.228	.207	.188	.172	.158	.147	.137	.129	.122	.117	.113
1.0*	.602	.528	.469	.419	.376	.339	.307	.280	.255	.234	.216	.200	.186	.174	.164	.155	.148	.142	.137
1.2*	.637	.570	.516	.469	.430	.395	.365	.338	.315	.294	.276	.259	.245	.232	.221	.212	.203	.196	.189
1.4*	.674	.615	.565	.524	.487	.455	.427	.401	.379	.359	.340	.324	.309	.296	.283	.272	.263	.253	.245
1.6*	.710	.659	.616	.579	.546	.517	.491	.467	.445	.425	.407	.390	.375	.361	.348	.335	.324	.314	.304
1.8*	.745	.702	.666	.633	.604	.578	.554	.531	.511	.491	.474	.457	.441	.426	.412	.398	.386	.374	.363
2.0*	.776	.741	.711	.684	.658	.635	.613	.593	.573	.555	.537	.520	.504	.488	.474	.460	.446	.433	.421
2.2*	.803	.776	.751	.728	.707	.687	.667	.649	.630	.613	.596	.579	.563	.547	.532	.518	.504	.490	.477
2.4*	.825	.804	.785	.767	.749	.732	.714	.698	.681	.665	.648	.632	.617	.601	.586	.572	.557	.544	.530
2.6*	.842	.827	.812	.798	.784	.769	.754	.740	.725	.710	.695	.680	.665	.650	.635	.621	.607	.593	.579
2.8*	.855	.844	.834	.822	.811	.799	.787	.774	.761	.748	.734	.720	.706	.693	.679	.665	.651	.638	.624
3.0*	.864	.857	.849	.841	.832	.823	.813	.802	.791	.779	.767	.755	.742	.730	.717	.704	.691	.678	.665
3.2*	.871	.866	.861	.855	.848	.840	.832	.824	.814	.805	.794	.783	.772	.761	.749	.737	.725	.713	.701
3.4*	.875	.872	.868	.864	.859	.854	.847	.840	.833	.825	.816	.807	.797	.787	.776	.766	.755	.744	.732
3.6*	.878	.876	.873	.870	.867	.863	.858	.853	.847	.840	.833	.825	.817	.808	.799	.790	.780	.770	.760
3.8*	.879	.878	.877	.875	.872	.869	.866	.862	.857	.852	.846	.840	.833	.826	.818	.810	.801	.792	.783
4.0*	.880	.880	.879	.877	.876	.874	.871	.868	.865	.861	.856	.851	.845	.839	.833	.826	.818	.810	.802
4.2*	.881	.880	.880	.879	.878	.877	.875	.873	.870	.867	.863	.859	.855	.850	.845	.839	.832	.826	.819
4.4*	.881	.881	.880	.880	.879	.878	.877	.876	.874	.872	.869	.866	.862	.858	.854	.849	.844	.838	.832
4.6*	.881	.881	.881	.881	.880	.880	.879	.878	.876	.875	.873	.870	.868	.865	.861	.857	.853	.848	.843
4.8*	.881	.881	.881	.881	.881	.880	.880	.879	.878	.877	.876	.874	.872	.869	.867	.863	.860	.856	.852
5.0*	.881	.881	.881	.881	.881	.881	.880	.880	.879	.879	.877	.876	.875	.873	.871	.868	.865	.862	.859
5.5*	.881	.881	.881	.881	.881	.881	.881	.881	.881	.880	.880	.879	.879	.878	.877	.876	.874	.872	.870
6.0*	.881	.881	.881	.881	.881	.881	.881	.881	.881	.881	.881	.881	.880	.880	.880	.879	.878	.877	.876
6.5*	.881	.881	.881	.881	.881	.881	.881	.881	.881	.881	.881	.881	.881	.881	.881	.880	.880	.880	.879
7.0*	.881	.881	.881	.881	.881	.881	.881	.881	.881	.881	.881	.881	.881	.881	.881	.881	.881	.881	.880

WEIGHT= .750

	.100	.150	.200	.250	.300	.350	.400	.450	.500	.550	.600	.650	.700	.750	.800	.850	.900	.950	1.000
.0*	.454	.371	.305	.254	.210	.173	.142	.115	.091	.072	.055	.041	.029	.020	.012	.007	.003	.001	.000
.1*	.455	.372	.307	.255	.211	.174	.143	.116	.093	.073	.056	.042	.031	.021	.014	.008	.004	.002	.001
.2*	.458	.375	.311	.259	.215	.179	.147	.120	.097	.077	.061	.047	.035	.025	.018	.012	.008	.006	.005
.3*	.462	.381	.317	.265	.222	.185	.154	.127	.104	.085	.068	.054	.042	.033	.025	.019	.015	.013	.012
.4*	.469	.388	.325	.274	.231	.195	.164	.137	.114	.094	.078	.064	.052	.042	.035	.029	.025	.023	.021
.5*	.477	.398	.336	.285	.243	.207	.176	.149	.127	.107	.090	.076	.065	.055	.047	.042	.037	.035	.033
.6*	.487	.410	.348	.298	.257	.221	.191	.164	.142	.122	.106	.092	.080	.070	.063	.057	.052	.049	.047
.7*	.499	.423	.363	.314	.273	.238	.208	.182	.159	.140	.124	.110	.098	.088	.080	.074	.069	.066	.063
.8*	.512	.438	.380	.332	.291	.257	.227	.201	.179	.160	.144	.130	.118	.108	.100	.093	.088	.084	.081
.9*	.526	.455	.398	.351	.312	.278	.249	.223	.202	.183	.166	.152	.140	.130	.122	.115	.109	.105	.102
1.0*	.541	.473	.418	.372	.334	.301	.272	.247	.226	.207	.191	.177	.165	.154	.146	.138	.132	.127	.123
1.2*	.574	.511	.461	.418	.382	.351	.324	.300	.279	.261	.245	.230	.218	.207	.198	.189	.182	.176	.171
1.4*	.609	.553	.507	.469	.435	.406	.380	.358	.337	.319	.303	.289	.276	.265	.254	.245	.236	.229	.222
1.6*	.643	.595	.555	.520	.490	.463	.439	.418	.398	.381	.365	.350	.336	.324	.313	.302	.293	.284	.275
1.8*	.677	.636	.601	.571	.544	.520	.498	.477	.459	.442	.426	.411	.397	.384	.372	.360	.349	.339	.330
2.0*	.707	.674	.645	.619	.595	.574	.553	.535	.517	.500	.485	.470	.456	.442	.429	.417	.405	.394	.383
2.2*	.733	.707	.684	.662	.642	.623	.605	.587	.571	.555	.540	.525	.511	.497	.483	.471	.458	.446	.435
2.4*	.755	.735	.716	.699	.682	.665	.650	.634	.619	.604	.589	.575	.561	.547	.534	.521	.508	.496	.484
2.6*	.772	.757	.743	.729	.715	.702	.688	.674	.660	.647	.633	.619	.606	.593	.580	.567	.554	.542	.530
2.8*	.784	.774	.763	.753	.742	.730	.719	.707	.695	.683	.670	.658	.645	.633	.620	.608	.595	.583	.572
3.0*	.794	.787	.779	.771	.763	.754	.744	.734	.724	.713	.702	.691	.679	.667	.656	.644	.632	.621	.609
3.2*	.801	.796	.790	.785	.778	.771	.763	.755	.746	.737	.728	.718	.707	.697	.686	.675	.665	.654	.643
3.4*	.805	.802	.798	.794	.789	.784	.778	.771	.764	.757	.748	.740	.731	.722	.712	.702	.692	.682	.672
3.6*	.808	.806	.803	.800	.797	.793	.788	.783	.778	.771	.765	.758	.750	.742	.733	.725	.716	.707	.697
3.8*	.809	.808	.807	.805	.802	.799	.796	.792	.788	.783	.777	.771	.765	.758	.751	.743	.736	.727	.719
4.0*	.810	.810	.809	.807	.806	.804	.801	.798	.795	.791	.787	.782	.777	.771	.765	.759	.752	.745	.737
4.2*	.811	.810	.810	.809	.808	.807	.805	.803	.800	.797	.794	.790	.786	.781	.776	.771	.765	.759	.753
4.4*	.811	.811	.810	.810	.809	.808	.807	.806	.804	.802	.799	.796	.793	.789	.785	.781	.776	.771	.765
4.6*	.811	.811	.811	.811	.810	.810	.809	.808	.807	.805	.803	.801	.798	.795	.792	.788	.784	.780	.775
4.8*	.811	.811	.811	.811	.811	.810	.810	.809	.808	.807	.806	.804	.802	.800	.797	.794	.791	.787	.783
5.0*	.811	.811	.811	.811	.811	.811	.810	.810	.809	.809	.808	.806	.805	.803	.801	.799	.796	.793	.790
5.5*	.811	.811	.811	.811	.811	.811	.811	.811	.811	.810	.810	.809	.809	.808	.807	.806	.804	.803	.801
6.0*	.811	.811	.811	.811	.811	.811	.811	.811	.811	.811	.811	.811	.810	.810	.810	.809	.808	.808	.807
6.5*	.811	.811	.811	.811	.811	.811	.811	.811	.811	.811	.811	.811	.811	.811	.811	.810	.810	.810	.809
7.0*	.811	.811	.811	.811	.811	.811	.811	.811	.811	.811	.811	.811	.811	.811	.811	.811	.811	.811	.810

WEIGHT= .800

	.100	.150	.200	.250	.300	.350	.400	.450	.500	.550	.600	.650	.700	.750	.800	.850	.900	.950	1.000
.0*	.389	.316	.260	.215	.177	.146	.119	.096	.077	.060	.046	.034	.025	.017	.010	.006	.002	.001	.000
.1*	.390	.317	.261	.216	.178	.147	.120	.098	.078	.062	.047	.036	.026	.018	.012	.007	.004	.002	.001
.2*	.392	.320	.264	.219	.182	.151	.124	.101	.082	.065	.051	.039	.029	.021	.015	.010	.007	.005	.005
.3*	.396	.325	.269	.224	.187	.156	.130	.107	.088	.071	.057	.045	.035	.027	.021	.016	.013	.011	.010
.4*	.402	.331	.276	.232	.195	.164	.138	.115	.096	.080	.065	.054	.044	.036	.030	.025	.021	.019	.018
.5*	.410	.340	.285	.242	.205	.174	.148	.126	.107	.090	.076	.064	.055	.047	.040	.035	.032	.029	.028
.6*	.419	.350	.296	.253	.217	.187	.161	.138	.120	.103	.089	.077	.068	.059	.053	.048	.044	.042	.040
.7*	.429	.362	.309	.267	.231	.201	.175	.153	.134	.118	.104	.093	.083	.074	.068	.063	.059	.056	.054
.8*	.441	.375	.324	.282	.247	.217	.192	.170	.151	.135	.121	.110	.100	.092	.085	.079	.075	.072	.070
.9*	.453	.390	.340	.299	.264	.235	.210	.189	.170	.154	.141	.129	.119	.111	.104	.098	.094	.090	.087
1.0*	.467	.406	.357	.317	.284	.255	.231	.209	.191	.175	.162	.150	.140	.131	.124	.118	.113	.109	.106
1.2*	.497	.440	.395	.358	.326	.299	.276	.255	.237	.222	.208	.196	.186	.177	.169	.162	.157	.152	.148
1.4*	.529	.478	.437	.402	.373	.347	.325	.305	.288	.273	.259	.247	.236	.227	.218	.211	.204	.198	.193
1.6*	.561	.516	.480	.449	.422	.398	.377	.358	.342	.327	.313	.301	.290	.279	.270	.262	.254	.246	.240
1.8*	.592	.554	.522	.495	.471	.449	.430	.412	.396	.381	.368	.355	.343	.333	.322	.313	.304	.296	.288
2.0*	.620	.589	.562	.539	.517	.498	.480	.464	.448	.434	.421	.408	.396	.384	.374	.363	.354	.344	.336
2.2*	.645	.620	.599	.579	.560	.543	.527	.512	.497	.484	.470	.458	.445	.434	.422	.412	.401	.391	.382
2.4*	.666	.647	.630	.613	.598	.583	.569	.555	.542	.528	.516	.503	.491	.479	.468	.457	.446	.436	.426
2.6*	.683	.669	.655	.642	.630	.617	.605	.592	.580	.568	.556	.544	.532	.521	.509	.498	.488	.477	.467
2.8*	.695	.686	.675	.665	.655	.644	.634	.623	.612	.601	.590	.579	.568	.557	.546	.536	.525	.515	.504
3.0*	.705	.698	.690	.683	.675	.666	.658	.649	.639	.629	.620	.609	.599	.589	.579	.569	.558	.548	.538
3.2*	.711	.706	.701	.696	.690	.683	.676	.668	.660	.652	.644	.635	.625	.616	.607	.597	.588	.578	.569
3.4*	.716	.712	.709	.705	.700	.695	.690	.684	.677	.670	.663	.655	.647	.639	.630	.622	.613	.604	.595
3.6*	.718	.716	.714	.711	.708	.704	.700	.695	.690	.684	.678	.672	.665	.658	.650	.642	.634	.626	.618
3.8*	.720	.719	.717	.715	.713	.710	.707	.704	.700	.695	.690	.685	.679	.673	.666	.659	.652	.645	.638
4.0*	.721	.720	.719	.718	.716	.715	.712	.710	.706	.703	.699	.695	.690	.685	.679	.673	.667	.661	.654
4.2*	.721	.721	.720	.720	.719	.717	.716	.714	.711	.709	.706	.702	.698	.694	.690	.685	.680	.674	.668
4.4*	.722	.721	.721	.721	.720	.719	.718	.717	.715	.713	.711	.708	.705	.702	.698	.694	.689	.685	.680
4.6*	.722	.722	.721	.721	.721	.720	.719	.718	.717	.716	.714	.712	.710	.707	.704	.701	.697	.693	.689
4.8*	.722	.722	.722	.722	.721	.721	.720	.720	.719	.718	.717	.715	.713	.711	.709	.706	.703	.700	.697
5.0*	.722	.722	.722	.722	.722	.721	.721	.721	.720	.719	.718	.717	.716	.714	.713	.710	.708	.705	.703
5.5*	.722	.722	.722	.722	.722	.722	.722	.722	.721	.721	.721	.720	.720	.719	.718	.717	.716	.714	.712
6.0*	.722	.722	.722	.722	.722	.722	.722	.722	.722	.722	.722	.721	.721	.721	.720	.720	.719	.719	.718
6.5*	.722	.722	.722	.722	.722	.722	.722	.722	.722	.722	.722	.722	.722	.722	.721	.721	.721	.721	.720
7.0*	.722	.722	.722	.722	.722	.722	.722	.722	.722	.722	.722	.722	.722	.722	.722	.722	.722	.721	.721

WEIGHT= .850

	.100	.150	.200	.250	.300	.350	.400	.450	.500	.550	.600	.650	.700	.750	.800	.850	.900	.950	1.000
.0*	.313	.252	.205	.170	.140	.115	.094	.076	.061	.047	.036	.027	.019	.013	.008	.004	.002	.000	.000
.1*	.313	.253	.207	.171	.141	.116	.095	.077	.061	.048	.037	.028	.020	.014	.009	.005	.003	.001	.001
.2*	.315	.255	.210	.173	.144	.119	.098	.080	.064	.051	.040	.031	.023	.017	.012	.008	.006	.004	.004
.3*	.319	.259	.214	.178	.148	.123	.102	.084	.069	.056	.045	.036	.028	.022	.017	.013	.010	.009	.008
.4*	.324	.265	.220	.184	.154	.130	.109	.091	.076	.063	.051	.042	.034	.028	.023	.020	.017	.015	.015
.5*	.330	.272	.227	.192	.162	.138	.117	.099	.084	.071	.060	.051	.043	.037	.032	.028	.025	.023	.023
.6*	.338	.280	.236	.201	.172	.148	.127	.109	.094	.081	.070	.061	.053	.047	.042	.038	.035	.033	.032
.7*	.346	.290	.247	.212	.183	.159	.139	.121	.106	.093	.082	.073	.065	.059	.054	.050	.047	.045	.044
.8*	.356	.301	.258	.224	.196	.172	.152	.134	.120	.107	.096	.087	.079	.073	.067	.063	.060	.058	.056
.9*	.367	.313	.272	.238	.210	.187	.167	.149	.135	.122	.111	.102	.094	.088	.082	.078	.075	.072	.070
1.0*	.379	.327	.285	.253	.226	.203	.183	.166	.151	.139	.128	.119	.111	.104	.099	.094	.091	.088	.086
1.2*	.405	.356	.318	.287	.261	.239	.219	.203	.189	.176	.166	.156	.148	.141	.135	.130	.126	.123	.120
1.4*	.432	.388	.353	.324	.300	.279	.260	.244	.230	.218	.208	.198	.190	.182	.176	.170	.165	.161	.157
1.6*	.461	.422	.390	.363	.341	.321	.304	.288	.275	.263	.252	.243	.234	.226	.219	.212	.207	.201	.196
1.8*	.488	.455	.427	.403	.383	.364	.348	.334	.321	.309	.298	.288	.279	.271	.263	.255	.249	.242	.237
2.0*	.514	.486	.462	.442	.423	.407	.392	.378	.366	.354	.343	.333	.324	.315	.306	.298	.291	.283	.277
2.2*	.537	.514	.495	.477	.461	.447	.433	.420	.408	.397	.386	.376	.366	.357	.348	.339	.331	.323	.316
2.4*	.557	.539	.523	.508	.495	.482	.470	.458	.447	.436	.426	.416	.406	.396	.387	.378	.370	.361	.353
2.6*	.572	.559	.547	.535	.524	.513	.502	.491	.481	.471	.461	.451	.441	.432	.423	.414	.405	.397	.388
2.8*	.584	.575	.565	.556	.547	.538	.528	.519	.510	.501	.491	.482	.473	.464	.455	.446	.437	.429	.421
3.0*	.593	.586	.580	.573	.565	.558	.550	.542	.534	.526	.517	.509	.500	.492	.483	.475	.466	.458	.450
3.2*	.599	.595	.590	.585	.579	.573	.567	.560	.554	.546	.539	.531	.524	.516	.508	.500	.492	.484	.476
3.4*	.604	.601	.597	.594	.589	.585	.580	.574	.569	.563	.556	.550	.543	.536	.529	.521	.514	.506	.499
3.6*	.606	.604	.602	.600	.597	.593	.589	.585	.580	.576	.570	.564	.558	.552	.546	.539	.533	.526	.519
3.8*	.608	.607	.605	.603	.601	.599	.596	.593	.589	.585	.581	.576	.571	.566	.560	.554	.548	.542	.536
4.0*	.609	.608	.607	.606	.605	.603	.601	.598	.595	.592	.589	.585	.581	.577	.572	.567	.562	.556	.551
4.2*	.609	.609	.608	.608	.607	.605	.604	.602	.600	.598	.595	.592	.589	.585	.581	.577	.572	.568	.563
4.4*	.609	.609	.609	.608	.608	.607	.606	.605	.603	.602	.599	.597	.594	.591	.588	.585	.581	.577	.573
4.6*	.610	.610	.609	.609	.609	.608	.608	.607	.606	.604	.603	.601	.599	.597	.594	.591	.588	.585	.581
4.8*	.610	.610	.610	.609	.609	.609	.608	.608	.607	.606	.605	.604	.602	.600	.598	.596	.593	.591	.587
5.0*	.610	.610	.610	.610	.610	.609	.609	.609	.608	.607	.607	.606	.604	.603	.601	.600	.598	.595	.593
5.5*	.610	.610	.610	.610	.610	.610	.610	.610	.609	.609	.609	.608	.608	.607	.606	.605	.604	.603	.601
6.0*	.610	.610	.610	.610	.610	.610	.610	.610	.610	.610	.610	.609	.609	.609	.608	.608	.607	.607	.606
6.5*	.610	.610	.610	.610	.610	.610	.610	.610	.610	.610	.610	.610	.610	.610	.609	.609	.609	.609	.608
7.0*	.610	.610	.610	.610	.610	.610	.610	.610	.610	.610	.610	.610	.610	.610	.610	.610	.609	.609	.609

WEIGHT= .900

	.100	.150	.200	.250	.300	.350	.400	.450	.500	.550	.600	.650	.700	.750	.800	.850	.900	.950	1.000
.0*	.224	.179	.146	.119	.098	.080	.066	.053	.042	.033	.025	.019	.013	.009	.006	.003	.001	.000	.000
.1*	.224	.179	.146	.120	.099	.081	.066	.054	.043	.034	.026	.020	.014	.010	.006	.004	.002	.001	.001
.2*	.226	.181	.143	.122	.101	.083	.068	.056	.045	.036	.028	.022	.016	.012	.008	.006	.004	.003	.003
.3*	.228	.184	.151	.125	.104	.086	.071	.059	.048	.039	.031	.025	.019	.015	.012	.009	.007	.006	.006
.4*	.232	.188	.155	.129	.108	.091	.076	.063	.053	.044	.036	.029	.024	.020	.016	.014	.012	.011	.010
.5*	.237	.193	.161	.135	.114	.097	.082	.069	.059	.050	.042	.035	.030	.026	.022	.020	.018	.017	.016
.6*	.243	.199	.167	.142	.121	.103	.089	.076	.066	.057	.049	.043	.037	.033	.029	.027	.025	.024	.023
.7*	.249	.207	.175	.150	.129	.112	.097	.085	.074	.065	.057	.051	.046	.041	.038	.035	.033	.032	.031
.8*	.257	.215	.184	.159	.138	.121	.107	.094	.084	.073	.067	.061	.055	.051	.047	.044	.042	.041	.040
.9*	.265	.224	.193	.169	.148	.132	.117	.105	.095	.086	.078	.072	.066	.062	.058	.055	.053	.051	.051
1.0*	.274	.234	.204	.180	.160	.143	.129	.117	.106	.098	.090	.084	.078	.074	.070	.067	.065	.063	.062
1.2*	.294	.257	.228	.205	.185	.169	.155	.144	.133	.125	.117	.111	.105	.100	.096	.093	.090	.088	.087
1.4*	.316	.282	.255	.233	.214	.199	.185	.174	.164	.155	.148	.141	.135	.130	.126	.122	.119	.117	.114
1.6*	.339	.308	.283	.263	.245	.231	.218	.207	.197	.188	.181	.174	.168	.163	.158	.154	.150	.147	.144
1.8*	.362	.334	.312	.294	.278	.264	.252	.241	.232	.223	.216	.209	.202	.197	.191	.187	.182	.178	.174
2.0*	.384	.360	.341	.324	.310	.297	.286	.276	.267	.258	.251	.243	.237	.231	.225	.219	.214	.209	.205
2.2*	.403	.384	.368	.353	.341	.329	.319	.309	.300	.292	.284	.277	.270	.263	.257	.251	.246	.240	.235
2.4*	.420	.405	.392	.380	.369	.359	.349	.340	.332	.324	.316	.308	.301	.294	.288	.282	.276	.270	.264
2.6*	.434	.423	.412	.402	.393	.384	.376	.368	.360	.352	.344	.337	.330	.323	.316	.310	.304	.297	.292
2.8*	.445	.437	.428	.421	.413	.405	.398	.391	.384	.376	.369	.362	.355	.349	.342	.336	.329	.323	.317
3.0*	.454	.447	.441	.435	.429	.423	.416	.410	.404	.397	.391	.384	.378	.371	.365	.359	.352	.346	.340
3.2*	.459	.455	.451	.446	.441	.436	.431	.426	.420	.415	.409	.403	.397	.391	.385	.379	.373	.367	.361
3.4*	.463	.460	.457	.454	.450	.447	.442	.438	.433	.428	.423	.418	.413	.407	.402	.396	.390	.385	.379
3.6*	.465	.464	.462	.459	.457	.454	.451	.447	.443	.439	.435	.430	.426	.421	.416	.411	.406	.401	.395
3.8*	.467	.466	.465	.463	.461	.459	.457	.454	.451	.448	.444	.440	.436	.432	.428	.423	.419	.414	.409
4.0*	.468	.467	.466	.465	.464	.463	.461	.459	.456	.454	.451	.448	.445	.441	.437	.433	.429	.425	.421
4.2*	.468	.468	.468	.467	.466	.465	.464	.462	.460	.459	.456	.454	.451	.448	.445	.442	.438	.434	.431
4.4*	.469	.468	.468	.468	.467	.467	.466	.465	.463	.462	.460	.458	.456	.454	.451	.448	.445	.442	.439
4.6*	.469	.469	.469	.468	.468	.468	.467	.466	.465	.464	.463	.461	.460	.458	.456	.453	.451	.448	.445
4.8*	.469	.469	.469	.469	.468	.468	.468	.467	.467	.466	.465	.464	.462	.461	.459	.457	.455	.453	.451
5.0*	.469	.469	.469	.469	.469	.468	.468	.468	.467	.467	.466	.465	.464	.463	.462	.460	.459	.457	.455
5.5*	.469	.469	.469	.469	.469	.469	.469	.469	.469	.468	.468	.468	.467	.467	.466	.465	.464	.463	.462
6.0*	.469	.469	.469	.469	.469	.469	.469	.469	.469	.469	.469	.469	.468	.468	.468	.467	.467	.466	.466
6.5*	.469	.469	.469	.469	.469	.469	.469	.469	.469	.469	.469	.469	.469	.469	.469	.468	.468	.468	.468
7.0*	.469	.469	.469	.469	.469	.469	.469	.469	.469	.469	.469	.469	.469	.469	.469	.469	.469	.469	.468

WEIGHT= .950

	.100	.150	.200	.250	.300	.350	.400	.450	.500	.550	.600	.650	.700	.750	.800	.850	.900	.950	1.000
.0*	.120	.095	.077	.063	.052	.042	.034	.028	.022	.017	.013	.010	.007	.005	.003	.002	.001	.000	.000
.1*	.121	.096	.077	.063	.052	.042	.035	.028	.022	.018	.014	.010	.007	.005	.003	.002	.001	.000	.000
.2*	.122	.097	.078	.064	.053	.044	.036	.029	.023	.019	.015	.011	.008	.006	.004	.003	.002	.002	.001
.3*	.123	.098	.080	.066	.055	.045	.037	.031	.025	.020	.016	.013	.010	.008	.006	.005	.004	.003	.003
.4*	.125	.100	.082	.068	.057	.048	.040	.033	.028	.023	.019	.015	.013	.010	.009	.007	.006	.003	.003
.5*	.128	.103	.085	.071	.060	.051	.043	.036	.031	.026	.022	.018	.016	.013	.012	.010	.009	.009	.005
.6*	.131	.107	.089	.075	.064	.054	.047	.040	.034	.030	.026	.022	.019	.017	.015	.014	.013	.012	.012
.7*	.135	.111	.093	.079	.068	.059	.051	.044	.039	.034	.030	.027	.024	.022	.020	.018	.017	.017	.017
.8*	.140	.115	.098	.084	.073	.064	.056	.049	.044	.039	.035	.032	.029	.027	.025	.024	.023	.022	.022
.9*	.145	.121	.103	.090	.079	.069	.062	.055	.050	.045	.041	.038	.035	.032	.031	.029	.028	.028	.027
1.0*	.150	.126	.109	.096	.085	.076	.068	.062	.056	.051	.047	.044	.041	.039	.037	.035	.034	.034	.033
1.2*	.162	.140	.123	.110	.099	.090	.083	.076	.071	.066	.062	.058	.056	.053	.052	.050	.049	.048	.047
1.4*	.176	.154	.138	.126	.115	.107	.099	.093	.087	.083	.079	.075	.073	.070	.068	.066	.065	.064	.063
1.6*	.190	.171	.155	.143	.133	.125	.118	.112	.106	.102	.098	.094	.091	.089	.086	.084	.083	.081	.080
1.8*	.206	.187	.174	.162	.153	.145	.138	.132	.126	.122	.118	.114	.111	.108	.106	.103	.101	.100	.098
2.0*	.220	.204	.192	.181	.173	.165	.158	.153	.147	.143	.139	.135	.132	.129	.126	.123	.121	.118	.116
2.2*	.234	.221	.210	.200	.192	.185	.179	.174	.168	.164	.160	.156	.152	.149	.146	.143	.140	.137	.135
2.4*	.247	.236	.227	.218	.211	.205	.199	.194	.189	.184	.180	.176	.172	.168	.165	.162	.158	.155	.152
2.6*	.258	.249	.241	.234	.228	.222	.217	.212	.207	.203	.199	.194	.190	.187	.183	.179	.176	.173	.170
2.8*	.266	.260	.254	.248	.243	.238	.233	.228	.224	.220	.215	.211	.207	.203	.200	.196	.193	.189	.186
3.0*	.273	.268	.264	.259	.255	.251	.246	.242	.238	.234	.230	.226	.222	.219	.215	.211	.208	.204	.201
3.2*	.278	.274	.271	.268	.264	.261	.257	.254	.250	.246	.243	.239	.235	.232	.228	.224	.221	.217	.214
3.4*	.281	.279	.276	.274	.271	.269	.266	.263	.260	.256	.253	.250	.246	.243	.240	.236	.233	.229	.226
3.6*	.283	.282	.280	.278	.276	.274	.272	.270	.267	.264	.261	.258	.255	.252	.249	.246	.243	.240	.237
3.8*	.285	.284	.283	.281	.280	.278	.277	.275	.273	.270	.268	.265	.263	.260	.257	.254	.252	.249	.246
4.0*	.285	.285	.284	.283	.282	.281	.280	.278	.277	.275	.273	.271	.269	.266	.264	.261	.259	.256	.254
4.2*	.286	.286	.285	.285	.284	.283	.282	.281	.280	.278	.277	.275	.273	.271	.269	.267	.265	.263	.260
4.4*	.286	.286	.286	.285	.285	.284	.284	.283	.282	.281	.280	.278	.277	.275	.273	.272	.270	.268	.266
4.6*	.286	.286	.286	.286	.286	.285	.285	.284	.283	.283	.282	.281	.280	.278	.277	.275	.274	.272	.270
4.8*	.286	.286	.286	.286	.286	.286	.285	.285	.285	.284	.283	.282	.282	.281	.279	.278	.277	.275	.274
5.0*	.286	.286	.286	.286	.286	.286	.286	.286	.285	.285	.284	.284	.283	.282	.281	.280	.279	.278	.277
5.5*	.286	.286	.286	.286	.286	.286	.286	.286	.286	.286	.286	.285	.285	.285	.284	.284	.283	.282	.282
6.0*	.286	.286	.286	.286	.286	.286	.286	.286	.286	.286	.286	.286	.286	.286	.286	.285	.285	.285	.284
6.5*	.286	.286	.286	.286	.286	.286	.286	.286	.286	.286	.286	.286	.286	.286	.286	.286	.286	.286	.285
7.0*	.286	.286	.286	.286	.286	.286	.286	.286	.286	.286	.286	.286	.286	.286	.286	.286	.286	.286	.286

WEIGHT= .980

	.100	.150	.200	.250	.300	.350	.400	.450	.500	.550	.600	.650	.700	.750	.800	.850	.900	.950	1.000
.0*	.050	.040	.032	.026	.021	.017	.014	.011	.009	.007	.005	.004	.003	.002	.001	.001	.000	.000	.000
.1*	.051	.040	.032	.026	.021	.017	.014	.011	.009	.007	.006	.004	.003	.002	.001	.000	.000	.000	.000
.2*	.051	.040	.032	.027	.022	.018	.015	.012	.010	.008	.006	.005	.003	.003	.002	.001	.001	.001	.000
.3*	.052	.041	.033	.027	.022	.019	.015	.013	.010	.008	.007	.005	.004	.003	.003	.002	.002	.001	.001
.4*	.053	.042	.034	.028	.023	.020	.016	.014	.011	.009	.008	.006	.005	.004	.002	.001	.001	.001	.001
.5*	.054	.043	.035	.029	.025	.021	.018	.015	.013	.011	.009	.008	.006	.006	.005	.004	.002	.002	.002
.6*	.055	.044	.037	.031	.026	.022	.019	.016	.014	.012	.010	.009	.008	.007	.006	.006	.002	.003	.003
.7*	.057	.046	.039	.033	.028	.024	.021	.018	.016	.014	.012	.011	.010	.009	.008	.008	.004	.004	.004
.8*	.059	.048	.041	.035	.030	.026	.023	.020	.018	.016	.014	.013	.012	.011	.010	.010	.005	.005	.005
.9*	.061	.051	.043	.037	.033	.029	.025	.023	.020	.019	.017	.015	.014	.013	.013	.012	.012	.007	.007
1.0*	.064	.053	.045	.040	.035	.031	.028	.025	.023	.021	.019	.018	.017	.016	.015	.015	.014	.008	.009
1.2*	.069	.059	.052	.046	.041	.037	.034	.032	.029	.027	.026	.024	.023	.022	.021	.021	.020	.020	.020
1.4*	.076	.066	.058	.053	.048	.045	.041	.039	.036	.034	.033	.031	.030	.029	.029	.028	.027	.027	.027
1.6*	.083	.073	.066	.061	.056	.053	.049	.047	.045	.043	.041	.040	.038	.037	.037	.036	.035	.035	.034
1.8*	.091	.082	.075	.069	.065	.061	.058	.056	.054	.052	.050	.049	.047	.046	.045	.045	.044	.043	.043
2.0*	.099	.090	.084	.079	.075	.071	.068	.066	.063	.061	.060	.058	.057	.056	.054	.054	.053	.052	.051
2.2*	.107	.099	.093	.088	.084	.081	.078	.076	.073	.071	.070	.068	.067	.065	.064	.063	.062	.061	.060
2.4*	.114	.107	.102	.098	.094	.091	.088	.086	.084	.082	.080	.078	.077	.075	.074	.073	.071	.070	.069
2.6*	.121	.115	.111	.107	.104	.101	.098	.096	.094	.092	.090	.088	.086	.085	.083	.082	.080	.079	.078
2.8*	.127	.122	.119	.115	.112	.110	.107	.105	.103	.101	.099	.097	.095	.094	.092	.090	.089	.087	.086
3.0*	.131	.128	.125	.122	.120	.117	.115	.113	.111	.109	.107	.105	.103	.102	.100	.098	.097	.095	.094
3.2*	.135	.132	.130	.128	.126	.124	.122	.120	.118	.116	.114	.113	.111	.109	.107	.106	.104	.102	.101
3.4*	.137	.136	.134	.132	.131	.129	.127	.126	.124	.122	.120	.119	.117	.115	.114	.112	.110	.109	.107
3.6*	.139	.138	.137	.135	.134	.133	.131	.130	.129	.127	.125	.124	.122	.121	.119	.118	.116	.115	.113
3.8*	.140	.139	.138	.138	.137	.136	.135	.133	.132	.131	.129	.128	.127	.125	.124	.123	.121	.120	.118
4.0*	.141	.140	.140	.139	.138	.138	.137	.136	.135	.134	.132	.132	.130	.129	.128	.127	.125	.124	.123
4.2*	.141	.141	.140	.140	.140	.139	.138	.138	.137	.136	.135	.134	.133	.132	.131	.130	.129	.127	.126
4.4*	.141	.141	.141	.141	.140	.140	.140	.139	.138	.137	.137	.136	.135	.134	.134	.133	.132	.130	.129
4.6*	.141	.141	.141	.141	.141	.141	.140	.140	.139	.139	.138	.138	.137	.136	.136	.135	.134	.133	.132
4.8*	.141	.141	.141	.141	.141	.141	.141	.140	.140	.140	.139	.139	.138	.138	.137	.136	.136	.135	.134
5.0*	.141	.141	.141	.141	.141	.141	.141	.141	.141	.140	.140	.140	.139	.139	.138	.138	.137	.136	.136
5.5*	.141	.141	.141	.141	.141	.141	.141	.141	.141	.141	.141	.141	.141	.140	.140	.140	.139	.139	.139
6.0*	.141	.141	.141	.141	.141	.141	.141	.141	.141	.141	.141	.141	.141	.141	.141	.141	.141	.140	.140
6.5*	.141	.141	.141	.141	.141	.141	.141	.141	.141	.141	.141	.141	.141	.141	.141	.141	.141	.141	.141
7.0*	.141	.141	.141	.141	.141	.141	.141	.141	.141	.141	.141	.141	.141	.141	.141	.141	.141	.141	.141

WEIGHT= .990

	.100	.150	.200	.250	.300	.350	.400	.450	.500	.550	.600	.650	.700	.750	.800	.850	.900	.950	1.000
.0*	.026	.020	.016	.013	.011	.009	.007	.006	.005	.004	.003	.002	.001	.001	.001	.000	.000	.000	.000
.1*	.026	.020	.016	.013	.011	.009	.007	.006	.005	.004	.003	.002	.002	.001	.000	.000	.000	.000	.000
.2*	.026	.020	.016	.013	.011	.009	.007	.006	.005	.004	.003	.002	.002	.001	.001	.000	.000	.000	.000
.3*	.026	.021	.017	.014	.011	.009	.008	.006	.005	.004	.003	.003	.002	.002	.001	.001	.001	.001	.000
.4*	.027	.021	.017	.014	.012	.010	.008	.007	.006	.005	.004	.002	.001	.001	.001	.000	.000	.000	.001
.5*	.027	.022	.018	.015	.012	.010	.009	.007	.006	.005	.004	.004	.003	.001	.001	.001	.001	.001	.001
.6*	.028	.023	.019	.016	.013	.011	.010	.008	.007	.006	.005	.005	.004	.004	.001	.001	.001	.001	.001
.7*	.029	.024	.020	.017	.014	.012	.011	.009	.008	.007	.006	.006	.005	.004	.002	.002	.002	.002	.002
.8*	.030	.025	.021	.018	.015	.013	.012	.010	.009	.008	.007	.007	.006	.006	.002	.002	.002	.003	.003
.9*	.031	.026	.022	.019	.016	.014	.013	.012	.010	.009	.008	.008	.007	.007	.003	.003	.003	.003	.004
1.0*	.033	.027	.023	.020	.018	.016	.014	.013	.012	.011	.010	.009	.009	.008	.004	.004	.004	.004	.005
1.2*	.036	.030	.025	.023	.021	.019	.017	.016	.015	.014	.013	.012	.012	.011	.006	.006	.006	.007	.007
1.4*	.039	.034	.030	.027	.025	.023	.021	.020	.018	.017	.017	.016	.015	.015	.015	.014	.014	.009	.010
1.6*	.043	.038	.034	.031	.029	.027	.025	.024	.023	.022	.021	.020	.020	.019	.019	.018	.018	.018	.018
1.8*	.047	.042	.038	.036	.033	.031	.030	.028	.027	.026	.026	.025	.024	.024	.023	.023	.023	.022	.022
2.0*	.052	.047	.043	.041	.038	.037	.035	.034	.032	.032	.031	.030	.029	.029	.028	.028	.027	.027	.027
2.2*	.056	.052	.049	.046	.044	.042	.040	.039	.038	.037	.036	.035	.035	.034	.034	.033	.033	.032	.032
2.4*	.061	.057	.054	.051	.049	.048	.046	.045	.044	.043	.042	.041	.040	.040	.039	.038	.038	.037	.037
2.6*	.065	.062	.059	.057	.055	.053	.052	.051	.049	.048	.047	.047	.046	.045	.044	.044	.043	.042	.042
2.8*	.069	.066	.064	.062	.060	.059	.057	.056	.055	.054	.053	.052	.051	.050	.049	.049	.048	.047	.047
3.0*	.073	.070	.068	.067	.065	.064	.062	.061	.060	.059	.058	.057	.056	.055	.054	.053	.053	.052	.051
3.2*	.075	.074	.072	.071	.069	.068	.067	.066	.065	.064	.063	.062	.061	.060	.059	.058	.057	.056	.055
3.4*	.077	.076	.075	.074	.073	.072	.071	.070	.069	.068	.066	.066	.065	.064	.063	.062	.061	.060	.059
3.6*	.079	.078	.077	.076	.075	.074	.073	.073	.072	.071	.070	.069	.068	.067	.066	.065	.064	.064	.063
3.8*	.080	.079	.078	.078	.077	.076	.076	.075	.074	.073	.073	.072	.071	.070	.069	.068	.068	.067	.066
4.0*	.080	.080	.079	.079	.078	.078	.077	.077	.076	.075	.075	.074	.073	.073	.072	.071	.070	.069	.069
4.2*	.080	.080	.080	.080	.079	.079	.079	.078	.078	.077	.076	.076	.075	.075	.074	.073	.072	.072	.071
4.4*	.081	.081	.080	.080	.080	.080	.079	.079	.079	.078	.078	.077	.077	.076	.075	.075	.074	.074	.073
4.6*	.081	.081	.081	.080	.080	.080	.080	.080	.079	.079	.079	.078	.078	.077	.077	.076	.076	.075	.075
4.8*	.081	.081	.081	.081	.081	.080	.080	.080	.080	.080	.079	.079	.079	.078	.078	.077	.077	.076	.076
5.0*	.081	.081	.081	.081	.081	.081	.081	.080	.080	.080	.080	.080	.079	.079	.079	.078	.078	.077	.077
5.5*	.081	.081	.081	.081	.081	.081	.081	.081	.081	.081	.081	.080	.080	.080	.080	.080	.079	.079	.079
6.0*	.081	.081	.081	.081	.081	.081	.081	.081	.081	.081	.081	.081	.081	.081	.080	.080	.080	.080	.080
6.5*	.081	.081	.081	.081	.081	.081	.081	.081	.081	.081	.081	.081	.081	.081	.081	.081	.081	.080	.080
7.0*	.081	.081	.081	.081	.081	.081	.081	.081	.081	.081	.081	.081	.081	.081	.081	.081	.081	.081	.081

APPENDIX 3 Computation of B_k

J. K. M. Moody and C. J. Jardine have devised an algorithm for computing $B_k(d)$ from d. The following program has been written by Moody in USASI FORTRAN. The program calls subroutines CLIQUE and UNPACK from subroutine CLQWTE; these are machine-code subroutines associated with the Moody-Hollis cluster-listing technique outlined in Appendix 5. No READ routine is given here because input requirements may vary; the READ routine is responsible for setting the highest required value of k, entering the DC values into section 1 of the array DIST, and setting L1 to the number of OTU's. The program comments explain details of the algorithm used, which is rather complicated.

```
 1   C           THIS PROGRAM PROVIDES THE BASIC REDUCTION FOR THE
 2   C     CLUSTER METHOD BK(D) OF JARDINE AND SIBSON. IT IS WRITTEN
 3   C     IN STANDARD ASA FORTRAN, AND IS DESIGNED TO ACCEPT INPUT AND
 4   C     LIST OUTPUT VIA SUBROUTINES, THE DATA BEING PASSED IN NAMED
 5   C     COMMON. THE PROGRAM IS WRITTEN IN SUCH A WAY AS TO EASE THE
 6   C     TASK OF THE COMPILER  -  SINGLE INDEXING IS USED FOR ARRAY
 7   C     SUBSCRIPTION WHERE POSSIBLE, AND TEMPORARY VARIABLES ARE
 8   C     TAKEN OUTSIDE LOOPS WHERE THIS WILL EFFECT A SAVING.
 9   C
10   C
11   C           THERE ARE TWO RESTRICTIONS ON THE USE OF THIS PROGRAM.
12   C     FIRSTLY, IT IS ASSUMED THAT NO ENTRIES IN THE DISSIMILARITY
13   C     MATRIX ARE NUMERICALLY GREATER THAN  1.0E10 .  SECONDLY, IT
14   C     IS ASSUMED THAT  1.1E10  IS A VALID FLOATING-POINT CONSTANT.
15   C           THESE VALUES ARE USED ESSENTIALLY AS FLAGS. PROVIDED
16   C     THAT THEIR USE IS PROPERLY UNDERSTOOD, THEY MAY BE REPLACED
17   C     BY ANY MORE SUITABLE VALUES.
18   C
19   C
20   C           THE ARRAY DIST IS USED TO HOLD TWO VERSIONS OF THE
21   C     DISSIMILARITY MATRIX, AS WE KNOW THAT THE MATRIX WILL BE
22   C     SYMMETRIC. IF   0 < I < J < L1+1   , L1 = (NUMBER OF OBJECTS)
23   C     IS THE CANONICAL FORM FOR THE INDICES, THEN STORAGE FOR
24   C     THE TWO VERSIONS MAY BE CONSIDERED TO BE   -
25   C
26   C           1.      DIST(I,J)
27   C           2.      DIST(J,I+1)
28   C
29   C           WE THUS HAVE THE FIRST COLUMN OF THE ARRAY  'DIST'
30   C     UNASSIGNED AND CAN USE THE STORAGE FOR A LINEAR INDEX.
31   C
32   C
33   C           THE BLOCKS OF NAMED COMMON ARE USED TO SHARE STORE
34   C     BETWEEN THE MAIN PROGRAM AND SUBROUTINES, AND IN PARTICULAR
35   C     TO PASS THE DISSIMILARITY MATRIX FOR I/O OPERATIONS.
36   C
37   C
38         DIMENSION ITEM(15),DIST(60,60),D(3540),KAD(15),INDEX(60),
39        1 NEWIND(60),LIST(60)
40         EQUIVALENCE (DIST(1,1),INDEX(1)),(DIST(1,2),D(1))
41         COMMON/BLANK/KB,L1,L2,N,DSTORE,DMAX0,DHALF0,DONE0,DMAX,DHALF,
42        1 DONE,DMAX1,DHALF1,DONE1
43         COMMON/DIST/DIST
44         COMMON/LIST/LIST
45         COMMON/NUM/NEWIND
46   C
```

```
 47    C              SUBROUTINE 'READ' SETS UP THE ORIGINAL DISSIMILARITY
 48    C       MATRIX IN THE SECTION  '1'  OF THE ARRAY  'DIST',
 49    C
 50    C              IT IS ALSO RESPONSIBLE FOR SETTING UP THE NUMBER OF
 51    C       OBJECTS TO BE CLASSIFIED IN THE VARIABLE 'L1' AND THE INITIAL
 52    C       LEVEL OF K IN THE VARIABLE 'KB' .
 53    C
 54            CALL READ
 55    C
 56    C
 57            L2=L1-1
 58            DMAX0=0.0
 59            DHALF0=0.0
 60            DONE0=0.0
 61            K1=0
 62            M1=0
 63            DO 2 I=1,L2
 64            I2=I+1
 65            M=M1+I
 66                   DO 5 J=I2,L1
 67                   K=K1+J
 68                   D(K)=D(M)
 69                   DONE0=DONE0+D(M)
 70                   DHALF0=DHALF0+D(M)*D(M)
 71                   IF (D(M)-DMAX0) 5,5,4
 72    4              DMAX0=D(M)
 73    5              M=M+60
 74            K1=K1+60
 75    2       M1=M1+60
 76            DHALF0=SQRT(DHALF0)
 77    C
 78    C       IF DESIRED A LISTING OF THE INPUT MAY BE OBTAINED HERE.
 79    C
 80            CALL WRITE1
 81    C
 82    C
 83    C              FOR EACH VALUE OF K DESIRED THE BASIC REDUCTION IS
 84    C       CARRIED OUT  -  THE OUTER LOOP MERELY INITIALISES A NUMBER OF
 85    C       VALUES REQUIRED FOR CONTROL THROUGH THE INNER LOOP. FOR EXAMPLE
 86    C       KMAX AND J2 ARE SET TO  (K-1)  AND (K+1)  RESPECTIVELY, AND
 87    C       N IS SET UP TO COUNT THE NUMBER OF LEVELS IN THE REDUCED
 88    C       DISSIMILARITY MATRIX.
 89    C
 90    3       LMAX=0
 91            KMAX=KB-1
 92            DSTORE=1.0E10
 93            J2=KB+1
 94            N=1
 95    C
 96    C       WE MAY HERE WRITE A HEADING FOR EACH VALUE OF  K .
 97    C
 98            CALL CLOONE
 99    C
100    C
101    C              THE METHOD DETERMINES SUCCESSIVELY THE LOWEST UNMARKED
102    C       EDGE. THIS EDGE IS MARKED, AND FOR EVERY NEW (K+1)-AD IN WHICH
103    C       EACH EDGE IS MARKED, AN APPROPRIATE REDUCTION OF UNMARKED
104    C       EDGES IS EFFECTED. ALL OPERATIONS ARE CARRIED OUT IN THE
105    C       SECTION  '1'  OF THE ARRAY DIST.
106    C
107    C              THE FULL REDUCTION HAS BEEN CARRIED OUT WHEN ALL THOSE
108    C       EDGES SO FAR UNMARKED ARE EQUAL  -  THE FIRST SECTION OF THE
109    C       INNER LOOP MERELY SELECTS THE LONGEST AND SHORTEST EDGE STILL
110    C       UNMARKED, MARKING CONSISTS SIMPLY IN SETTING THE SIGN OF THE
111    C       APPROPRIATE EDGE NEGATIVE  -  IN THE CASE OF A NON-DEFINITE
112    C       DISSIMILARITY MATRIX ZERO DISTANCES ARE SET TO  -1.1E10 .
113    C              THE VERTICES AT EACH END OF THE LEAST EDGE AT A GIVEN
114    C       RUN ARE RECORDED IN IMIN AND JMIN, WHERE  IMIN < JMIN .
115    C
116    1       DMIN=1.0E10
117            DMAX=0.0
118            M1=0
119            DO 8 I=1,L2
120            I2=I+1
121            M=M1+I
```

```
122                    DO 7 J=I2,L1
123                    DD=D(M)
124                    IF (DD) 7,9,9
125     9              IF (DD-DMAX) 10,10,11
126    11              DMAX=DD
127    10              IF (DD-DMIN) 6,7,7
128     6              DMIN=DD
129                    IMIN=I
130                    JMIN=J
131     7              M=M+60
132     8        M1=M1+60
133     C
134     C              A CHECK IS MADE FOR EVERY DISTANCE MARKED TO SEE IF
135     C        THERE IS A CHANGE IN THE 'LEVEL' , THAT IS THE LENGTH OF THE
136     C        CURRENT LEAST DISTANCE. IF SO A CALL IS MADE TO SUBROUTINE
137     C        'CLOWTE' WHICH CAN BE USED TO WRITE OUT ALL THE CLUSTERS AT
138     C        THE GIVEN LEVEL. MACHINE CODE SUBROUTINES FOR THIS PURPOSE
139     C        HAVE BEEN WRITTEN FOR THE  ATLAS II  AND IBM/360  COMPUTERS.
140     C
141              IF (DSTORE-DMIN) 16,17,17
142    16        N=N+1
143              CALL CLOWTE
144    17        DSTORE=DMIN
145     C
146     C        IF ALL UNMARKED DISTANCES ARE EQUAL, THAT'S ALL FOR THIS K.
147     C
148              IF (DMIN-DMAX) 14,75,75
149     C
150     C              IF NOT, WE MUST MARK THE LEAST DISTANCE AND CARRY OUT
151     C        THE REDUCTION. FOR THIS REASON WE MAINTAIN TWO INTEGER ARRAYS -
152     C              1.       AN 'INDEX' OF ALL VERTICES ON A MARKED EDGE.
153     C              2.       A 'LIST' OF ALL VERTICES JOINED BY MARKED EDGES
154     C        TO BOTH IMIN AND JMIN  -  THIS LIST CONTAINS ALL THE VERTICES
155     C        WHICH CAN BE MEMBERS OF NEW 'COMPLETE' (K+1)-ADS.
156     C
157     C              BOTH 'INDEX' AND 'LIST' ARE HELD IN ASCENDING ORDER
158     C        TO EASE PROCESSING.
159     C
160    14        DIST(IMIN,JMIN)=-DMIN
161              IF (DMIN) 12,12,13
162    12        DIST(IMIN,JMIN)=-1.1E10
163    13        NIND=0
164              NLIST=0
165              IF (LMAX) 34,34,15
166    15        IJON=-1
167              DO 23 I=1,LMAX
168              M=INDEX(I)
169              IF (M-IMIN) 18,19,20
170    18        IF (DIST(M,IMIN)) 21,22,22
171    21        IF (DIST(M,JMIN)) 24,22,22
172    19        IJON=IJON+1
173              GO TO 22
174    20        IF (IJON) 25,26,26
175    25        IJON=0
176              NIND=NIND+1
177              NEWIND(NIND)=IMIN
178    26        IF (M-JMIN) 27,19,29
179    27        IF (DIST(IMIN,M)) 21,22,22
180    29        IF (IJON) 30,30,31
181    30        IJON=1
182              NIND=NIND+1
183              NEWIND(NIND)=JMIN
184    31        IF (DIST(IMIN,M)) 32,22,22
185    32        IF (DIST(JMIN,M)) 24,22,22
186    24        NLIST=NLIST+1
187              LIST(NLIST)=M
188    22        NIND=NIND+1
189              NEWIND(NIND)=M
190    23        CONTINUE
191     C
192     C              WE HAVE ESTABLISHED A NEW INDEX (WITH THE POSSIBLE
193     C        EXCEPTION OF IMIN AND JMIN) IN 'NEWIND'. WE NOW COPY IT BACK
194     C        TO 'INDEX' AND ADD IMIN AND JMIN IF NECESSARY.
195     C
196              DO 33 I=1,NIND
197    33        INDEX(I)=NEWIND(I)
```

```
198          IF (IJJN) 34,35,36
199    34    NIND=NIND+1
200          INDEX(NIND)=IMIN
201    35    NIND=NIND+1
202          INDEX(NIND)=JMIN
203    C
204    C
205    C             WE RECORD THE UPDATED NUMBER OF VERTICES IN THE INDEX,
206    C     I.E. THOSE WHICH LIE ON AT LEAST ONE MARKED EDGE, WE ALSO
207    C     CHECK BEFORE CONTINUING THAT THERE ARE AT LEAST (K-1)
208    C     VERTICES IN THE LIST OF THOSE JOINED TO BOTH IMIN AND JMIN,
209    C
210    36    LMAX=NIND
211          IF (KMAX-NLIST) 38,38,1
212    C
213    C
214    C             WE NOW HAVE AN UPDATED INDEX OF ALL THOSE VERTICES
215    C     WHICH LIE ON A MARKED EDGE, AND A LIST OF CANDIDATES FOR
216    C     THE INNER LOOP (K+1)-ADS, WE GENERATE APPROPRIATE (K-1)-ADS
217    C     TO ADD TO IMIN, JMIN,  BY CONSTRUCTING ORDERED SEQUENCES
218    C
219    C             0 < L1 < L2 < L3.... < LN < NLIST+1  ,      N=K-1
220    C
221    C
222    C             WE MAINTAIN AN ORDERED INDEX OF VERTICES  KAD(I),
223    C     I=1,(K-1), SELECTED FROM THE LIST, WE ADD VERTICES BY
224    C     QUERYING SUCCESSIVELY  -
225    C
226    C     1.      Q, HAVE WE ENOUGH VERTICES LEFT IN THE LIST,
227    C             A)      YES.     THEN ASK QUESTION 2,
228    C             B)      NO,      Q, ARE WE AT VERTEX  I=1 ,
229    C                     1)       YES,     THEN WE HAVE FOUND THEM ALL,
230    C                     2)       NO,      UPDATE  KAD(I-1) AND ASK  '1'
231    C
232    C     2.      Q, ARE JOINS TO ALL PREVIOUS VERTICES MARKED,
233    C             A)      YES,     Q, ARE WE AT VERTEX (K-1),
234    C                     1)       YES,     THEN WE MAY PROCESS ,
235    C                     2)       NO,      THEN LOOK AT VERTEX (I+1),
236    C             B)      NO,      UPDATE   KAD(I) AND ASK  '1'
237    C
238    38    K=0
239          I=0
240          KAD(KB)=IMIN
241          KAD(J2)=JMIN
242          IF (KMAX) 46,46,47
243    47    K=K+1
244    45    I=I+1
245          IF (NLIST-I-KMAX+K) 41,42,42
246    42    ITEM(K)=I
247          I1=LIST(I)
248          KAD(K)=I1
249          K1=K-1
250          IF (K1) 43,43,37
251    37    DO 44 J=1,K1
252          J1=KAD(J)
253          IF (DIST(J1,I1)) 44,45,45
254    44    CONTINUE
255    43    IF (KMAX-K) 46,46,47
256    41    IF (K-1) 1,1,49
257    49    K=K-1
258          I=ITEM(K)
259          GO TO 45
260    C
261    C             THE ARRAY 'KAD' NOW CONTAINS A NEW COMPLETE (K+1)-AD,
262    C     FOR ALL POINTS 'V' WHICH ARE NOT VERTICES OF THIS (K+1)-AD,
263    C     WE SET THE LONGEST UNMARKED EDGE JOINING 'V' TO A  VERTEX
264    C     OF THE (K+1)-AD TO THE LONGER OF
265    C
266    C             1.      THE MOST RECENTLY MARKED DISTANCE,
267    C             2.      THE NEXT LONGEST EDGE JOINING 'V' TO A KAD(I),
268    C
269    46    DO 50 I1=1,L1
270                  DO 63 J=1,J2
271                  IF (I1-KAD(J)) 63,50,63
272    63            CONTINUE
273          DMAX=0.0
```

```
274           DNEXT=0.0
275                   DO 48 J=1,J2
276                   J1=KAD(J)
277                   IF (I1-J1) 51,51,52
278     51            DD=DIST(I1,J1)
279                   GO TO 53
280     52            DD=DIST(J1,I1)
281     53            IF (DD) 48,48,62
282     62            IF (DD-DMAX) 54,54,55
283     55            DNEXT=DMAX
284                   DMAX=DD
285                   JMAX=J1
286                   GO TO 48
287     54            IF (DD-DNEXT) 48,48,56
288     56            DNEXT=DD
289     48            CONTINUE
290           IF (DMAX) 50,50,57
291     57    IF (DNEXT) 58,58,59
292     58    DNEXT=DMIN
293     59    IF (I1-JMAX) 64,64,61
294     64    DIST(I1,JMAX)=DNEXT
295           GO TO 50
296     61    DIST(JMAX,I1)=DNEXT
297     50    CONTINUE
298           GO TO 45
299     C
300     C             WHEN THE DATA IS COMPLETELY REDUCED WE CALCULATE THE
301     C     DISTORTION FROM THE ORIGINAL DATA, WHICH HAS BEEN PRESERVED
302     C     IN SECTION  '2'  OF THE ARRAY  'DIST'.
303     C
304     75    DMAX=0.0
305           DHALF=0.0
306           DONE=0.0
307           K1=0
308           M1=0
309           DO 66 I=1,L2
310           I2=I+1
311           M=M1+I
312                   DO 65 J=I2,L1
313                   K=K1+J
314                   IF (D(M)+1.0E10) 71,71,72
315     71            D(M)=0.0
316                   GO TO 68
317     72            IF (D(M)) 67,67,68
318     67            D(M)=-D(M)
319     68            DD=D(K)-D(M)
320                   DONE=DONE+DD
321                   DHALF=DHALF+DD*DD
322                   IF (DD-DMAX) 65,65,69
323     69            DMAX=DD
324     65            M=M+60
325           K1=K1+60
326     66    M1=M1+60
327           DHALF=SQRT(DHALF)
328           DMAX1=DMAX/DMAX0
329           DHALF1=DHALF/DHALF0
330           DONE1=DONE/DONE0
331     C
332     C             WE ARE NOW AT THE FINAL LEVEL AND CAN GIVE THE TOTAL
333     C     NUMBER OF LEVELS FOR STATISTICAL PURPOSES.
334     C
335           CALL CLOTWO
336     C
337     C             'WRITE2' IS NOW CALLED AND MAY BE USED TO WRITE OUT THE
338     C     MODIFIED DISSIMILARITY MATRIX AND THE VARIOUS DISTORTION
339     C     MEASURES.
340     C
341           CALL WRITE2
342     C
343     C             WE NOW DECREASE THE VALUE OF K BY ONE AND TEST WHETHER
344     C     THE NEW VALUE OF  K  IS ZERO. IF NOT, WE CARRY OUT THE
345     C     REDUCTION BK(D) FOR THE NEW VALUE OF K, USING THE PARTIALLY
346     C     PROCESSED DATA IN SECTION  '1'  OF 'DIST'. THIS IS VALID AS
347     C     THE METHODS BK(D) ARE NESTED AS K DECREASES.
348     C
349           KB=KB-1
```

```
350            IF (KB) 70,70,3
351    70      STOP
352            END
353    C
354    C
355            SUBROUTINE READ
356            DIMENSION  DIST(60,60)
357            COMMON/DIST/DIST
358            COMMON/BLANK/KB,L1,L2,N,DSTORE,DMAX0,DHALF0,DONE0,DMAX,DHALF,
359          1 DONE,DMAX1,DHALF1,DONE1
360            CALL SCLOCK
361            READ (5,2) KB,L1
362    2       FORMAT (2I3)
363            DO  6  J=2,L1
364            J1=J-1
365            READ (5,3) (DIST(I,J),I=1,J1)
366    3       FORMAT (10F7.3)
367    6       CONTINUE
368            RETURN
369            END
370    C
371    C
372            SUBROUTINE WRITE1
373            DIMENSION NUMBER(60),DIST(60,60)
374            COMMON/BLANK/KB,L1,L2,N,DSTORE,DMAX0,DHALF0,DONE0,DMAX,DHALF,
375          1 DONE,DMAX1,DHALF1,DONE1
376            COMMON/DIST/DIST
377            COMMON/NUM/NUMBER
378            DO 16 J=1,L2
379    16      NUMBER(J)=J
380            I1=1
381            I2=11
382            IJON=0
383    20      IF (I2-L2) 22,22,21
384    21      I2=L2
385    22      J2=I1+1
386            IF (IJON) 26,27,26
387    27      J3=I1+22
388            GO TO 28
389    26      J3=I1+11
390    28      IF (J3-L1) 24,24,25
391    25      J3=L1
392    24      WRITE (6,1)
393    1       FORMAT(1H1,48X,12HINITIAL DATA,////)
394            IF (L1-12) 29,29,30
395    30      IF (J2-J3) 38,39,38
396    38      WRITE (6,23) I1,I2,J2,J3
397    23      FORMAT(1H0,30X,7HCOLUMNS,I4,4H  TO,I4,20X,
398          1 4HROWS,I4,4H  TO,I4///)
399            GO TO 29
400    39      IF (I1-I2) 44,41,44
401    44      WRITE (6,42) I1,I2,J2
402    42      FORMAT(1H0,30X,7HCOLUMNS,I4,4H  TO,I4,25X,3HROW,I4)
403            GO TO 29
404    41      WRITE (6,43) I1,J2
405    43      FORMAT(1H0,35X,6HCOLUMN,I4,29X,3HROW,I4)
406    29      IF (IJON) 31,32,31
407    31      IF(J3-I1-11) 33,33,32
408    33      WRITE (6,17)
409    17      FORMAT(1H0/////////////////////)
410    32      DO 5 J=J2,J3
411            IF (I2-J) 18,19,19
412    18      J1=I2
413            GO TO 5
414    19      J1=J-1
415    5       WRITE (6,2) J,(DIST(I,J),I=I1,J1)
416    2       FORMAT (1H0,3X,I3,3X,11F10.4)
417            WRITE (6,3) (NUMBER(I),I=I1,I2)
418    3       FORMAT (//1H0,7X,11(7X,I3))
419            IF (J3-L1)  34,35,35
420    34      J2=J3+1
421            J3=J3+22
422            GO TO 28
423    35      IF(I2-L2) 36,37,37
424    36      IJON=1-IJON
```

```
425            I1=I1+11
426            I2=I2+11
427            GO TO 20
428    37      IF (L1-12) 14,14,13
429    13      WRITE (6,9)
430    9       FORMAT(1H1,36X,27HSTATISTICS FOR INITIAL DATA)
431    14      WRITE (6,4) DMAX0
432    4       FORMAT(////1H0,32X,13HSIZE-ZERO    =,F11.4)
433            WRITE (6,6) DHALF0
434    6       FORMAT(1H0,32X,13HSIZE-HALF    =,F11.4)
435            WRITE (6,7) DONE0
436    7       FORMAT(1H0,32X,13HSIZE-ONE     =,F11.4)
437            CALL RCLOCK(T)
438            WRITE (6,8) T
439    8       FORMAT(//1H0,65X,14HLOCAL TIME IS ,F7.2,5H SECS)
440            RETURN
441            END
442    C
443    C
444            SUBROUTINE WRITE2
445            DIMENSION NUMBER(60),DIST(60,60)
446            COMMON/BLANK/KB,L1,L2,N,DSTORE,DMAX0,DHALF0,DONE0,DMAX,DHALF,
447          1 DONE,DMAX1,DHALF1,DONE1
448            COMMON/DIST/DIST
449            COMMON/NUM/NUMBER
450            DO 16 J=1,L2
451    16      NUMBER(J)=J
452            I1=1
453            I2=11
454            IJON=0
455    20      IF (I2-L2) 22,22,21
456    21      I2=L2
457    22      J2=I1+1
458            IF (IJON) 26,27,26
459    27      J3=I1+22
460            GO TO 28
461    26      J3=I1+11
462    28      IF (J3-L1) 24,24,25
463    25      J3=L1
464    24      WRITE (6,1) KB
465    1       FORMAT(1H1,51X,3HK =,I2////)
466            IF (L1-12) 29,29,30
467    30      IF (J2-J3) 38,39,38
468    38      WRITE (6,23) I1,I2,J2,J3
469    23      FORMAT(1H0,30X,7HCOLUMNS,I4,4H  TO,I4,20X,
470          1 4HROWS,I4,4H  TO,I4///)
471            GO TO 29
472    39      IF (I1-I2) 44,41,44
473    44      WRITE (6,42) I1,I2,J2
474    42      FORMAT(1H0,30X,7HCOLUMNS,I4,4H  TO,I4,25X,3HROW,I4)
475            GO TO 29
476    41      WRITE (6,43) I1,J2
477    43      FORMAT(1H0,35X,6HCOLUMN,I4,29X,3HROW,I4)
478    29      IF (IJON) 31,32,31
479    31      IF (J3-I1-11) 33,33,32
480    33      WRITE (6,17)
481    17      FORMAT(1H0//////////////////)
482    32      DO 5 J=J2,J3
483            IF (I2-J) 18,19,19
484    18      J1=I2
485            GO TO 5
486    19      J1=J-1
487    5       WRITE (6,2) J,(DIST(I,J),I=I1,J1)
488    2       FORMAT(1H0,3X,I3,3X,11F10.4)
489            WRITE (6,3) (NUMBER(I),I=I1,I2)
490    3       FORMAT(//1H0,7X,11(7X,I3))
491            IF (J3-L1) 34,35,35
492    34      J2=J3+1
493            J3=J3+22
494            GO TO 28
495    35      IF (I2-L2) 36,37,37
496    36      IJON=1-IJON
497            I1=I1+11
498            I2=I2+11
499            GO TO 20
```

```
500   37      IF (L1-12) 14,14,13
501   13      WRITE (6,9) KB
502   9       FORMAT(1H1,39X,19HSTATISTICS FOR  K =,I2)
503   14      WRITE (6,4) DMAX,DMAX1
504   4       FORMAT(////1H0,17X,14HDELTA-ZERO    =,F11.4,10X,
505         1 18HDELTA-ZERO-HAT    =,F10.4)
506           WRITE (6,15) DHALF,DHALF1
507   15      FORMAT(1H0,17X,14HDELTA-HALF    =,F11.4,10X,
508         1 18HDELTA-HALF-HAT    =,F10.4)
509           WRITE (6,6) DONE,DONE1
510   6       FORMAT(1H0,17X,14HDELTA-ONE     =,F11.4,10X,
511         1 18HDELTA-ONE-HAT     =,F10.4)
512           CALL RCLOCK(T)
513           WRITE (6,8) T
514   8       FORMAT(//1H0,65X,14HLOCAL TIME IS ,F7.2,5H SECS)
515           IF (KB-1) 10,10,11
516   10      WRITE (6,12)
517   12      FORMAT(///1H0,7X,16HK = 1  COMPLETED/1H1)
518   11      RETURN
519           END
520   C
521   C
522           SUBROUTINE CLOWTE
523           DIMENSION DIST(60,60),NUMBER(60),LIST(60)
524           COMMON/LIST/LIST
525           COMMON/DIST/DIST
526           COMMON/NUM/NUMBER
527           COMMON/BLANK/KB,L1,L2,N,DSTORE,DMAX0,DHALF0,DONE0,DMAX,DHALF,
528         1 DONE,DMAX1,DHALF1,DONE1
529           CALL CLIQUES(DIST,60,L1,DSTORE)
530           WRITE (6,5) DSTORE
531   5       FORMAT(//1H0,42X,19HCURRENT LEVEL IS    ,F7.3/)
532           L=0
533           IND=0
534   1       CALL UNPACK(LIST,IND)
535           K=0
536           L=L+1
537           DO 8 J=1,L1
538           IF (LIST(J)) 9,8,9
539   9       K=K+1
540           NUMBER(K)=J
541   8       CONTINUE
542           IF (K.GT.10) GO TO 7
543           WRITE (6,4) (NUMBER(J),J=1,K)
544   4       FORMAT(5X,33HA MAXIMAL COMPLETE SUBGRAPH IS  -,10I8)
545           GO TO 3
546   7       WRITE (6,4) (NUMBER(J),J=1,10)
547           WRITE (6,10) (NUMBER(J),J=11,K)
548   10      FORMAT(38X,10I8)
549   3       IF (IND) 2,1,1
550   2       WRITE (6,6) L,DSTORE
551   6       FORMAT(1H0,6X,I4,19H  M.C.S. AT LEVEL   ,F7.3)
552           RETURN
553           END
554   C
555   C
556           SUBROUTINE CLQONE
557           COMMON/BLANK/KB,L1,L2,N,DSTORE,DMAX0,DHALF0,DONE0,DMAX,DHALF,
558         1 DONE,DMAX1,DHALF1,DONE1
559           WRITE(6,16) KB
560   16      FORMAT(1H1,32X,44HMAXIMAL COMPLETE SUBGRAPHS BY LEVEL FOR  K =,
561         1 I2)
562           RETURN
563           END
564   C
565   C
566           SUBROUTINE CLQTWO
567           COMMON/BLANK/KB,L1,L2,N,DSTORE,DMAX0,DHALF0,DONE0,DMAX,DHALF,
568         1 DONE,DMAX1,DHALF1,DONE1
569           WRITE(6,7) DSTORE,N
570   7       FORMAT(////20X,17HFINAL LEVEL IS    ,F7.3,25X,I4,
571         1 14H LEVELS IN ALL)
572           RETURN
573           END
```

APPENDIX 4 Computation of C_u

We pointed out in Chapter 8.7 that a DC d satisfies the u-diametric inequality if and only if it satisfies the following four-point condition for all 4-tuples A, B, C_1, C_2 of elements of P.

$$d(C_1, C_2) > ul \Rightarrow d(A, B) \leqslant l$$

where

$$l = \max\{d(C_1, C_2), d(C_1, A), d(C_2, A), d(C_1, B), d(C_2, B)\}.$$

The value of u lies in [0, 1]. Clearly if X_1, X_2, X_3, X_4 is a 4-tuple of elements from P, the only pair A, B for which this condition might fail for an arbitrary DC d is a pair such that $d(A, B)$ is the maximal value of d on the 4-tuple. Since C_u is the subdominant method associated with u-diametric DC's, to find $C_u(d)$ from d we may proceed by checking each 4-tuple in turn, and if necessary reducing the largest value of the partly transformed DC on the 4-tuple to the next largest value, until the entire list of 4-tuples can be run through without further changes to the DC.

L. F. Meintjes and R. Sibson have written a program in USASI FORTRAN to implement this algorithm. The central routine which transforms d to $C_u(d)$ in the above-diagonal section of the array DIST is given below. I/O routines, not given here, are similar to those for the B_k program in Appendix 3. Since the algorithm for C_u, unlike that for B_k, does not operate on successively increasing levels, the Moody-Hollis technique for listing the ML-sets has to be applied level by level to the completely transformed DC; it is not possible to apply it concurrently with the algorithm in the way that can be done for B_k. Comments on the program contain details of notation. The vector D is of dimension 6, and DIST is a square array handled as in the program for B_k described in Appendix 3.

```
 1   C
 2   C
 3   C
 4         SUBROUTINE CU(NOBJ,U)
 5         DIMENSION D(6)
 6         COMMON/DIST/DIST(60,60)
 7         COMMON/NFLAG/NFLAG
 8         COMMON/ZU/ZU
 9   C
10   C   THIS ROUTINE IMPLEMENTS SIBSON'S CLUSTER METHOD CU, WHICH CONSTRUCTS
11   C   CLUSTERS WHOSE OVERLAP DIAMETER IN TERMS OF THE TRANSFORMED DC IS AT
12   C   MOST U*(CURRENT LEVEL).  NOBJ IS THE NUMBER OF OTU'S.  THE DC IS
13   C   STORED IN THE ABOVE-DIAGONAL REGION OF THE ARRAY 'DIST', WHICH
14   C   IS PASSED IN NAMED COMMON.
15   C
16         ZU = 1.1
17         NOBJ3 = NOBJ-3
18         NOBJ2 = NOBJ-2
19         NOBJ1 = NOBJ-1
20   C
21   C   N4 IS THE NUMBER OF 4-TUPLES.
22   C
23         N4 = NOBJ*NOBJ1*NOBJ2*NOBJ3/24
24   C
25   C   NSCORE IS A CONTROL VARIABLE WHICH COUNTS THE NUMBER OF SUCCESSIVE
26   C   4-TUPLES ON WHICH NO CHANGE IS MADE.  WHEN NSCORE REACHES N4 WE LEAVE
27   C   THE DO-LOOP NEST AND RETURN TO THE CALLING PROGRAM.
28   C
29         NSCORE = 0
30   C
31   C   NFLAG IS A FLAG VARIABLE USED TO CHECK POTENTIAL ROUNDING ERRORS.
32   C
33         NFLAG = 1
34   C
35   C   THE 4-DEEP NEST OF DO-LOOPS TOGETHER WITH THE TRANSFER INSTRUCTION
36   C   'GO TO 99' AFTER '10...CONTINUE' CAUSES 4-TUPLES TO BE CONSIDERED
37   C   IN LEXICOGRAPHIC ORDER, STARTING AGAIN WITH 1234 EACH TIME THE LAST
38   C   ONE IS REACHED.
39   C
40      99 DO 10  I1 = 1,NOBJ3
41         I11 = I1+1
42         DO 10  I2 = I11,NOBJ2
43         I22 = I2+1
44         DO 10  I3 = I22,NOBJ1
45         I33 = I3+1
46         DO 10  I4 = I33,NOBJ
47   C
48   C   VALUES OF DIST(I1,I2),...,DIST(I3,I4) ARE ENTERED INTO THE VECTOR 'D'
49   C   IN SUCH A WAY THAT I AND 7-I LABEL OPPOSITE EDGES.
50   C
51         D(1) = DIST(I1,I2)
52         D(2) = DIST(I1,I3)
53         D(3) = DIST(I1,I4)
54         D(4) = DIST(I2,I3)
55         D(5) = DIST(I2,I4)
56         D(6) = DIST(I3,I4)
57   C
58   C   WE FIND THE LARGEST VALUE 'DTOP' OF D, AND THE VALUE 'ITOP' OF I FOR
59   C   WHICH IT OCCURS.
60   C
61         DTOP = 0.0
62         DO 2  I = 1,6
63         IF(D(I)-DTOP) 2,4,4
64       4 DTOP = D(I)
65         ITOP = I
66       2 CONTINUE
67   C
68   C   'DNEXT' IS THE LARGEST VALUE OF THE D(I) EXCLUDING I = ITOP.
69   C
70         DNEXT = 0.0
71         DO 3  I = 1,6
72         IF(I-ITOP) 5,3,5
73       5 IF (D(I)-DNEXT) 3,3,55
74      55 DNEXT = D(I)
75       3 CONTINUE
76   C
```

```
 77  C      IF DTOP = DNEXT, NO CHANGE IS MADE ON THE 4-TUPLE.  1 IS ADDED TO
 78  C      NSCORE AND IT IS CHECKED AGAINST N4, AS A RESULT OF WHICH WE EITHER
 79  C      LEAVE THE MAIN DO-LOOP NEST OR GO TO THE END OF THIS LOOP.
 80  C
 81           IF(DTOP-DNEXT) 6,6,7
 82        6  NSCORE = NSCORE+1
 83           IF(N4-NSCORE) 11,11,10
 84  C
 85  C      THE OPPOSITE EDGE TO THAT LABELLED ITOP IS THAT LABELLED ICROSS.
 86  C
 87        7  ICROSS = 7-ITOP
 88  C
 89  C      U*DNEXT IS CHECKED AGAINST D(ICROSS).  IF U*DNEXT IS GREATER THAN OR
 90  C      EQUAL TO D(ICROSS) THEN NO CHANGE IS MADE ON THE 4-TUPLE, AND WE
 91  C      PROCEED AS FOR DTOP = DNEXT.  IF U*DNEXT IS LESS THAN D(ICROSS) WE
 92  C      REDUCE THE LARGEST VALUE OF DIST ON THE 4-TUPLE FROM DTOP TO DNEXT
 93  C      AND RESET NSCORE TO ZERO.  IF THE DIFFERENCE BETWEEN U*DNEXT AND
 94  C      D(ICROSS) IS SMALL, NFLAG IS SET TO ZERO, OTHERWISE IT IS LEFT AS
 95  C      IT WAS ON ENTRY TO THIS LOOP, THE INITIAL VALUE BEING 1.  THE FINAL
 96  C      VALUE OF NFLAG IS PASSED BACK TO THE CALLING PROGRAM, AND MAY BE
 97  C      USED TO CAUSE PRINTING OF A WARNING MESSAGE IF DESIRED.
 98  C
 99           TEST = U*DNEXT-D(ICROSS)
100           UU = D(ICROSS)/DNEXT
101           IF(UU-ZU) 15, 14, 14
102       15  ZU = UU
103       14  CONTINUE
104           IF(0.0000001-ABS(TEST)) 12,12,13
105       13  NFLAG = 0
106       12  IF(TEST) 8,6,6
107        8  NSCORE = 0
108           GO TO (61,62,63,64,65,66),ITOP
109       61  DIST(I1,I2) = DNEXT
110           GO TO 10
111       62  DIST(I1,I3) = DNEXT
112           GO TO 10
113       63  DIST(I1,I4) = DNEXT
114           GO TO 10
115       64  DIST(I2,I3) = DNEXT
116           GO TO 10
117       65  DIST(I2,I4) = DNEXT
118           GO TO 10
119       66  DIST(I3,I4) = DNEXT
120       10  CONTINUE
121           GO TO 99
122       11  CONTINUE
123           RETURN
124           END
125  C
126  C
127  C
```

APPENDIX 5 A Cluster-listing Method

We have pointed out earlier that although in the case of an ultrametric DC d describing a dendrogram it is very simple to pass between the DC d and the lists of clusters at each level which are equivalent to it, the general case presents much more difficulty. Indeed, it is no exaggeration to say that the possibility of fruitfully applying sophisticated cluster methods such as B_k and C_u is crucially dependent on the availability of an efficient cluster listing method. J. K. M. Moody and J. Hollis have recently devised such a method based on a suggestion by Dr R. M. Needham and we are indebted to Dr Moody for his assistance in preparing the following account of it. The method operates one level at a time, and to list the ML-sets for a DC d it needs to be applied separately at each splitting level. The present account is given in terms of finding maximal complete subgraphs in a graph; this problem is trivially equivalent to that of finding ML-sets for a relation.

Let P be the vertex set for the graph Γ. As usual, take $|P| = p > 0$. A maximal complete subgraph for Γ is a subgraph in which every vertex is linked to every other vertex, and which is maximal with respect to this property. Conventionally, every vertex is assumed to be linked to itself. We set up mutually inverse 1-1 correspondences σ, V between binary vectors of length p and subsets of P as follows. Order the elements of P as $A_1, \ldots, A_p$.

If $S \subseteq P$, define $V(S)$ by

$$[V(S)]_i = 1 \text{ if } A_i \in S$$
$$= 0 \text{ if } A_i \notin S$$

If v is a binary vector, define $\sigma(v) \subseteq P$ by

$$A_i \in \sigma(v) \text{ if } v_i = 1$$
$$A_i \notin \sigma(v) \text{ if } v_i = 0$$

Let M be the binary $p \times p$ matrix such that $M_{ij} = 1$ if and only if A_i and A_j are joined by an edge; thus $M_{ij} = M_{ji}$ and $M_{ii} = 1$. The i'th row of M will be denoted by ρ_i. M is commonly known as an *incidence matrix*. δ_i will denote the binary vector with a 0 in the i'th place and 1's elsewhere. $\mathbf{1}$ will denote the binary vector of 1's. If v_1, v_2 are binary vectors, $v_1 \wedge v_2$ will denote the vector v which has a 1 in those places where both v_1 and v_2 have 1's, and has 0's elsewhere; thus, $v_1 \wedge v_2$ is the logical AND of the two vectors.

The correspondences σ, V identify binary vectors with subsets. The vector $\mathbf{1}$ is identified with the subset $\sigma(\mathbf{1}) = P$. δ_i corresponds to $\sigma(\delta_i) = P - \{A_i\}$. ρ_i

corresponds to the set $\sigma(\rho_i)$ of elements of P linked to A_i. Suppose that $Q \subseteq P$ is the vertex set of a maximal complete subgraph. Then for each vertex $A_i \in P$, either $A_i \notin Q$, or A_i is linked to every member of Q. We may express this in terms of vectors by saying that either $\delta_i \wedge V(Q) = V(Q)$ or $\rho_i \wedge V(Q) = V(Q)$. The algorithm operates by using this property for each row of the matrix in turn to construct at each stage a set $C^{(i)}$ of vectors $c_j^{(i)}$ such that every maximal complete subgraph is contained in some $\delta(c_j^{(i)})$. The set $C^{(p)}$ obtained as the final outcome of the process describes exactly all the maximal complete subgraphs.

The algorithm proceeds as follows. $C^{(0)}$ is taken to be $\{1\}$. $\sigma(1) = P$, so $C^{(0)}$ certainly has the property that every maximal complete subgraph is in a $\sigma(c_j^{(i)})$. Now consider the first vertex A_1. Every maximal complete subgraph either contains A_1, in which case it lies in some $\sigma(1 \wedge \rho_1) = \sigma(\rho_1)$, or it does not, in which case it lies in $\sigma(1 \wedge \delta_1) = \sigma(\delta_1)$. So we construct $C^{(1)} = \{\rho_1, \delta_1\}$. Similarly, for A_2, every maximal complete subgraph lies in one of $\rho_1 \wedge \rho_2, \rho_1 \wedge \delta_2, \delta_1 \wedge \rho_2, \delta_1 \wedge \delta_2$. However, these may not all be distinct, we eliminate duplicates, and also any vectors v_1 such that $\sigma(v_1) \subsetneqq \sigma(v_2)$ where v_1, v_2 are vectors in the set we have just constructed. The resultant reduced set is $C^{(2)}$. We go on to form $C^{(3)}$ by generating new vectors by the operations $\wedge\, \rho_3$ and $\wedge\, \delta_3$ and performing the same elimination processes. Eventually we reach a set $C^{(p)}$ of vectors which still have the property that every maximal complete subgraph is contained in some $\sigma(c_j^{(p)})$. It is not hard to prove that in fact the $\sigma(c_j^{(p)})$ actually are the maximal complete subgraphs.

This completes the definition of the fundamental algorithm. Its computational efficiency depends on a systematic method of checking duplication and containment in the sets $\sigma(c_j^{(i)})$. This is done by dividing the vectors in $C^{(i)}$ into three subsets $D^{(i)}$, $E^{(i)}$, $F^{(i)}$ as they are constructed. $D^{(i)}$ consists of those vectors $c_j^{(i-1)}$ such that $A_i \notin \sigma(c_j^{(i-1)})$. $E^{(i)}$ consists of those vectors of the form $c_j^{(i-1)} \wedge \rho_i$ where $A_i \in \sigma(c_j^{(i-1)})$, and $F^{(i)}$ of those of the form $c_j^{(i-1)} \wedge \delta_i$ where $A_i \in \sigma(c_j^{(i-1)})$. It can be shown that, in the obvious notation, the only possible relations between vectors which would be put into these three sets are $\sigma(e) \subseteq \sigma(e')$, $\sigma(f) \subseteq \sigma(e)$, and $\sigma(f) \subseteq \sigma(d)$, and the systematic use of a checking procedure based on this makes it possible to construct $C^{(i)}$ efficiently from $C^{(i-1)}$.

It is obvious that the algorithm is very ill-suited to being programmed in a language such as FORTRAN. Moody and Hollis have implemented it in machine-code on the Titan computer in the Cambridge University Mathematical Laboratory. A BCPL program implementing a slightly improved version of the algorithm has been written by Dr M. Richards and a version in a general macro language is being prepared.

APPENDIX 6 Fast Hierarchic Clustering

We are indebted to C. J. van Rijsbergen for the following account of his fast large-scale algorithm for the single-link cluster method.

The algorithm which implements the cluster methods B_k (see Appendix 3) produces a numerical representation of the clusterings. Clusters are listed by the method described in Appendix 5. When only a specification of the single-link (B_1) clusters and their levels is required faster and more direct algorithms are available. Gower and Ross (1969) describe an algorithm which does not produce the clusters directly but relies on initially generating a minimum spanning tree of the object points. A faster algorithm designed for large data sets is given by Wishart (1969b); it proceeds agglomeratively, but involves successive transformations of the input matrix as fusion of clusters proceeds. The algorithm below, which has been described in van Rijsbergen (1970b), operates directly on the input matrix.

The subprogram to be described generates the dendrogram specified by single-link. The clusters are generated level by level starting at the lowest level, so the algorithm is of the agglomerative type. The levels considered by the subroutine are the splitting levels in the dendrogram. The program also evaluates $\hat{\Delta}_1$, a normalized distortion measure. For the initialization of parameters and size of arrays see the comments in the program. The dissimilarity matrix is passed to the subroutine in sections. Each section is stored in the vectors IX and IY.

LINK is called for each distinct value of the dissimilarity coefficient. It then proceeds to operate on the objects stored in the vectors IX and IY. These vectors contain the pairs (I, J) such that d(I, J) = h where h is the current level. Each cluster generated by LINK at the current level is uniquely defined by a number NCL; that is, OB(I, = NCL ⇔ object I B cluster NCL Single-element clusters are ignored. Vector IT is maintained to point to the second member of a cluster. All clusters are stored in LIST. This means that if IT(NCL) = K1 then K1 ∈ cluster NCL and LIST(K1) = K2 is the second member of the cluster NCL, LIST(K2) = K3 is the third, etc. The cluster is terminated by LIST(KN) = – 1. Thus the cluster NCL = {K1, K2, . . ., KN}.

For each linked pair (I, J) in the list exactly one of the following conditions is satisfied:

(1) neither I nor J belong to a cluster.
(2) I belongs to a cluster and J belongs to a different cluster.

(3) I and J belong to the same cluster.
(4) I (or J) belongs to a cluster and J (or I) does not belong to a cluster.

Note again that single-element clusters are ignored. The subsequent action taken by LINK depends on the condition satisfied. For condition (1) a new cluster is created consisting of I and J. For condition (4) the already existing cluster is extended. In both cases the relevant modifications to vectors IT, OB and LIST are carried out immediately. For condition (3) no action is taken and the next pair in line is considered. For condition (2) no immediate action is taken and the two clusters are tagged. When IX and IY have been exhausted the two tagged clusters, which in the meantime may have acquired new elements, are fused. This again is achieved by modifying IT, OB and LIST.

Subroutine CLOUT unpacks the clusters from LIST at the current level providing that at least one cluster has been generated which is different from the cluster at the previous level.

The algorithm can readily be adapted to find the clusters satisfying the L and L* conditions described in Chapter 12. It is necessary only to select single-link clusters which satisfy the appropriate condition. An algorithm suggested by McQuitty (1963) could readily be modified to find directly all clusters which satisfy the L-condition. The rank ordering of dissimilarity values between each object and all the remaining objects is first determined. The algorithm then operates on the set of such rank-orderings. Van Rijsbergen (1970a) has described an algorithm which finds directly all clusters satisfying the L*-condition. For large data sets both of these direct algorithms are slower than the indirect process of first performing single-link cluster analysis and then examining clusters.

Subroutine LINK has been tested on the Titan computer at the Cambridge University Mathematical Laboratory. For 20 objects the program takes 2 seconds to generate and output the clusters and levels. It has been run against the Cranfield collection of 200 documents. This it took about 70 seconds to cluster. In each case DEL, the sum of the entries in the upper diagonal matrix representing the ultrametric dissimilarity coefficient characterizing the dendrogram, was calculated. From DEL the distortion $\hat{\Delta}_1$ is simply calculated. The computation time is a function of the number of levels. The program will run in 12K words of store for 1000 objects.

```
      SUBROUTINE LINK(MO,OB,CL,NBUF,NCL,IX,IY,JM,DEL,LIST,IT)
C     THE ARRAYS OB,LIST,IT MUST BE INITIALISED TO ZERC BY THE
C     CALLING PROGRAM. THE DIMENSION OF OB,IT AND LIST MUST
C     EQUAL JM.THE DIMENSION OF IX AND IY EQUALS THE MAXIMUM
C     EXPECTED VALUE OF NBUF OR JM WHICH EVER IS THE LARGER.
C     ON ENTRY TO LINK IX AND IY WILL CONTAIN THE PAIRS
C     (IX(K),IY(K)),K=1,NBUF SUCH THAT Z(K)=Z(K+1),K=1,NBUF-1
C     WHERE D(IX(K),IY(K))=Z(K) AND D IS THE DISSIMILARITY
C     COEFFICIENT. CL IS THE CURRENT LEVEL AT WHICH CLUSTERING
C     IS TAKING PLACE. OTHER VARIABLES TO BE INITIALISED;
C      JM - NO. OF OBJECTS TO BE CLUSTERED.
C      DEL- INITIAL VALUE 0.
C      MO - NO. OF OUTPUT STREAM.
C      NCL- INITIAL VALUE 1 .
      DIMENSION LIST(JM),IT(JM),OB(JM),IX(NBUF),IY(NBUF)
      INTEGER OB,OBI,OBJ,OBIJ
      CH=0.
      NN=0
      DO 8 M=1,NBUF
      I=IX(M)
      J=IY(M)
      OBI=OB(I)
      OBJ=OB(J)
      OBIJ=OBI+OBJ
      IF(OBIJ)3,3,1
 1    IF(OBI)4,4,2
 2    IF(OBJ)5,5,6
 3    OB(I)=NCL
      OB(J)=NCL
      IT(NCL)=J
      NCL=NCL+1
      LIST(I)=-1
      LIST(J)=I
      DEL=DEL+CL
      GOTO 7
 4    OB(I)=OBJ
      LIST(I)=IT(OBJ)
      IT(OBJ)=I
      DEL=DEL+(COUNT(LIST,I,JM)-1.)*CL
      GOTO 7
 5    OB(J)=OBI
      LIST(J)=IT(OBI)
      IT(OBI)=J
      DEL=DEL+(COUNT(LIST,J,JM)-1.)*CL
      GOTO 7
 6    IF(OBI.EQ.OBJ)GOTO 8
      NN=NN+1
      IX(NN)=OBI
      IY(NN)=OBJ
 7    CH=2.
 8    CONTINUE
      IF(CH)17,17,9
 9    IF(NN)16,16,10
 10   DO 15 J=1,NN
      II=IX(J)
      JJ=IY(J)
      IF(II.EQ.JJ)GOTO 15
      IB1=IT(II)
      JB1=IT(JJ)
      XXX=COUNT(LIST,IB1,JM)*COUNT(LIST,JB1,JM)
      DEL=DEL+XXX*CL
      DO 12 L=1,JM
      INOB=OB(L)
      IF(INOB.NE.II)GOTO 12
      OB(L)=JJ
      IF(LIST(L))11,12,12
 11   LIST(L)=IT(JJ)
      IT(JJ)=IT(II)
      IT(II)=0
 12   CONTINUE
      IF(J.EQ.NN)GOTO 16
      JDO=J+1
      DO 14 LL=JDO,NN
      IF(IY(LL).EQ.II)IY(LL)=JJ
      IF(IX(LL).EQ.JJ)GOTO 13
      IF(IX(LL).NE.II)GOTO 14
 13   IX(LL)=IY(LL)
      IY(LL)=JJ
 14   CONTINUE
```

```
15        CONTINUE
16        CALL CLOUT(MO,NCL,LIST,CL,IT,IX,JM)
17        RETURN
          END
C
          SUBROUTINE CLOUT(MO,NCL,LIST,CL,IT,IX,JM)
C         THIS SUBR. UNPACKS THE VECTOR LIST AT LEVEL CL.THE VECTOR
C         CONTAINS ALL THE CLUSTERS OF MORE THAN ONE ELEMENT AT LEVEL CL
C         THE VECTOR IT POINTS TO THE SECOND ELEMENT OF THE CLUSTER
C         IDENTIFIED BY I. IT(I) IS THE FIRST ELEMENT OF CLUSTER
C         I. IF IT(I) IS 0 IT IS IGNORED.
          DIMENSION LIST(JM),IT(JM),IX(JM)
          WRITE(MO,1)CL
1         FORMAT(//30X,14HCURRENT LEVEL=,F10.3)
          DO 7 I=1,NCL
          ICL=IT(I)
          IF(ICL)7,7,2
2         J=1
          IX(J)=ICL
3         J=J+1
          ICL=LIST(ICL)
          IF(ICL)5,5,4
4         IX(J)=ICL
          GOTO 3
5         WRITE(MO,6)(IX(II),II=1,J-1)
6         FORMAT(9H0CLUSTER=,10(I5,1H,))
7         CONTINUE
          RETURN
          END
C
          FUNCTION COUNT(LIST,I,JM)
C         THIS SUBPROGRAM COUNTS THE NUMBER OF ELEMENTS IN A GIVEN
C         CLUSTER. THIS ROUTINE IS ONLY NECESSARY TO CALCULATE DEL
          DIMENSION LIST(JM)
          II=I
          COUNT=0.
1         COUNT=COUNT+1.
          II=LIST(II)
          IF(II)2,2,1
2         RETURN


          END
```

APPENDIX 7 Illustrative Applications

The two examples given here show how some of the methods described in the book may be used to attack specific taxonomic problems. Both examples are fragments of larger scale taxonomic studies.

Morphological differentiation in the genus Silene

The data used in this example was collected by Dr E. A. Bari in the course of a study of the experimental taxonomy of annual European *Silene*, carried out under the supervision of Dr S. M. Walters, and described in Bari (1969). The genus *Silene* is a large genus of the family Caryophyllaceae of flowering plants. The sectioning of the genus adopted by Chowdhuri (1957), and by Chater and Walters (1964) may, at least in part, be an artificial one. One of

Table 11. *Silene* species tested in the study.

OTU's	Section in *Flora Europaea*
1 *S. sedoides*	*Atocion*
2 *S. sericea*	*Scorpioideae*
3 *S. coeli-rosa*	*Eudianthe*
4 *S. nocturna* var. *brachypetala*	*Scorpioideae*
5 *S. pendula*	*Erectorefractae*
6 *S. graeca*	*Behenantha*
7 *S. conoidea*	*Conomorpha*
8 *S. trinervia*	*Lasiocalycinae*
9 *S. cerastoides*	*Silene*
10 *S. bellidifolia*	*Silene*
11 *S. gallica*	*Silene*
12 *S. dichotoma*	*Dichotomae*
13 *S. scabriflora*	*Scorpioideae*
14 *S. inaperta*	*Rigidulae*
15 *S. portensis*	*Rigidulae*
16 *S. nocturna*	*Scorpioideae*
17 *S. nicaeensis*	*Scorpioideae*
18 *S. colorata*	*Dipterospermae*
19 *S. psammitis*	*Erectorefractae*
20 *S. littorea*	*Erectorefractae*
21 *S. cretica*	*Behenantha*
22 *S. rubella*	*Atocion*
23 *S. conica* ssp. *subconica*	*Conomorpha*
24 *S. conica* ssp. *conica*	*Conomorpha*
25 *S. conica* ssp. *sartorii*	*Conomorpha*
26 *S. fuscata*	*Atocion*
27 *S. behen*	*Behenantha*
28 *S. muscipula*	*Behenantha*
29 *S. colorata* var. *decumbens*	*Dipterospermae*
30 *S. apetala*	*Dipterospermae*
31 *S. laeta*	*Eudianthe*

the aims of the work described here was to find out which, if any, of the sections recognized by these authors correspond to phenetic clusters obtained using a large selection of morphological attributes. Bari studied the seed morphology of the annual species in detail; a second aim of the work described here was to find out whether a phenetic classification based on seed attributes alone gives a clustering which fits well the clustering obtained from the whole set of attributes. A final aim was to find out whether or not different kinds of attributes give highly discordant discriminations of the species studied.

The selection of OTU's was largely determined by the availability of material. The thirty-one OTU's selected correspond to taxa of specific or subspecific rank in Chater and Walters' (1964) account in *Flora Europaea.*

Table 12. Character states for some attributes of the seed of *Silene sedoides.* Where character states of conditionally defined attributes are used, the weight is indicated to the left of the attribute name. When the weight is zero, as for *Back type 2,* character states cannot be recorded. For *Seed size* the mean and variance of the logarithms of the values are used.

Attribute		Character states
Seed size		$\log [\mathcal{N}(-.3506, .0655)]$
Crest		crested (0) not crested (1)
Beard		bearded (0) not bearded (1)
Tubercles		tuberculate (.27) not tuberculate (.73)
	if *tuberculate*	
Weight (.27)	Tubercle type	granulate (0) low rounded (0) cylin. rounded (0) cylin. truncate (0) umbonate (1)
Back type 1		convex (0) concave (1) canaliculate (0)
	if *canaliculate*	
Weight (0)	Back type 2	winged (–) not winged (–)
Back width		broad (1) narrow (0)
Face type		convex (.50) flat (.50) shallow concave (0) deep concave (0) excavate (0)
Hylar zone		distinct (0) not distinct (1)
	if *distinct*	
Weight (0)	Hylar zone type	protruding (–) inset (–)
Type of zonation		hatched (0) irregular (0) transverse (.50) concentric (.50) longitudinal (0)
Cell boundary type		serrate (0) straight (1)
Seed shape		globular (0) broad reniform (0) reniform (1)

The representation of the various sections of *Silene* is uneven. The OTU's selected are listed with their *Flora Europaea* sections in Table 11.

As pointed out in Chapter 15, it is inadvisable to estimate character states from one or two specimens, even for attributes thought to be constant within OTU's; where quantitative attributes are used larger samples are obviously needed. In this case fifteen specimens from each OTU were examined. Sixty-nine attributes were used. The way in which character states were recorded is illustrated in Table 12 which shows the information derived from attributes of the seed in *Silene sedoides.*

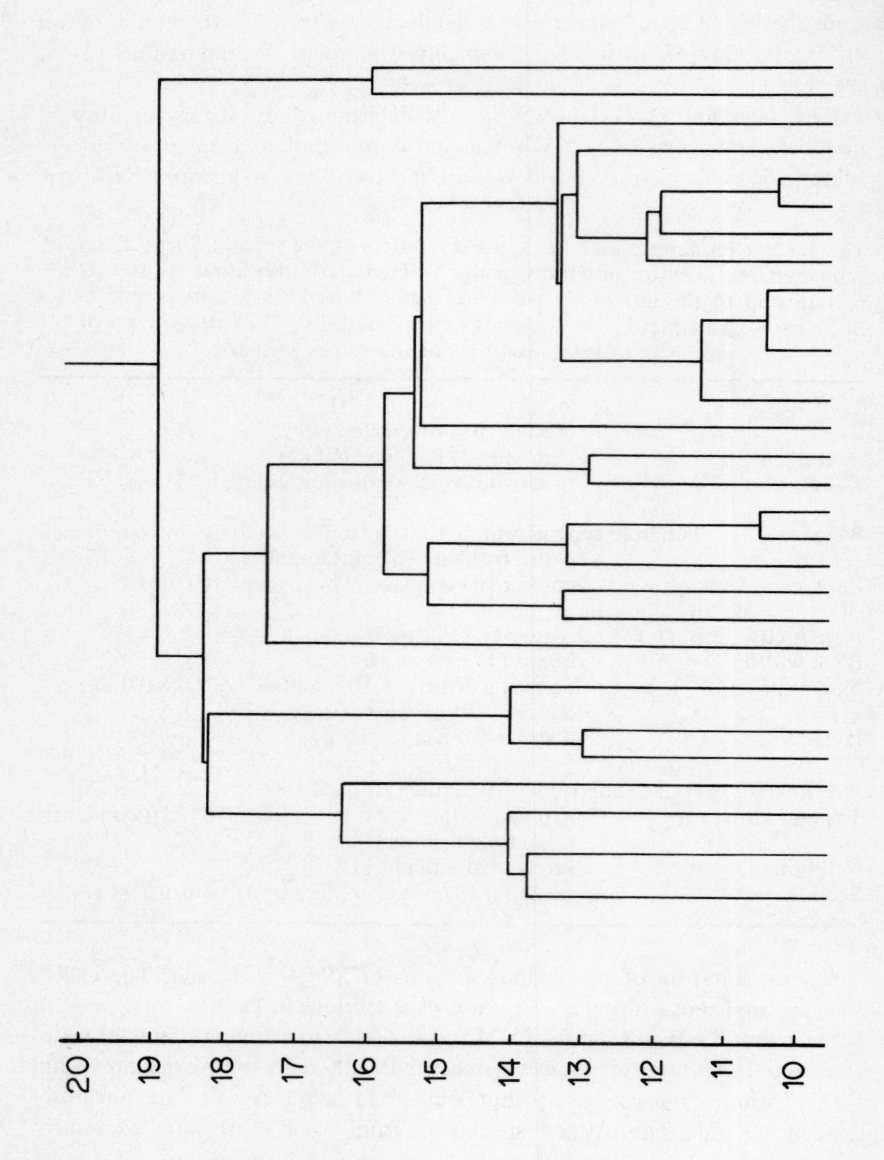
20
19
18
17
16
15
14
13
12
11
10

31 3 10 14 16 11 9 4 30 29 18 2 22 19 13 8 20 17 5 26 1 28 27 21 12 6 15 25 24 23 7
9 8 7 6 5 4 3 2 1 0

Figure 1.

K-dissimilarity coefficients on the thirty-one OTU's were then calculated both for the entire set of attributes and for particular selections of attributes. The single-link clustering for the entire set of attributes is shown in Figure 1. The clusterings obtained by B_2 and B_3 did not produce any marked drop in distortion and yielded little further information.

The single-link clustering yields information which confirms conclusions reached independently by Bari and previous workers from intuitive assessment of morphology. OTU's 3 (*S. coeli-rosa*) and 31 (*S. laeta*) which are placed in the section *Eudianthe* by Chater and Walters (1964) are highly distinct both from each other and from the rest of the genus. Bari (1969) suggested that the former should be allocated to the allied genus *Agrostemma* and that the latter should be allocated to the allied genus *Lychnis*. OTU's 7, 23, 24, and 25 form a fairly isolated cluster. They all belong to the section *Conomorpha* which differs from all the other annual *Silene* in chromosome number and which previous authors have consistently recognized as a natural group. It has sometimes been treated as a distinct genus or as a distinct subgenus of *Silene*. OTU's 5, 19, and 20, the representatives of the section *Erectorefractae*, form a moderately isolated cluster. OTU's 6, 21, and 27 which belong to the section *Behenantha* fall in the same cluster as 12 which belongs to the section *Dichotomae*, and are separated from 28 which belongs to *Behenantha*. The low-level clusters {18, 29} and {23, 24} each represent a pair of subtaxa of the same species. There is no correspondence between clusters in the single-link clustering or between clusters in the clustering obtained by B_2 and other sections of the genus recognized by Chater and Walters (1964) or by earlier authors. The conclusion is that *Conomorpha* and *Erectorefractae* may be natural groups, that *Behenantha* may be in part a natural group, and that other sections are artificial. It appears that in the species referred to other sections the pattern of differentiation is reticulate.

These conclusions are tentative. A wider selection of OTU's would be needed to confirm them, and collection of further data is in progress. In so large a genus sectioning, however artificial, is useful for diagnostic purposes and these results do not provide any grounds for drastic taxonomic revision.

In order to test whether different kinds of attributes give significantly different patterns of discrimination of the OTU's *K*-dissimilarities were calculated separately for selections of fourteen seed attributes, fourteen flower and fruit attributes, fourteen habit and branching attributes, and fourteen calyx and leaf attributes. In Figure 2 the discordances between these dissimilarity coefficients are compared with the distribution of discordances between sixty pairs of dissimilarity coefficients based on randomly selected

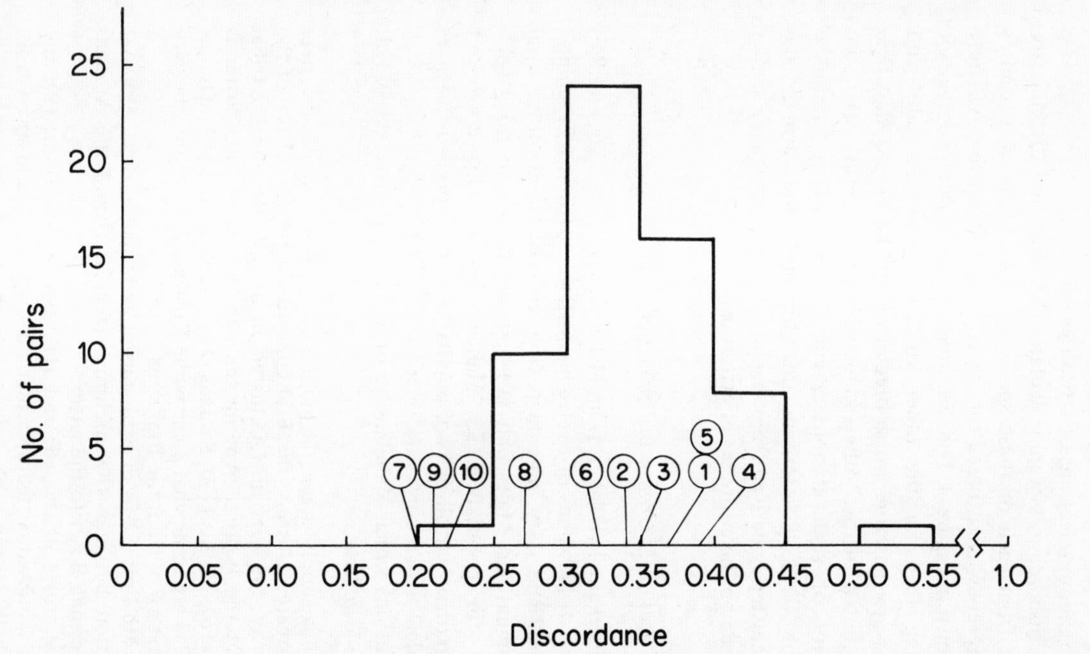

Figure 2. Distribution of the $\Delta_1{}^*$ discordance measure (see Chapter 11.3) between sixty pairs of dissimilarity coefficients based on randomly selected sets of fourteen attributes. The discordances between dissimilarity coefficients based on pairs of fourteen attributes from different functional complexes are indicated as follows. (1) Seed: Habit and Inflorescence. (2) Seed: Fruit and Flower. (3) Seed: Calyx and Leaf. (4) Habit and Inflorescence: Fruit and Flower. (5) Habit and Inflorescence: Calyx and Leaf. (6) Fruit and Flower: Calyx and Leaf. The discordances between the dissimilarity coefficients based on fourteen attributes from each functional complex and the dissimilarity coefficient based on the entire set of attributes are indicated as follows. (7) Seed: Total. (8) Habit and Inflorescence: Total. (9) Fruit and Flower. Total. (10) Calyx and Leaf: Total.

sets of fourteen attributes. These results indicate that sets of attributes from different 'functional complexes' are not, in this case, significantly more discordant than are random selections of attributes.

The normalized discordances were calculated between the dissimilarity coefficients derived from each of these specially selected sets of attributes and the dissimilarity coefficient based on the entire set of attributes. These discordances are shown in Figure 2. The discordance is substantially larger for habit and branching than for the other kinds of attribute. Habit and branching attributes are therefore worse predictors for the overall classification than are attributes of the other kinds. This may result from the phenotypic plasticity of such attributes, and perhaps partly from the difficulty in finding ways of describing unambiguously the form of the inflorescence. The clustering derived from seed attributes alone is very similar to the clustering based on the entire set of attributes.

Differentiation of Local Caste Groups in Upper Bengal

The data used in this example is taken from Mahalanobis, Majumdar, and Rao (1949). They recorded the values of twelve quantitative attributes in samples of about one hundred and sixty individuals from each of twenty-three local caste and tribal populations. Males only were scored for populations 1-22, and females only for population 23. Mahalanobis and his coworkers calculated the D^2 statistic for all pairs of populations, using pooled variances and covariances to deal with inconstant dispersion.

The main conclusions which Mahalanobis drew from intuitive assessment of the D^2 values are as follows.

> The D^2-values thus supply a general picture of the following kind. There are three well-demarcated clusters, the Brahmins (B-cluster) at the top of the Hindu social hierarchy; the Artisans (A-cluster) in the middle; and the tribal groups (T-cluster) at the bottom. Among the tribals, a small sub-cluster is formed by Cheros (16), Majhis (17), Panikas (18) and Kharwars (19); Oraons (20) and Rajwars (21) are somewhat close while Korwas (22) occupy the very bottom place furthest away from the Brahmins.
>
> Chattris (4) and Muslims (5) form a sub-cluster and together with Agharias (3) occupy a position between the Brahmins (B-cluster) and the Artisans (A-cluster) but somewhat closer to the latter.
>
> The two criminal tribes Bhatus (6) and Habrus (7), and Doms (9), and Bhils (8) have highly individual features of their own, and stand apart from the other groups which form a main sequence. Each of these four groups has in fact a distinct position of its own, and cannot be fitted into the general scheme determined by the three basic clusters (B), (A), and (T). Tharu (14) has also a place of its own somewhere on the border line of the T-cluster but quite distinct from the tribal groups. Finally, Chamar (15) is also a border-

line case separating the higher castes and tribals but nearer to the Artisans. (Quoted from Mahalanobis, Majumdar, and Rao (1949); our reference numbers.)

He summarized these conclusions together with conclusions based on language, burial and marriage customs, and geographical disposition in the diagram shown in Figure 3.

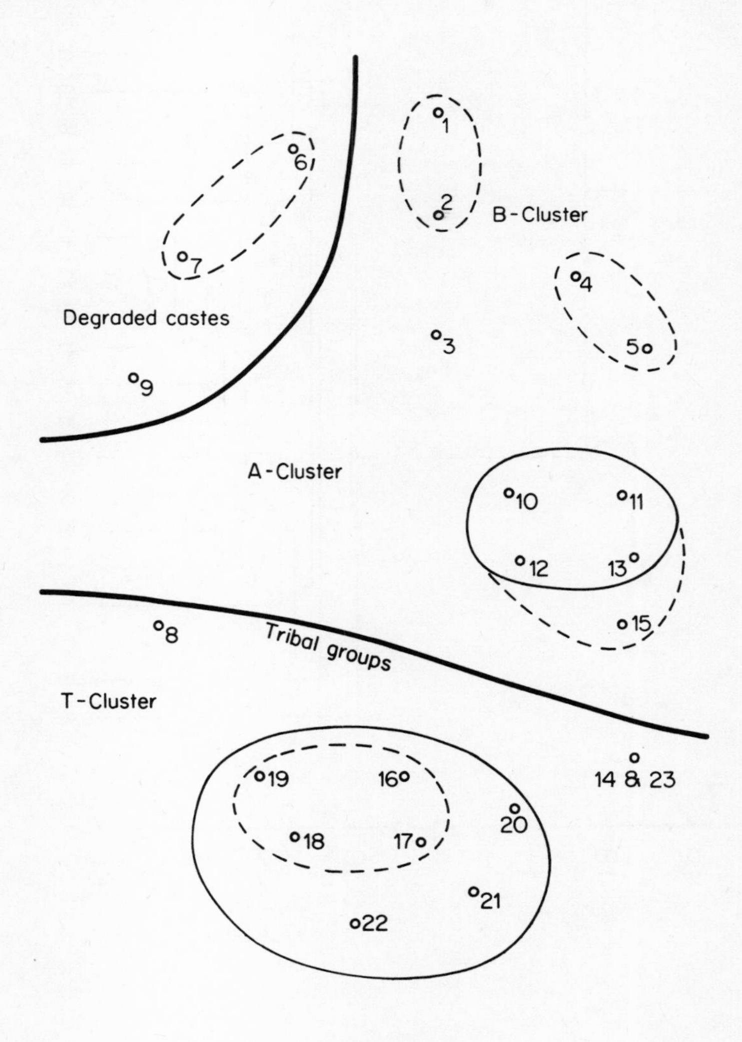

Figure 3.

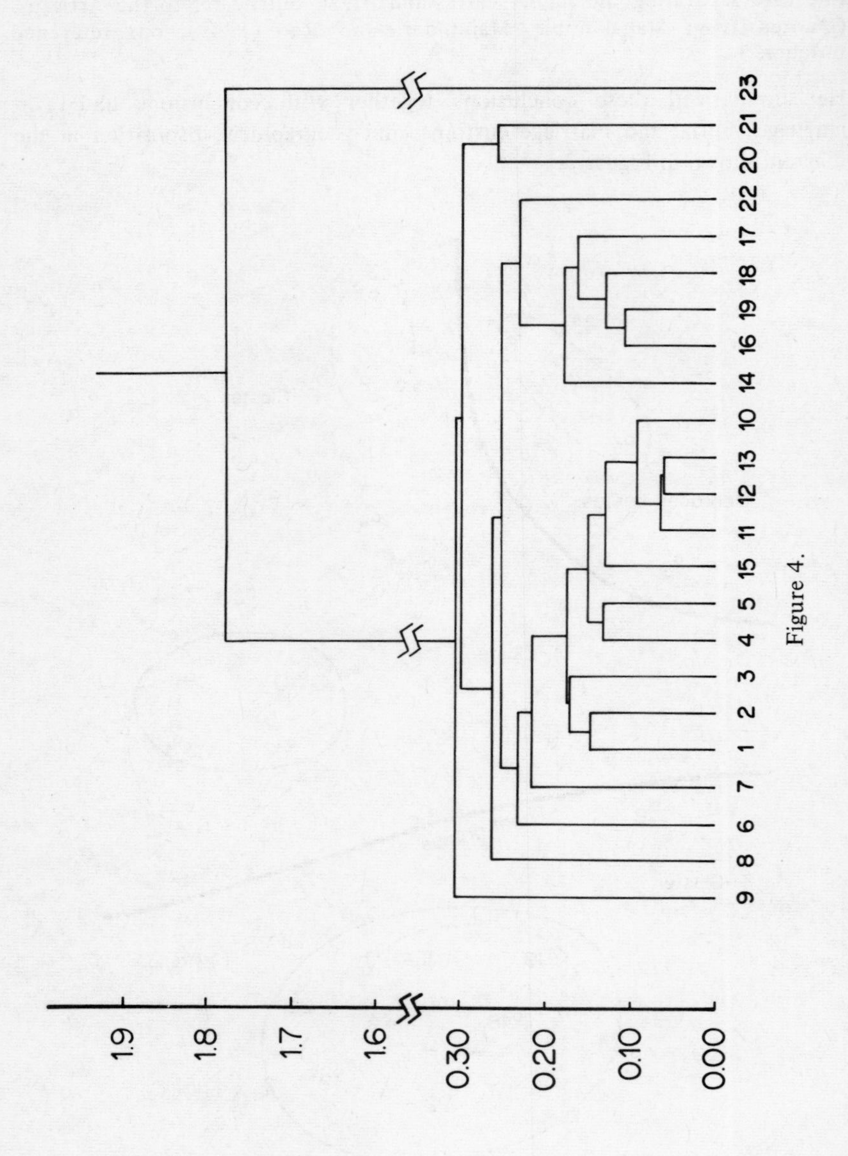
1.9
1.8
1.7
1.6
0.30
0.20
0.10
0.00
9
8
6
7
1
2
3
4
5
15
11
12
13
10
14
16
19
18
17
22
20
21
23

Figure 4.

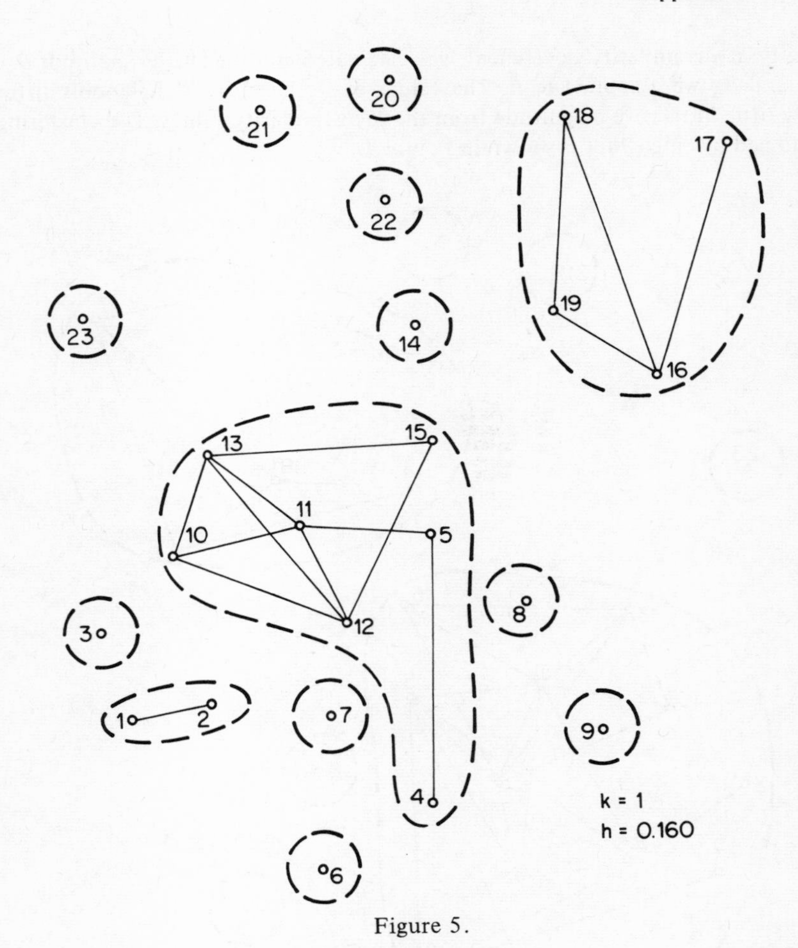

Figure 5.

KEY TO FIGURES 4-10

1 Brahmin Basti
2 Brahmin Others
3 Agharia
4 Chattri
5 Muslim
6 Bhatu
7 Habru
8 Bhil
9 Dom
10 Ahir
11 Kurmi
12 Artisan Others
13 Kahar
14 Tharu
15 Chamar
16 Chero
17 Majhi
18 Panika
19 Kharwar
20 Oraon
21 Rajwar
22 Korwa
23 Tharu Females

The *K*-dissimilarity coefficient was calculated and the cluster methods B_1, B_2 and B_3 were applied to it. The values of D^2 given by Mahalanobis differ very little in relative magnitude from the *K*-dissimilarity values. The clustering obtained by single-link is shown in Figure 4.

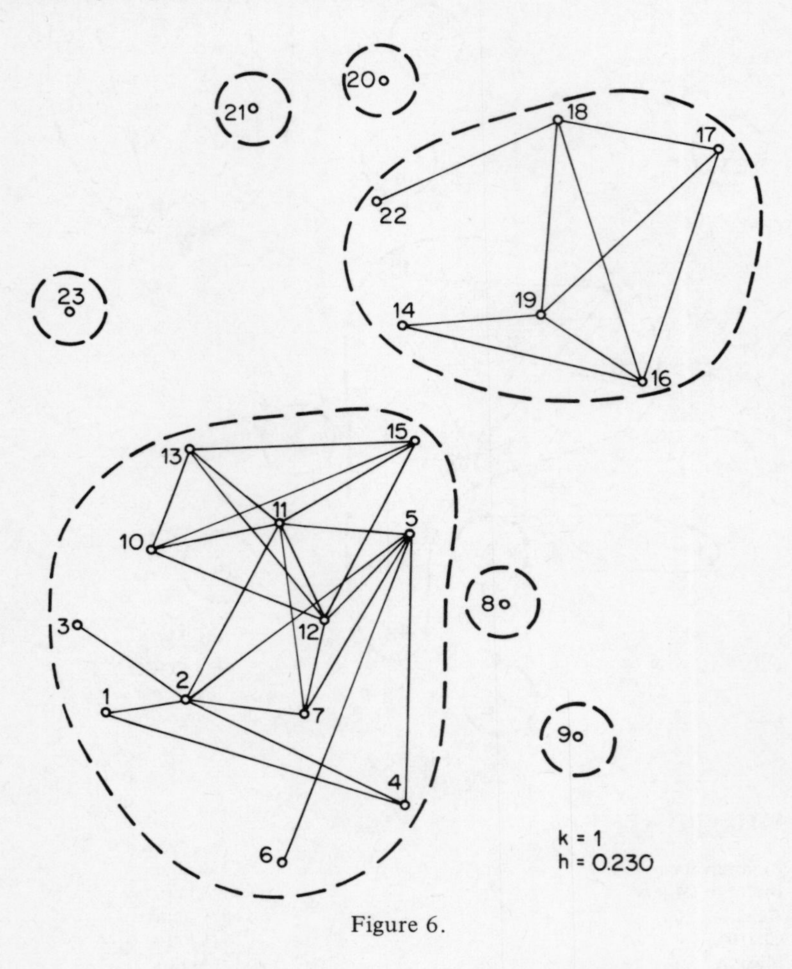

Figure 6.

The clusters obtained at levels $h = 0.160$, 0.230, and 0.300 are shown in Figures 5-7 on the graph which represents these values of the *K*-dissimilarity coefficient. The disposition of vertices representing the populations is based on the two-dimensional representation obtained by non-metric multi-dimensional scaling of the dissimilarity coefficient output by B_1.

The single-link clustering imposes a relatively high distortion on the input dissimilarity coefficient; much higher for example than the distortion induced by single-link in the *Silene* study. Such high distortion is to be expected when

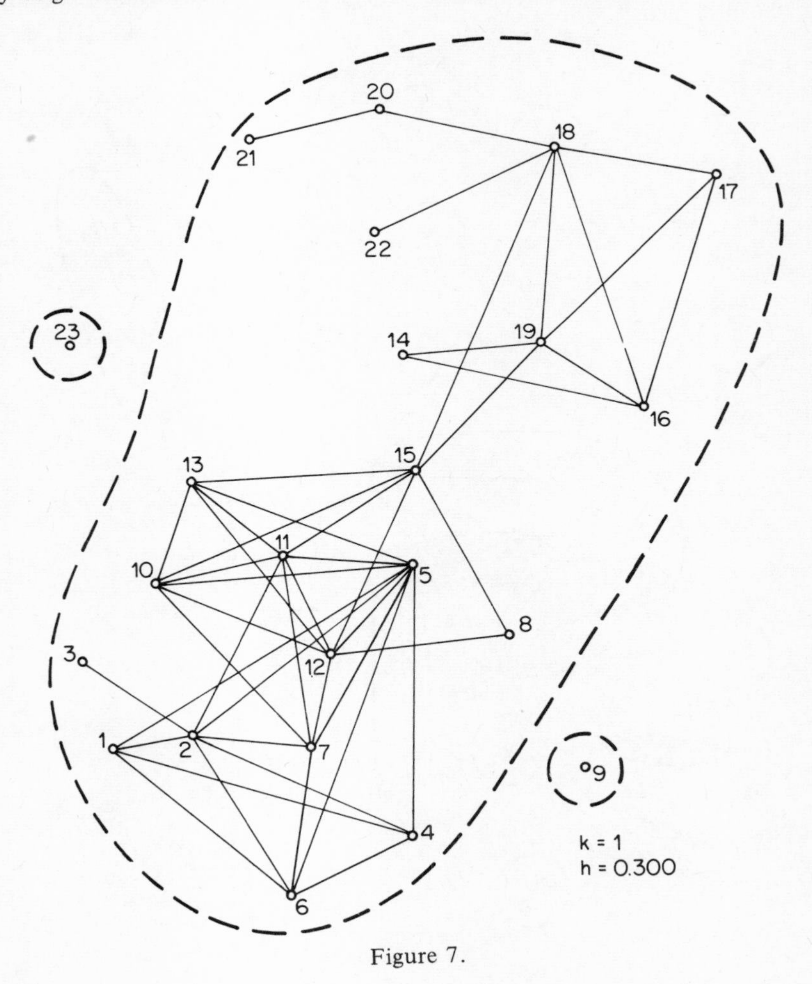

Figure 7.

microgeographic differentiation within a species is studied. Nevertheless the hierarchic clustering vindicates several of Mahalanobis' conclusions. Population 23 which included only females is appropriately isolated. Bhatus (6), Habrus (7), Bhils (8), and Doms (9) are likewise relatively isolated. Mahalanobis' A and B clusters are confirmed, and his suggested subcluster

{16, 17, 18, 19} of his T-cluster is confirmed. Likewise Chattris (4) and Muslims (5) cluster together.

In Figures 8-10 the clusters obtained by B_2 at levels 0.160, 0.230 and 0.300 are shown.

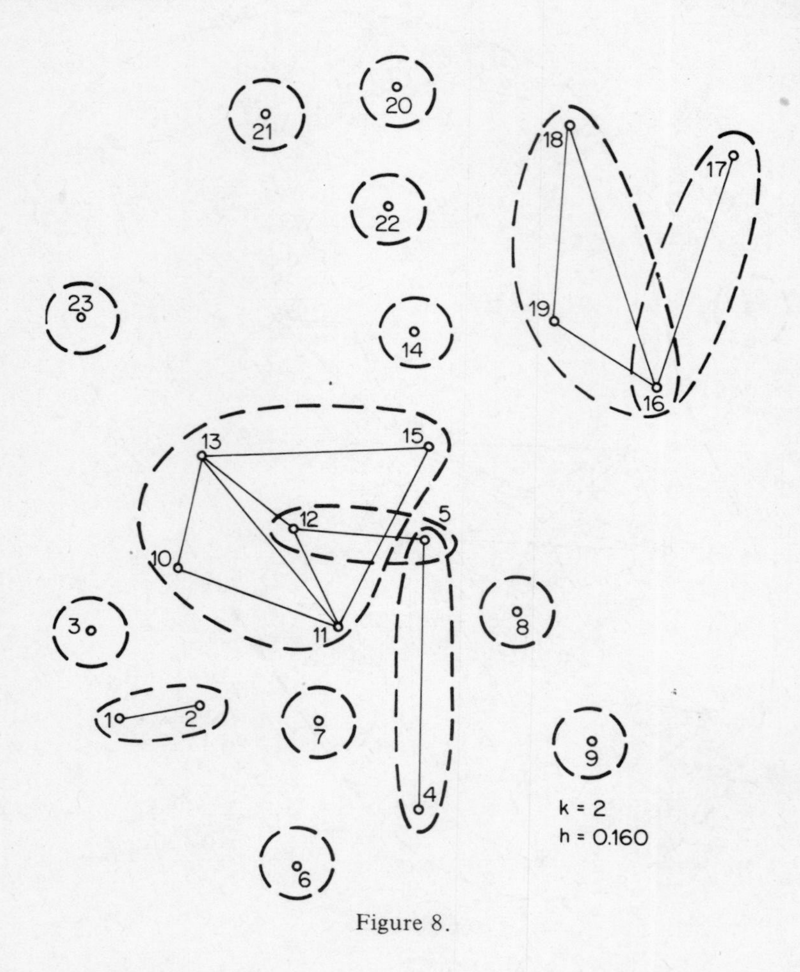

Figure 8.

The Chamars (15) lie in the overlap between two clusters which represent the 'tribal group' and the 'higher castes' recognized by Mahalanobis. The only conclusion drawn by Mahalanobis which is not substantiated is the supposition that the Brahmins (1 and 2) are well isolated from the A-cluster and that Chattris (4), Muslims (5) and Agharias (3) occupy an intermediate position. Mahalanobis' implication that there is some kind of sequence in physical differentiation corresponding to social position in the caste system, with Brahmins (1 and 2) at the 'top' and Korwas (22) at the 'bottom' is not vindicated by the results of non-metric multidimensional scaling. The results

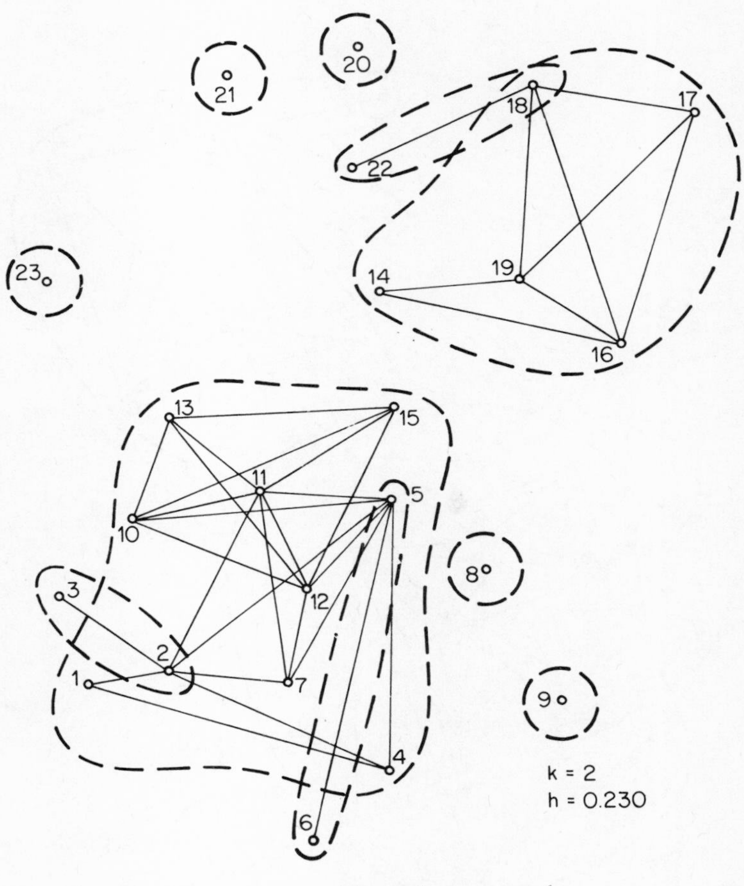

Figure 9.

do, however, suggest an element of topoclinal variation between the hill-dwelling tribes and the plains-dwelling groups.

No unequivocal interpretation of these results can be given. The amounts of differentiation between groups are small. As would be expected they are much smaller than the differentiation between males and females of the same tribal or caste group. Some correlation between caste and physical differentiation is indicated. It is, however, unclear to what extent differentiation has arisen historically from partial genetic isolation imposed by the caste system, and to what extent the caste system has preserved pre-existing racial differentiation.

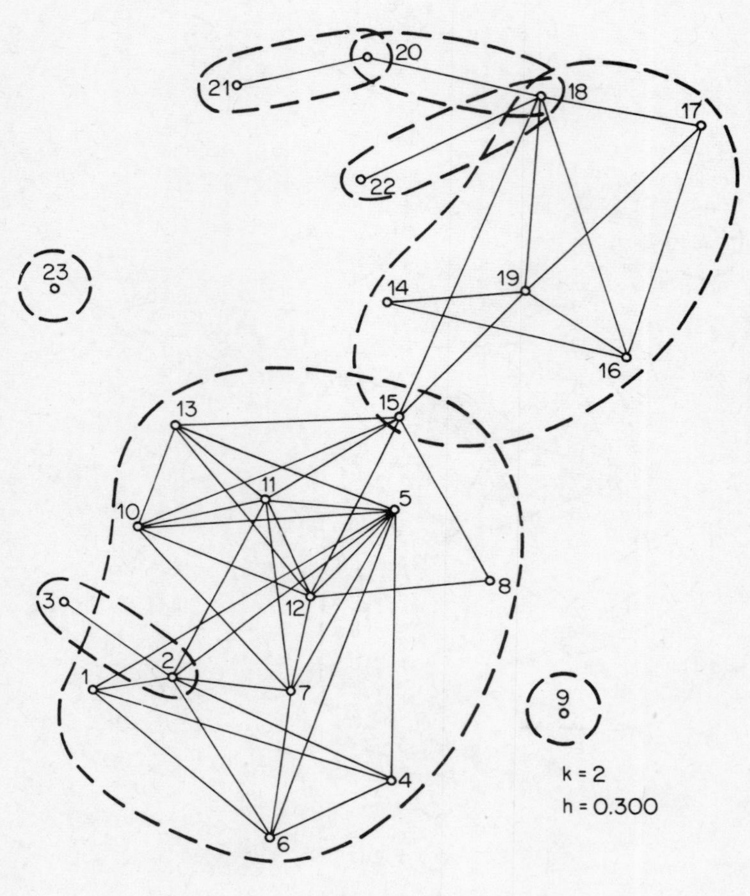

Figure 10.

APPENDIX 8 Bibliographic Sources

The publications listed here include some of the many surveys and bibliographies of methods of automatic classification and their application in biological taxonomy and allied fields.

Ball, G. H. (1966). Data analysis in the social sciences: what about the details? *Proc. Fall Joint Comput. Conf.*, **27**, 533-599.

Cole, A. J. (Ed.) (1969). *Numerical Taxonomy.* Proceedings of the colloquium on numerical taxonomy held in the University of St Andrew's, September 1968. Academic Press, London and New York.

Dagnelie, P. A. (1966). A propos des différentes méthodes de classification numérique. *Revue Statist. appl.*, **14**, 55-75.

Driver, H. E. (1965). Survey of numerical classification in anthropology. In D. Hymes (ed.), *The Use of Computers in Anthropology,* Mouton, The Hague. pp. 301-356.

Harrison, I. (1968). Cluster analysis. *Metra,* **7**, 513-528.

Heywood, V. H., and J. McNeill (Eds.) (1964), *Phenetic and Phylogenetic Classification,* Systematics Association, Publ. No. 6, London.

Hill, L. R. (1962). *Taxometrics.* A newsletter dealing with mathematical and statistical aspects of classification. Issued from the National Collection of Type Cultures, Central Public Health Laboratory, Colindale Avenue, London N.W.9, England.

Johnson, L. A. S. (1968). Rainbow's end: the quest for an optimal taxonomy. *Proc. Linn. Soc. N.S.W.*, **93**, 8-45.

Maisel, H., and L. R. Hill (1969), A KWIC index of publications in numerical taxonomy in the period 1948-1968. Available from Dr H. Maisel, Director, Computation Center, Georgetown University, 37th and 0 Sts., N.W., Washington, D.C. 20007, United States of America.

Sokal, R. R., and P. H. A. Sneath (1963). *Principles of Numerical Taxonomy,* Freeman, San Francisco and London.

Williams, W. T., and M. B. Dale (1965). Fundamental problems in numerical taxonomy. *Adv. bot. Res.*, **2**, 35-68.

The following are some of the journals which have published papers on automatic classification and its applications in biological taxonomy and related fields. An asterisk indicates journals which have carried a substantial number of such articles. The journal abbreviation, which is given alongside in each case, is from the *World List of Scientific Periodicals.*

American Journal of Botany	*Am. J. Bot.*
Australian Journal of Botany	*Aust. J. Bot.*

Biological Journal of the Linnean Society (formerly Proceedings of the Linnean Society of London)	*Biol. J. Linn. Soc. (Proc. Linn. Soc. (Biol.))*
**Biometrics*	*Biometrics*
Biometrika	*Biometrika*
**Classification Society Bulletin*	*Classifn Soc. Bull.*
Computer Journal	*Comput. J.*
Current Anthropology	*Current Anthrop.*
Educational and Psychological Measurement	*Educ. Psychol. Measmt.*
Evolution	*Evolution, Lancaster, Pa.*
IBM Journal of Research and Development	*IBM Jl. Res. Dev.*
Journal of the American Statistical Association	*J. Am. statist. Ass.*
Journal of Applied Bacteriology	*J. appl. Bact.*
Journal of the Association for Computing Machinery	*J. Ass. comput. Mach.*
Journal of Ecology	*J. Ecol.*
**Journal of General Microbiology*	*J. gen. Microbiol.*
Journal of the Operations Research Society of America	*J. Ops Res. Soc. Am.*
Kansas University Science Bulletin	*Kans. Univ. Sci. Bull.*
Mathematical Biosciences	*Math. Biosci.*
Metra	*Metra*
Nature	*Nature, Lond.*
New Phytologist	*New Phytol.*
Proceedings of the Zoological Society of London	*Proc. zool. Soc. Lond.*
**Psychometrika*	*Psychometrika*
Science	*Science, N.Y.*
Sociometry	*Sociometry*
**Systematic Zoology*	*Syst. Zool.*
Taxon	*Taxon*

REFERENCES

Bari, E. A. (1969). Experimental taxonomy of annual European species of *Silene* L., Ph.D. thesis, University of Cambridge, England.

Chater, A. O., and S. M. Walters (1964). *Silene.* In T. G. Tutin and others (Eds.), *Flora Europaea, Vol. 1.* Cambridge University Press. pp. 158-181.

Chowdhuri, P. K. (1957). Studies in the genus *Silene*. *Notes R. bot. Gdn Edinb.*, **44**, 221-278.

Gower, J. C., and G. J. S. Ross (1969). Minimum spanning trees and single-linkage cluster analysis. *Appl. Statist.*, **18**, 54-64.

McQuitty, L. L. (1963). Rank order typal analysis. *Educ. Psychol. Measmt*, **23**, 55-61.

Mahalanobis, P. C., D. N. Majumdar, and C. R. Rao (1949). Anthropometric survey of the United Provinces. *Sankhyā*, **9**, 89-324.

Van Rijsbergen, C. J. (1970a). A clustering algorithm. *Comput. J.*, **13**, 113-115 (Algorithm 47).

Van Rijsbergen, C. J. (1970b). A fast hierarchic clustering algorithm. *Comput. J.*, **13**, 324-326.

Wishart, D. (1969b). An algorithm for hierarchical classifications. *Biometrics*, **25**, 165-170.

GLOSSARY

GLOSSARY

Agglomerative: A type of clustering algorithm which operates by successive grouping together of objects.

Allopolyploid: See **Polyploid**.

a posteriori: Based on evidence or observation.

a priori: Used in this book to describe statements or rules which are not based upon evidence or factual considerations.

Artificial classification: Classification based on successive division or pooling according to single criteria. *cf.* **Monothetic**.

A-similarity: The concept of similarity associated with possession of common properties.

Association measure: A measure of resemblance between clusters. See Chapter 10.3.

Attribute: A set of mutually exclusive attribute states. See **Attribute state**.

Attribute state: A descriptive term applicable to individuals or to parts of individuals; for example, *petals red, leaf length 2 cm*.

Autogamous: Self-fertilized.

Average-link cluster method: A hierarchic cluster method. See Chapter 7.

Axioms for cluster methods: See Chapters 9 and 10.

Backcrossing: Interbreeding between hybrid offspring and one of the parental populations.

Beckner-Simpson criterion: The requirement that every taxon of supra-specific rank should be minimally monophyletic. See **Monophyly, minimal**.

Biological definition of species: The delimitation of taxa of specific rank as groups of organisms genetically isolated from all other groups, but able to exchange genes with each other. We maintain in Chapter 13 that this is only one of several criteria which can be used to delimit taxa of specific rank, and hence that it is not definitory. *cf.* **Morphological definition of species, Species**.

B_k: The cluster method B_k is a non-hierarchic cluster method which allows clusters at the same level to overlap by up to $k-1$ objects, and which makes optimal use of this overlap. B_k is the method known as **(fine) *k*-clustering**. See Chapter 8.5.

B_k^c : B_k^c is the coarse *k*-clustering method, which allows clusters at the same level to overlap by up to $k-1$ objects, but makes less efficient use of this overlap than the method B_k.

Campylotropous: An ovule which is recurved so that the apex is near the point of attachment.

Category, taxonomic: See **Hierarchy**.

Character: Generally used by taxonomists for the items of information used as a basis for classification. Used in this book in a restricted sense for sets of probability measures over states of an attribute. See **Character state, Attribute, Attribute state.**

Character state: Used in this book as a technical term for a probability measure over states of an attribute. See **Attribute, Attribute state.**

Chromosomal inversion: Reversal of a segment of a chromosome and resultant reversal of the order of the genetic loci.

Chromosome: Thread-like organelles within the cell which carry much of the genetic information.

Chromosome pairing: The pairing of genomes in meiosis. In organisms of hybrid origin where the genomes are dissimilar pairing may not occur or be only partial. Extent of chromosome pairing has sometimes been used as a measure of dissimilarity between genomes. See **Genome, Meiosis.**

Cladism: The view of taxonomy which maintains that the sole purpose of a classification is to convey information about the sequence in which populations have diverged in evolution.

Cladistic analysis: A term sometimes used for methods for guessing evolutionary branching sequences from information about present-day populations.

Classification: Used in this book to describe the processes by which classificatory systems are constructed. *cf,* **Diagnosis.**

Clinal variation: A pattern of infraspecific variation in which there is a trend in one or more attributes correlated with some external factor such as geographical location **(topoclinal variation)**, soil-type, altitude, or climate **(ecoclinal variation).**

Cline: See **Clinal variation.**

Cluster: Informally, a set of objects characterized by the properties of isolation and coherence. Technical discussion is given in Part II; see in particular Chapter 9.

Clustering: Simple, partitional, and stratified clusterings are defined in the Introduction. See also **Numerically stratified clustering.**

Cluster listing method: A method for listing the ML-sets of a relation. See Appendix 5.

Coarse *k*-clustering: See B_k^c.

Coarse *k*-dendrogram: The type of generalized dendrogram produced by the cluster method B_k^c. See Chapter 8.6.

Complete-link cluster method: A hierarchic cluster method. See Chapter 7.

Complex: Used as a non-committal term for a group of closely related taxa.

Conditional definition: The phenomenon that an attribute is only meaningful when some other attribute or attributes take certain states is known as conditional definition. For example, the attribute *PETAL COLOUR* is conditionally defined, being conditional on the state *present* of the attribute *PETAL PRESENCE*.

Congruence: The term used in this book for the extent to which two partitions coincide.

Conservatism: Here used for the view of the aims of classification which maintains that they should be based as far as possible on existing classifications.

Conservative attribute state: Used in this book for an attribute state which persists in all the descendants of the population in which it arose.

Convergence, evolutionary: Convergence is said to have occurred when distinct phyletic lines become more similar with respect to some attribute or set of attributes.

C_u: The cluster method C_u is a non-hierarchic cluster method which controls cluster overlap by relating overlap diameter to the current level. It is known as u-diametric clustering.

Cytology, cytological: Relating to the structure and biochemistry of cells and their parts.

Data set: The set of possible data DC's.

DC: See **Dissimilarity coefficient.**

Metric inequality: See **Dissimilarity coefficient.**

Ultrametric inequality: See **Dissimilarity coefficient.**

***D*-dissimilarity**: The concept of dissimilarity associated with information gain on identification of populations.

Definite: See **Dissimilarity coefficient** and **Numerically stratified clustering.**

Deme: A neutral stem which denotes a population of organisms. 'Deme' can be variously prefixed according to the criteria used to delimit the population. Thus **ecodeme** denotes a population from a single kind of habitat, **topodeme** a population from a single contiguous locality, **cytodeme** a population homogeneous with respect to some cytological feature such as chromosome number, **gamodeme** a population within which unrestricted gene-flow can occur. Gamodemes are sometimes called **Mendelian** populations. See **Experimental category**.

Dendrogram: see **Numerically stratified clustering.**

Diagnosis: The process by which individuals are assigned to classes.

Discordant variation: A pattern of variation in which the relative extents to which populations are differentiated do not coincide for different sets of attributes. The product of mosaic evolution. See **Mosaic evolution.**

Discriminant analysis: Methods whereby linear combinations of quantitative attributes which give optimal discrimination of populations are calculated from the joint distributions over the attributes of each population.

Dissimilarity coefficient: A function defined on pairs of objects and taking real values which represent the pairwise dissimilarities of the objects. A **similarity coefficient** resembles a dissimilarity coefficient but has the numerical scale running in the opposite direction. Formally, a dissimilarity coefficient (DC) is a function $d : P \times P \to \mathscr{R}$ (where P is the set of objects) satisfying the following conditions.

DC1 $d(A, B) \geqslant 0$ for all A, $B \in P$

DC2 $d(A, A) = 0$ for all $A \in P$

DC3 $d(A, B) = d(B, A)$ for all A, $B \in P$

Extra conditions which a DC may satisfy include the **definiteness** condition,

DCD $d(A, B) = 0 \Rightarrow A = B$,

the **metric inequality**

DCM $d(A, C) + d(C, B) \geqslant d(A, B)$ for all A, B, $C \in P$,

and the **ultrametric inequality**,

DCU $\max\{d(A, C), d(C, B)\} \geqslant d(A, B)$ for all A, B, $C \in P$.

Distortion: The term used in this book for measures of the accuracy with which a classification represents a dissimilarity coefficient on a set of objects. Where the classification is hierarchic this is called the **hierarchic classifiability** of the dissimilarity coefficient.

Divisive: A type of clustering algorithm which operates by successive subdivision of the set of objects.

DNA: Deoxyribonucleic acid. The molecule which carries the genetic information in the chromosomes of every cell of an organism. The basic components are nucleotides which are linked to form two polynucleotide chains twined in a double helix. There are four different nucleotides which are specifically paired, Adenine with Thymine and Cytosine with Guanine. The sequence of base-pairing carries the genetic information. **DNA-hybridization**, the extent to which single polynucleotide strands will link together by base-pairing, is a measure of overall similarity between DNA for different organisms. **Nucleotide-pair frequency** is an alternative and more precise measure of similarity.

DNA hybridization: See **DNA**.

DNA nucleotide-pair: See **DNA**.

Domination: If d, d' are DC's, d dominates $d'(d \geqslant d')$ if

$$d(A, B) \geqslant d'(A, B) \text{ for all } A, B \in P$$

Ecoclinal variation: See **Clinal variation**.

Ecodeme: See **Deme**.

Ecotype; A morphological or physiological variant within a species which occurs in a particular kind of habitat.

Edaphic: Relating to the composition of the soil.

Experimental category: Groups of populations delimited by genetic, cytological, geographical or ecological criteria, or by any other special-purpose criterion. Experimental categories usually cover populations within a single species or complex of related species. The **deme** terminology is a widely used system for naming experimental categories. Experimental categories do not in general correspond to taxa although sometimes the range of an experimental category may coincide with that of a taxon.

Fine k-clustering: See B_k.

Fine k-dendrogram: The type of generalized dendrogram produced by the cluster method B_k. See Chapter 8.5.

Flat cluster method: A cluster method which satisfies certain basic axioms and which commutes with suitable monotone transformations of the dissimilarity scale.

Functional correlation: The relation between attributes of parts which are involved in performance of the same or closely related functions by an organism.

Gamodeme: See **Deme**.

Gene: A heritable unit which occupies a fixed place on a chromosome. This crude definition is defective, but any more accurate definition would be lengthy.

Genecology: The study of the genetic bases for differentiation of populations from the same species which occupy different kinds of habitat.

Gene-flow: The exchange of genetic material between populations due to interbreeding.

Genome: A complete set of chromosomes. The sexual cells, gametes, usually contain a single genome; the somatic cells contain a double set.

Genotype: The genetic constitution of an organism.

Hierarchic classifiability: See **Distortion.**

Hierarchic, non-hierarchic: See **Numerically stratified clustering.**

Hierarchy, taxonomic or Linnaean: A nested sequence of partitions in which each partition is assigned a rank. This is the classificatory system which taxonomists have used for the last two hundred years. The elements of the partitions are called **taxa** and each taxon is assigned the **taxonomic rank** of the partition to which it belongs. The taxa of a given rank constitute a **taxonomic category**. Thus categories which are always used are called the **principal categories**; they are *species, genus, family, order,* and *class.* Often such **subordinate categories** as *subspecies, subgenus,* and *tribe* are used also.

Homology: The relation between parts of organisms which are regarded as the same. A basic criterion of homology is correspondence in relative position. Secondary criteria are similarity of composition and similarity of embryological origin. Homology is often defined as holding between parts descended from a single part in a common ancestor. Grounds for rejection of this definition are given in Chapter 15.

Identification: See **Diagnosis.**

***I*-distinguishability**: The concept of dissimilarity associated with diagnosis of individuals.

Indicator family: A set of relations satisfying certain conditions. See Chapter 10.2.

Information: All technical material on information theory is dealt with in Chapter 2, to which reference should be made.

Information factor: One of the two factors making up a K-term in K-dissimilarity. See Chapter 3.3.

Introgression, Introgressive hybridization: Flow of genes from members of one taxon to members of another through past or present hybridization.

Jaccard's coefficient: Jaccard's coefficient is a similarity coefficient for objects described by a fixed number of binary attributes each of which has a preferred state. If A, B are objects, then $J(A, B)$ is the number of attributes in which both A and B are described by the preferred state, divided by the number of attributes in which either A or B is described by the preferred state. $1 - J(A, B)$ is a metric DC.

K-dissimilarity: See **Simple matching coefficient** and Chapter 3.

K-term: One of the terms whose sum constitutes K-dissimilarity. There is one K-term associated with each attribute. See Chapter 3.3.

Macrotaxonomy: See **Supraspecific classification.**

Mahalanobis' D^2 statistic: A divergence measure for pairs of multivariate normal distributions with the same covariance matrix Σ. If β_1 and β_2 are the mean vectors, D^2 is given by

$$D^2 = (\beta_1 - \beta_2)^{\mathrm{T}} \Sigma^{-1} (\beta_1 - \beta_2)$$

Mahalanobis' postulate: The hypothesis that as the number of attributes considered is increased so the values of dissimilarity measures between populations, or the values of their ratios, will approach some limiting value.

Meiosis: The process in which the chromosome number is halved in the production of the sexual cells, the gametes.

Mendelian population: See **Deme.**

Meristic variation: Variation in numbers of parts.

Microspecies: Taxa of specific rank which are discriminated by small morphological differences.

ML-set: An ML-set of a symmetric reflexive relation r on a set P is a subset S of P which is maximal subject to the condition that $r|S$ is the universal relation on S.

Monophyly, minimal: A taxon is said to be minimally monophyletic if all its component populations have direct descent from populations referred to one and only one taxon of the same rank.

Monophyly, strict: A set of populations is said to be strictly monophyletic when it comprises all and only descendants of a single ancestral population.

Monothetic class: A class defined by a single attribute state or logical compound of attribute states.

Monotypic taxon: A taxon is said to be monotypic if it includes only one taxon of the next lower rank. Thus the plant genus *Ginkgo* is monotypic because it contains only a single species *Ginkgo biloba* (the maidenhair tree).

Morphology, morphological: Relating to the sizes, shapes and dispositions of parts of organisms.

Morphological definition of species: The delimitation of taxa of specific rank as groups of organisms isolated from all other organisms by some particular amount of dissimilarity in size, shape and disposition of parts. *cf.* **Biological definition of species.**

Mosaic evolution: A pattern of evolution in which the rates of divergence of populations with respect to different attributes or complexes of attributes differ widely. This gives rise to discordant variation in groups of organisms. See **Discordant variation.**

Multidimensional scaling: Simplified representation of a similarity or dissimilarity measure on a set of objects by a disposition of points representing objects in a euclidean space of some dimension.

Multivariate analysis: A generic name for statistical methods which handle many descriptive variables simultaneously.

Nexus hypothesis: The supposition that each attribute is generally determined by many genes, and conversely that each gene generally affects many attributes.

NSC: See **Numerically stratified clustering.**

Numerically stratified clustering: A numerically stratified clustering (NSC) is a mathematical object which defines a numerically nested system of (possibly overlapping) clusters. It is the basic output of a cluster method. Formally, it is a function $c : [0, \infty) \to \Sigma(P)$, where $\Sigma(P)$ denotes the set of symmetric reflexive relations on P; c must satisfy the following conditions.

NSC1 $h \leqslant h' \Rightarrow c(h) \subseteq c(h')$
NSC2 $c(h)$ is eventually $P \times P$.
NSC3 Given h, there exists $\delta > 0$ such that $c(h + \delta) = c(h)$.

An NSC may also satisfy the **definiteness** condition

NSCD $c(0)$ is the equality relation

and the **hierarchic** condition

NSCH $c(h) \subseteq \mathrm{E}(P)$

(where E(P) denotes the set of equivalence relations on P). A **hierarchic** NSC is called a **dendrogram.** Other kinds of NSC are called **non-hierarchic.**

Object: This term is frequently used to describe the elements of the set P being classified. See also **OTU.**

Occurence factor: One of the two factors making up a K-term in K-dissimilarity. See Chapter 3.3.

Ontogeny: The process of development of an adult organism.

Order-isomorphism: Order-preserving invertible transformation.

Orthotropous: An ovule which is erect so that the apex is opposite to the point of attachment.

OTU: This term is frequently used to describe the elements of the set P being classified when these are biological populations.

Parallel evolution: The development of similar parts or structures in the evolution of populations from a common ancestor which did not possess the parts or structures.

Pheneticism: The view of taxonomy which maintains that the sole purpose of a classification is to express the relative similarities or dissimilarities of the populations classified.

Phenon level: A term sometimes used for the levels in a hierarchic dendrogram at which all clusters are recognized as taxa of a given rank.

Phenotype, phenotypic: The totality of the attributes of an organism; the product of the interaction of the genetic constitution and environment of an organism.

Phylogenetic: Relating to the evolutionary history of organisms.

Polyploid: Having more than two sets of comparable or homologous chromosomes. In **allopolyploids** the chromosome sets are derived from different taxa following hybridization.

Polythetic class: A class which is not defined by any single attribute state or logical compound of attribute states.

Population: Used in this book for a morphologically homogeneous group of organisms from a single locality or from a single kind of habitat. Populations in this sense are considered as the basis for all classification. The terms 'ancestral population' and 'derived population' are used for groups of organisms considered as segments of an evolutionary tree.

Primitive attribute state: Used in this book for an attribute state which occurred in the common ancestor of a group of populations.

Principal category: See **Hierarchy**.

Rank, taxonomic: See **Hierarchy**.

Reticulate variation: A pattern of differentiation in which populations show significant differences which cannot be represented adequately by a hierarchic classification and in which clinal variation is not involved.

Serology: The use of techniques in which the extent of reaction between a protein and the antibodies produced by animals when a related protein is injected is used as a measure of overall chemical similarity between the proteins.

Sibling species: A term used by zoological taxonomists to describe taxa accorded specific rank on grounds of intersterility or difference in chromosome number, but not or only partially discriminable by morphological criteria. Botanists do not normally accord specific rank to such taxa.

Similarity coefficient: See **Dissimilarity coefficient.**

Simple matching coefficient: The simple matching coefficient is a similarity coefficient for objects described by a fixed number n of binary attributes. If A, B are objects, then $S(A, B)$ is the number of attributes in which A and B agree, divided by n. $1 - S(A, B)$ is a metric DC, and $n(1 - S(A,B))$, which is also a metric DC, is a special case of **K-dissimilarity.**

Single-link cluster method: A hierarchic cluster method. See Chapter 7.

Species: The least category which is consistently used in taxonomic hierarchies. Taxa of specific rank often coincide with genetically isolated populations within which there are no barriers to gene-flow.

Stability: A method of cluster analysis is said to be stable if small changes in the data lead to commensurately small changes in the result. A classification is said to be stable under extension of range when introduction of new objects does not drastically alter it. A classification is said to be stable under change or increase of attributes when addition of further relevant attributes or change in selection of attributes does not substantially alter it. Methods for investigating these two kinds of stability of classifications are given in Chapter 14.

Strong k-transitivity: The condition on relations associated with the cluster method B_k^c.

Strong k-ultrametric inequality: The inequality associated with the cluster method B_k^c. See Chapter 8.6.

Subdominant method: A type of cluster method characterized by its target set. See Chapter 8.3.

Subordinate category: See **Hierarchy.**

Subspecies: Morphologically differentiated groups of populations within a species, in which the differentiation is correlated with habitat, geographical location or both.

Supraspecific classification: Grouping of taxa of specific rank into taxa of higher rank. Sometimes called **macrotaxonomy.**

Target set: The set of DC's representing output NSC's.

Taxon: Taxa are the components of taxonomic hierarchies and are named according to the conventions laid down in the international rules of zoological and botanical nomenclature. *cf*, **Experimental category.** See **Hierarchy.**

Topoclinal variation: See **Clinal variation.**

Topodeme: See **Deme.**

Type A cluster methods: The type A cluster methods are the hierarchic methods, that is, those which lead to a dendrogram. See Chapter 7.3.

Type B cluster methods: The type B cluster methods are the non-hierarchic methods based on absolute constraints. See Chapter 8.

Type C cluster methods: The type C cluster methods are the non-hierarchic methods based on internal constraints. See Chapter 8.

***u*-Diametric clustering**: See C_u.

***u*-Diametric inequality**: The inequality associated with the cluster method C_u. See Chapter 8.7.

Uniform cluster method: A method whose operation at a fixed level is determined wholly by which DC values equal or exceed that level.

Weak *k*-transitivity: The condition on relations associated with the cluster method B_k.

Weak *k*-ultrametric inequality: The inequality associated with the cluster method B_k. See Chapter 8.5.

Index of References

Numbers in italic indicate the page on which the full reference is given.

Subject Index

Page numbers given in italics indicate where an entry is listed in the Glossary